Glyconanotechnology

Glyconanotechnology

Nanoscale Approach for Novel Glycan Analysis and Their Medical Use

edited by
Tomáš Bertók

JENNY STANFORD
PUBLISHING

Published by

Jenny Stanford Publishing Pte. Ltd.
Level 34, Centennial Tower
3 Temasek Avenue
Singapore 039190

Email: editorial@jennystanford.com
Web: www.jennystanford.com

British Library Cataloguing-in-Publication Data
A catalogue record for this book is available from the British Library.

Glyconanotechnology: Nanoscale Approach for Novel Glycan Analysis and Their Medical Use

ISBN 978-981-4800-16-7 (Hardcover)
ISBN 978-0-429-44849-2 (eBook)

Contents

Preface

Although the term nanotechnology was first used in 1974 by Professor Nario Taniguchi, the concept behind this scientific field was theoretically formulated by Professor Richard Feynman a decade earlier in his famous talk "There´s Plenty of Room at the Bottom", when discussing processes allowing the manipulation of single atoms. This exciting vision was fully realised in 1981, when scanning tunnelling microscope was applied not only for the visualisation of atoms but also for precise positioning of individual atoms on the substrate. At that time nanotechnology as a scientific discipline was born. Why is nanotechnology so fascinating? It is not only controlled manipulation of single atoms, but rather fascinating properties of nanomaterials (i.e. material having at least one dimension within the range of 1 to 100 nm) so dramatically distinctive from their macroscopic counterparts what makes this field so exciting. For example, pure gold nanoparticles can have colour from red to violet, while gold we all have indulgence for has a different and so familiar colour and shine. Nanomaterials are full of surprises. When scientists tried to prepare a single layer of carbon atoms, better known as graphene, they speculated if such a nanomaterial can be stable when exposed to air, since some thought that such a thin material will be so reactive it would ignite immediately upon exposure to air. The other surprise came with the process graphene was firstly prepared—using ordinary Scotch tape. The tape was used to peel off graphite layer by layer until a single layer of carbon remained. Discovery of graphene in 2004 proved not only its stability in air but so many extraordinary properties with myriads of different possible applications.

Application of nanotechnology in glycomics, a scientific discipline that studies the biological functions of complex carbohydrates (glycans) in living organisms, made real revolution. Nanotechnology not only helped to control spatial orientation of glycans, but also the density of glycans on surfaces allowing to mimic cellular surfaces with a degree of control unimaginable without nanotechnology. The other significant aspect is that nanotechnology offered unprecedented possibilities in glycan analysis making glycan analysis highly robust, automated, and quick leading to fascinating discoveries about the role of glycans in our "ordinary" lives, but also about significant role of glycans in diseases. It is the combination of these two worlds that helps us to better understand the huge universe within us and around us. This book is an effort of the authors to present the most recent and fascinating advances in the area of carbohydrate nanotechnology and applications mainly in medicine and clinical fields. We sincerely hope that the readers will enjoy reading this book.

Chapter 1

Introductory Nanotechnology: What Is So Special about the Nanoscale?

Tomáš Bertók

Glycanostics, Ltd., and Department of Glycobiotechnology, Institute of Chemistry, Slovak Academy of Sciences, Dubravska cesta 9, 84538 Bratislava, Slovak Republic

tomas.bertok@savba.sk

This chapter is for all the research scientists and students out there without any knowledge of nanoscale physics and chemistry, or just wanting to improve their elementary knowledge on this topic. It covers the history, classification, synthesis and characterization techniques of different nanomaterials, as well as some examples of their common use in research and industry. Some basic mathematical explanations are involved as well for deeper understanding; however, these are not necessary for further reading. At the end, fundamentals of nanobiotechnology as a young discipline are described.

Glyconanotechnology: Nanoscale Approach for Novel Glycan Analysis and Their Medical Use
Edited by Tomáš Bertók

ISBN 978-981-4800-16-7 (Hardcover), 978-0-429-44849-2 (eBook)
www.jennystanford.com

1.1 Introduction, *id est* Motivation

According to the definition, nanoscience is the study of the materials properties in nanoscale. The properties often differ compared to their macroscopic counterparts. For example, macroscopic and colloidal gold have different optical properties (shiny metal vs. red solution of gold nanoparticles), as well as electrochemical and catalytic ones (gold nanoparticles can catalyze the oxidation of carbon monoxide (CO) to carbon dioxide (CO_2), but only when they're between 4–10 nm in diameter) [1]. Scientists have been doing research about the nature of small length scales for a long time, and even longer mankind was using them without even knowing. One of the most famous example is Michael Faraday's conclusion (1857) that his quite unusual colors of gold salts solutions are due to the presence of gold particles—small aggregates reduced with phosphorus [2].

Nanotechnology (according to Norio Taniguchi's definition, who was the first person to use this term) "consists of the processing of separation, consolidation and deformation of materials by one atom or one molecule." Nanotechnology is the manipulation of a single nanoscale object—such as nanoscale electronic elements [3, 4]. Nanotechnology occurred in modern art, too. Some of it was a part of famous movies, such as *Minority Report* (see Fig. 1.1),

Figure 1.1 Picture from the movie *Minority Report*, starring Tom Cruise (2002). News full of recent advances in the nanotechnology research.

and there was even a short movie created purely by manipulating individual atoms—in 2013 released on YouTube by IBM Research, called *Boy and His Atom*. Two-atom carbon monoxide molecules (CO) were magnified 100 million times using a scanning tunneling microscope (STM) and moved to make a small boy dance and play with a single atom. John Hart, a mechanical engineer from the University of Michigan, created a portrait of US President Barack Obama (named Nanobama), sculpted from millions of carbon nanotubes and measuring about half a millimeter in diameter. The world's smallest book ever created is *Teeny Ted from Turnip Town*, produced at Simon Fraser University in Vancouver, Canada. The book measures up to 100 micrometers and requires an electron microscope to read it. Moreover, one of the 100 copies of this book costs about USD 15,000. Nanomaterials became quite popular with the Nobel Prize in Chemistry (1996) awarded to Harry Kroto, Richard Smalley, and Robert Curl for their discovery of fullerenes and then Nobel Prize in Physics (2010) awarded to Andre Geim and Konstantin Novoselov [5] for their groundbreaking experiments with two-dimensional nanomaterial graphene (discovered in 2004, until then it was believed to be thermodynamically unstable) [6]. Together with carbon nanotubes, these three nanomaterials are an example of three different topologies with hexagonally bonded carbon. It was already mentioned that different nanomaterials were used by mankind for centuries; still, we probably can't talk about nanotechnology, since this was not the manipulation with individual nanoscale objects. The most common example of using these objects, namely gold nanoparticles in this case, is the Lycurgus cup (a 4th century Roman glass cup prepared from dichroic glass) located in British Museum in London. This cup (Fig. 1.2) shows a mythical king Lycurgus, who tried to kill Ambrosia—follower of Bacchus (Dionysus in Greek), a god of wine [7]. It shows different colors depending whether the light is passing through the glass or not. Due to the presence of metal nanoparticles, many glass windows in medieval cathedrals are stained (e.g., ruby color caused by the presence of gold nanoparticles).

The whole concept of nanoparticles for medical applications (by definition particles with at least one dimension being in the range of 1–100 nanometers) was inspired by the visit of

Figure 1.2 Photo of Lycurgus cup under different conditions, showing its glowing red appearance when light is passing through, caused by the presence of gold nanoparticles.

Karl Maria von Weber's opera by Paul Ehrlich. In this opera called "Der Freischütz", so-called "Freikugeln" played the central role. These were something like bullets able to make a direct hit, even if the target was out of reach or if the shooter didn't aim properly. This was the idea of targeted drug delivery for therapeutic purposes [8, 9].

In the fascinating world of nanoscience and nanotechnology, we can find a great pleasure of learning new things, of discovery of new materials with novel and unique properties (just like MXenes, a new family of two-dimensional nanomaterials described later in the text, synthesized only recently in 2011) [10] and challenge of searching for new applications. To do this, we must understand the concept of matter, to understand the particle and wave-like properties of matter and some basic characteristics of nanomaterials described and used through this whole book.

1.2 Basic Quantum Mechanics and Material Chemistry

You're probably familiar with Bohr planetary model of an atom (1913). Despite the fact it is already well known that this model doesn't mirror the real situation of an electron occupying atomic and molecular orbitals, we can easily imagine the structure of an atom. Nucleus, being about 1/100,000th of the whole atom despite the fact that 99% of the mass of an atom is contained here (now we know what Nobel Prize laureate Richard Feynman meant by his famous lecture "There is a Plenty of Room at the Bottom" in 1959) [11], consists of protons and neutrons [12]. According to standard model of particle physics, those are baryons (triquarks, subatomic particles that are quark-based), and participate in strong interactions, whereas leptons (e.g., electrons) don't. Most of the visible matter consists of these two baryons. Leptons, quarks and their composite particles (made of odd number of these, such as protons and neutrons) are called fermions, named after Enrico Fermi ("architect of the atomic bomb"). There are other, much more exotic particles, such as pentaquarks; however, these are not important for our understanding of nanoscience. For every single (nano)material, it is the size that matters and determines the material's properties. Those properties can emerge with a material's size small enough for the atomic properties to become prevalent. A surface to volume ratio (R) is a parameter that helps us to understand this feature. For the materials in the nanoscale, more and more atoms come from the bulk to the surface—as can be seen on Fig. 1.3. This phenomenon is common for all different structures, such as plane, cube, sphere or cylinder. Considering a plane ($S = 2l^2$, $V = al^2$, l being length and a being thickness), if $a << l$ (or in extreme case, if the nanomaterial is just one atom layer thick just like graphene), the relative number of the surface atoms is still the same (100%). For cube ($S = 6l^2$, $V = l^3$) and sphere ($S = 4\pi r^2$, $V = 4\pi r^3/3$) the ratio is $6/l$ and $3/r$, respectively, with l being the length and r being the sphere diameter. According to the figure below, it is clear the ratio R increases with the decreasing size. In other words, a larger fraction of the underlying atoms is now present on the surface. A disadvantage may be due to more "visible" defects on the surface, but at the same time, advantages emerge such as that more efficient catalysts arise, since the most effective catalysts are those with high surface area. In case of nanostructures, the area may be also used for the immobilization of some diagnostic biomolecules or therapeutic drugs. For nanostructured biosensors, enhanced area means higher surface density of biorecognition elements, thus enhanced performance (in most cases, but certainly not always, considering the sterical hindrance for a biorecognition event if the density is too high). More on this topic will be discussed in next chapters. Another way how to increase the surface area for enhanced applications in catalysis or biorecognition is to use porous nanomaterials (such as nanoporous carbon nanofibers or nanoporous hybrids; nanoporous means poresize ≤ 100 nm, mesoporous 2–50 nm) [13].

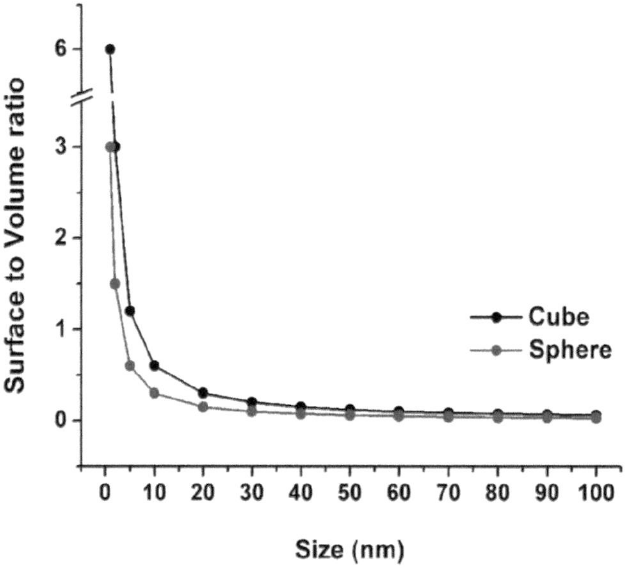

Figure 1.3 Surface to Volume ratio dependence on the material size. The smaller a particle become, the more atoms get to the surface. This phenomenon is shape-dependent, but valid in general for all other shapes mentioned above.

As for different types of nanomaterials, they can be divided according to their dimensionality (degree of freedom: 0D, 1D and 2D; all macroscopic objects are considered to be 3D), or materials used for their preparation (e.g., metallic particles, including magnetic nanoparticles, carbon-based materials, polymer-based nanostructures and dendrimers or hybrid nanomaterials). For 0D nanostructures, there are three degrees of confinement (e.g., a particle, such as electron, is confined in all three directions). These structures are called quantum dots. The other types of structures involve quantum wires and quantum wells (Fig. 1.4).

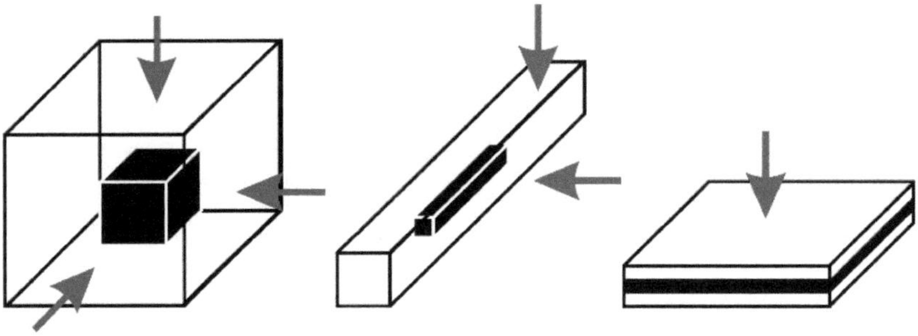

Figure 1.4 Schematic presentation of (from left to right) 0D quantum dots, 1D quantum wires and 2D quantum wells with their corresponding degrees of confinement (red arrows). Dimensionality in this case represents the degree of freedom, which for bulk material is 3 (3D macroscopic objects).

But what exactly can be considered a matter? Is it an atom, or even an electron? The truth is, every single object possesses particle- and wave-like properties. This particle-wave duality of all objects (not just photons, as originally proposed for its wave-like properties by Huygens or particle-like properties by Newton) was first introduced by Louis Victor Pierre Raymond de Broglie. De Broglie's contribution had implications beyond light. Generally, everything possesses both properties of a wave and those of a particle. The wave properties of matter become apparent only when we're dealing with really small objects. Even though macroscopic objects also have wave-like properties, they are generally better described by Newton's laws. For small objects like electrons and photons, so-called wave functions (Ψ) are used to describe the (quantum) state of the system. Perhaps the most common example is the mathematical function called atomic orbital, used for the calculation of the probability of finding an electron in some region around the nucleus—not in the center, although there is a strong electrostatic attraction between protons and electrons, since this would break what we call Heisenberg's uncertainty (indeterminacy) principle. According to Werner Heisenberg (1927), the more precisely a momentum of a particle is known, the less precisely its position can be determined, and vice versa (this can also be applied to energy and time). Formally:

$$\sigma_x \sigma_p \geq \frac{\hbar}{2},$$

where $\hbar$ is the reduced Planck constant ($h/2\pi$) and σ_x and σ_p is the standard deviation of position and momentum, respectively.

When an electron is described by a wave function, we can say it occupies that orbital. An overlap of orbitals then produces bands. And finally by equation:

$$\int_{-\infty}^{\infty} |\Psi(V,t)|^2 dV = 1,$$

where the wave function Ψ and its complex conjugate Ψ^* (complex number multiplied by its complex conjugate yields a real number, i.e., $\Psi\ \Psi^* = |\Psi|^2$) can be used to calculate the probability of finding the particle (it HAS to be somewhere, that's why there's 1 on the right side of the equation, i.e., we are sure to eventually find the particle). This is so-called probabilistic interpretation of wave functions, which can look like this:

$$\Psi = e^{ikx}$$

for a free particle (e.g., an electron), or (for a particle confined in one dimension, i.e., quantum well) like this:

$$\Psi = \sqrt{\frac{2}{a}} \sin \frac{n\pi x}{a},$$

where a is the length of the region which the particle is confined to, and n is an integer. But how do we know the wave function looks like this? We get this form of Ψ by solving what is called a Schrödinger equation (developed by Erwin Schrödinger in 1924 during his Christmas holidays in Alps). It is the central equation in quantum mechanics with a general form

$$i\hbar \frac{\partial}{\partial t} \Psi(\mathbf{r},t) = \hat{\mathrm{H}} \Psi(\mathbf{r},t),$$

where $\hat{\mathrm{H}}$ is Hamiltonian. Although this equation is time-dependent, it can be shown there is also a time-independent form, simply by assuming that the potential energy operator V in the next form of Schrödinger's equation is time-independent ($V(\mathbf{r},t) \rightarrow V(\mathbf{r})$; m is the particle's mass):

$$i\hbar \frac{\partial}{\partial t}\Psi(\boldsymbol{r},t) = \left[-\frac{\hbar^2}{2m}\nabla^2 + V(\boldsymbol{r},t) \right]\Psi(\boldsymbol{r},t)$$

∇^2 being a Laplacian—in Cartesian coordinate system:

$$\nabla^2 = \frac{\partial^2}{\partial x^2} + \frac{\partial^2}{\partial y^2} + \frac{\partial^2}{\partial z^2}$$

What is important for electrical and optical properties of matter are electrons and complementary holes (hole is an absence of an electron), together forming an exciton. According to de Broglie, the wavelength associated with every particle or object is described as:

$$\lambda = \frac{h}{p},$$

where h is Planck constant ($h = 6.62 \times 10^{-34}$ J s) and p is the object's momentum. It becomes clear that the larger the object the smaller the de Broglie wavelength. Another length scale that defines nano-objects is called bulk exciton Bohr radius. Bohr radius is in physics textbooks defined as the closest stable orbit of an electron to the nucleus in Bohr's planetary model of an atom. For an electron, the Bohr radius a_0 is:

$$a_0 = \frac{4\pi\varepsilon_0\hbar^2}{m_0 q^2},$$

where ε_0 is the vacuum permittivity ($\varepsilon_0 = 8.854 \times 10^{-12}$ F/m), $\hbar$ is the reduced Planck's constant ($\hbar = h/2\pi$), m_0 is the mass of an electron ($m_0 = 9.11 \times 10^{-31}$ kg) and q is the elementary charge ($q = 1.602 \times 10^{-19}$ C). For an exciton (electrostatically bound electron–hole pair formed upon light absorption, which promotes the electron from its valence to conduction band), the equation becomes:

$$a_B = \frac{4\pi\varepsilon\varepsilon_0\hbar^2}{\mu_0 q^2},$$

where μ is the reduced mass of the electron–hole pair. The bulk exciton Bohr radius a_B is a characteristics of every semiconductor.

For a semiconductor nanoparticles sufficiently small (e.g., a semiconductor quantum dot), the nanoparticle's size affect the a_B value—this is the effect of so-called quantum confinement. According to Brus equation, the size of the band gap between conduction and valence band increases with smaller size of semiconducting nanocrystals. This results in shift of the absorption spectra to lower wavelengths (so-called blue shift). It is thus possible to tune optical properties of quantum dots by changing the crystal size [13, 14].

1.2.1 "Crystal" Clear Properties of Matter

What is currently considered to be a matter must have both rest mass and volume. It exists in phases like gas, liquid or solid, but there are more, like plasma or Bose–Einstein condensate. The structure of atoms or molecules in objects can be either amorphous (lacks any order, e.g., in glass) or crystalline (highly oriented and arranged in regular and periodic manner). Somewhere between are polycrystalline structures. Nobel Prize in chemistry for the discovery of so-called quasi-crystals (containing regular pattern following mathematical rules, but those patterns never repeat themselves) was awarded to Dan Shechtman in 2011 [15].

There are seven basic crystal systems, participating in formation of a single unit cell of a crystal, namely:

- Triclinic
- Monoclinic
- Orthorhombic
- Tetragonal
- Rhombohedral
- Cubic
- Hexagonal

By repeating the unit cell, we can construct a full crystal. An important feature of each unit cell is to define the angles, cell lengths and atoms per unit (this allows us to calculate the amount of atoms in a nanocrystal). These atoms can be localized inside, on the corner, edge or face of a unit cell.

To denote different planes, directions and equivalent families, so-called Miller indices are often used, e.g., to describe nanowire growth or to define crystal faces. To obtain the Miller indices for a plane of atoms, we follow several rules:

- We choose a plane and determine where the intersection with *x*, *y* and *z* axes of the unit cell is
- We calculate the reciprocal of each of these intersection points. These values are associated with *x*, *y* and *z* directions and are generally denoted as *h*, *k* and *l*. In case there is no intersection, we may assume the intersection lies in the infinity, and its reciprocal is 0.
- Furthermore, in case the intersection occurs on the other side of the axis, the negative value is denoted by placing a bar over an index.

The Miller notation currently used is:

- (*hkl*) for a plane of atoms in a crystal lattice (see Fig. 1.5)
- {*hkl*} for a family of equivalent planes of atoms in a crystal
- [*hkl*] for a crystallographic direction
- <*hkl*> for a family of equivalent crystallographic directions

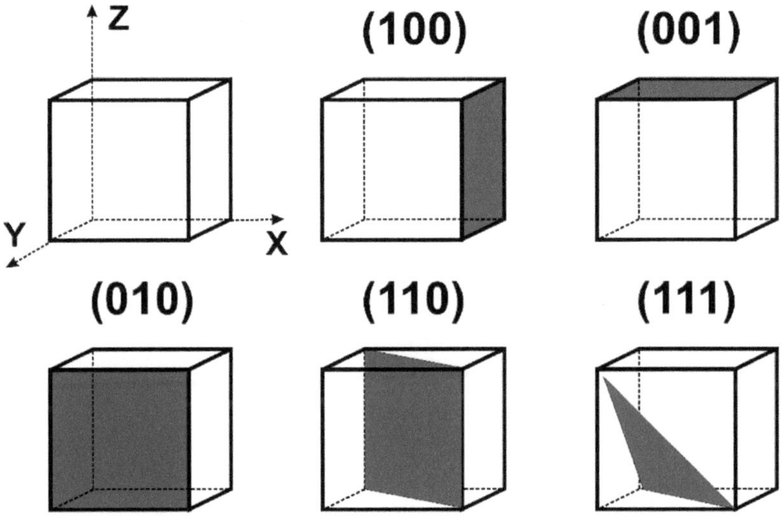

Figure 1.5 Schematic presentation of different atom planes and their notation using Miller indices. Number 1 denotes an intersection at the end of a unit cell axis, 0 means there is no intersection (or that it lies in ∞). Individual numbers in triplets in the brackets are *h*, *k* and *l* and correspond to *x*, *y* and *z* directions.

1.3 Production of Most Common Nanomaterials

There are two basic approaches for nanomaterials synthesis. Since they are on a half way from atoms to macroscopic objects, we can start with individual atoms and proceed with their "aggregation" to larger objects, or start with a macroscopic object and to isolate individual structures it is composed of. This is known as a bottom-up synthesis and top-down manufacturing. Some of the most commonly used nanoparticles and their synthesis will be briefly described in the following text.

1.3.1 Synthesis of Gold Nanoparticles

A chemist's approach for nanoparticles synthesis is colloidal growth (bottom-up synthesis). There are many "recipes" for preparation of every possible nanomaterial, and generally speaking, no description is valid for all protocols. We can say the common feature for all of these methods is precipitation-based approach starting with a nucleation and growth of these nanoparticles in a solution. For gold nanoparticles, which are very often used in biosensing applications and can be easily modified using thiol chemistry (preparation of so-called self-assembled monolayers) [16] by small molecules containing different functionalities, such as polar carboxylic group for covalent immobilization of proteins, two basic approaches can be used, namely Turkevich (1951) and Brust (1994) approach [13].

In both cases, the precursor is chloroauric acid ($HAuCl_4$) dissolved in appropriate solvent. Gold atoms are present in the solution in oxidized state (Au^{III}) and only after addition of a reducing agent, the gold is reduced to Au^0 and a formation of small clusters of Au atoms begins. The Turkevich approach uses high temperature (the solution of $HAuCl_4$ in deionized water is boiling) and citrate solution as a reducing agent, which at the same time passivates the surface of newly formed gold nanoparticles (AuNPs), giving them their negative charge, thus protecting them from further agglomeration (due to electrostatic repulsion). In the latter protocol suggested by Brust, gold ions are extracted from the water phase (from their water solution) to the organic phase (tetra-*n*-octylammonium bromide in toluene). After the water phase becomes clear, sodium borohydride is added to the mixture and nucleation and reduction of gold atoms to nanoparticles begins. The stabilizing agent in this method is trioctylphosphine, which acts pretty much the same way as the citrate in Turkevich method [17].

1.3.2 Synthesis of Quantum Dots and Wires

An interesting approach for the synthesis of quantum dots (QDs), even though it has one huge drawback, is the use of so-called reverse micelles. Basically, using surfactants with aliphatic tail and polar head (providing hydrophilic interior of the micelles in nonpolar solvents) such as sodium dodecylsulphate (SDS), cetyltrimethylammonium bromide (CTAB) or sodium bis(2-ethylhex yl)sulfosuccinate, allows us to control the growth of the nanocrystal inside the micelle, controlling the micelle size by the ratio of water to surfactant (ω parameter). The drawback is that despite an elegant approach of the size control, the room temperature during the preparation results in quite a bit of surface defects on QDs.

To produce high-quality QDs (e.g., CdSe QDs), colloidal synthesis is often applied. Here, the temperature is a critical factor, as well as subsequent size-selective precipitation, allowing us to collect nanoparticles with narrow size distribution. In case of QDs, during the synthesis small aliquots of reaction mixture can be collected and their absorption spectra are measured to stop the reaction at a desired nanocrystal size. Another approach can be synthesis by nonthermal plasma. An expensive and high-tech bottom-up method for the QDs and quantum wells syntheses is molecular beam epitaxy (MBE). It uses a high-vacuum chamber and a beam of atoms, which are "shooted" towards a substrate in the chamber. Because of the high vacuum, the atoms don't experience any collision on their way and by controlling temperature and timing, we are able to prepare structures with atomically precise thicknesses. QDs synthesized by MBE method are often referred to as epitaxial QDs. There are several types of different QDs currently available commercially, namely core type (e.g., CdSe), core-shell type (e.g., CdSe/ZnS) or alloyed QDs [18].

A similar method compared to MBE (however, using different experimental setup) for the synthesis of nanowires is chemical vapor deposition (CVD) of organometallic precursors. In 1964,

Ellis and Wagner from Bell Laboratories discovered the phenomenon of small metal nanoparticle catalysts, which could promote the asymmetric growth of semiconductors from the thermal lysis of precursors in gas phase. This has been subsequently used for vapor-liquid-solid (VLS) growth of nanowires. Small nanometer-sized metal droplets induce the nanowire growth upon (i) their formation using high temperatures (these droplets typically have higher diameter, i.e., several hundreds of nanometers, and this therefore results in nanowires with greater diameter), or (ii) their preparation using colloidal syntheses protocols (described above). In each case, small molten droplets (formed upon heating the substrate to high temperatures after nanoparticles formation) solubilize and incorporate elements from a vapor-phase precursors, starting a nucleation between a liquid droplet and solid substrate phase. Similar approach is called solution-liquid-solid (SLS) growth, using temperatures below 400°C, so the metal nanocatalysts used must be made of elements with lower melting points [13, 19].

1.3.3 Preparation of Carbon Nanomaterials

We already mentioned fullerenes—carbon-composed nanomaterials (hollow sphere-shaped) named after an American architect Richard Buckminster-Fuller [20], who designed a dome in the form of a football in 1967. It consists of 20 (smallest one), 26, 28, 32, 50, 60 (most common form), 70 and 84 (some authors claim even more, up to 100) carbon atoms, being formed in nature by lightning discharges in the atmosphere, found commonly in coal. C_{60} has a nucleus to nucleus diameter of 0.71 nm and van der Waals diameter of 1.1 nm (a hard sphere model used to describe the atom distances, representing the half of the internuclear separation of two non-bonded atoms of the same element on their closest possible approach; named after Johannes Diderik van der Waals, Nobel Prize in physics in 1910) [21].

In 1991, carbon nanotubes (CNTs) were synthesized for the first time. Carbon nanotubes can be considered a single graphite sheet (called graphene, with thickness of exactly one carbon atom) rolled up into a tube. The tube can be made of a single sheet or more of them. That's the reason why there are so-called single-walled (SWCNTs) and multi-walled carbon nanotubes (MWCNTs). Even though graphene sheets are made of sp^2 bonded carbon atoms (honeycomb structure), some amount of sp^3 bonded carbon is still present in nanotubes as a result of the nanotube curvature. Although these nanotubes can be relatively long, their diameter is about 1.2–1.4 nm for single-walled and up to 40 nm for multi-walled CNTs, respectively (individual layers are about 0.34 nm distant). As can be seen on Fig. 1.6, CNTs can have different chiral vectors, defined by n and m integers, determining their properties. A chiral vector C_h is a linear combination of the unit vectors a_1 and a_2 ($C_h = na_1 + ma_2$, where n and m are integers). Chiral vector defines the propagation vector, which represents the periodicity of the CNT's structure along with its axis. There are three different types of carbon nanotubes defined by their n and m, possessing different electrical properties, namely [14, 22]:

- Armchair ($n = m$), with metallic properties
- Zigzag (n or $m = 0$)
  - $n - m = 3i$, with metallic properties
  - $n - m \neq 3i$, with semiconductor properties (i is an integer)
- Chiral
  - $n - m = 3i$, with metallic properties
  - $n - m \neq 3i$, with semiconductor properties

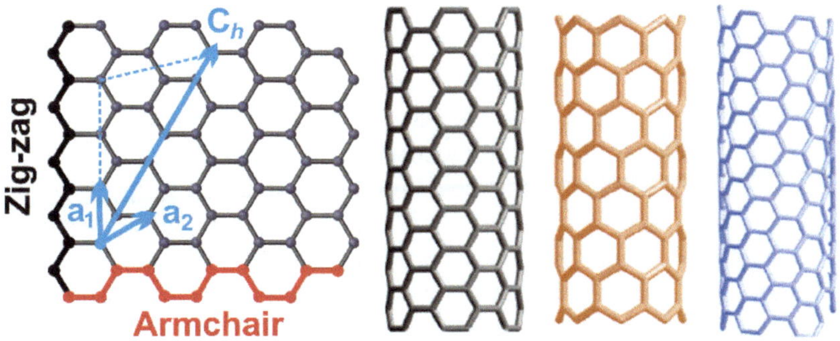

Figure 1.6 Schematic presentation of chirality in hexagonal carbon structures, such as graphene and carbon nanotubes (left). Three structures determining the CNT's unit periodicity and conductivity (right), namely (from left to right) armchair (red), zigzag (black) and chiral (blue) are shown.

Graphene does not conduct electric current, except for certain directions. This phenomenon is called anisotropy—it exhibits different electronic properties in different directions, namely direction parallel and perpendicular to C–C bond in its structure.

> **Anisotropy** (opposed to isotropy) is a property of matter, which is directionally dependent, i.e., different properties are observed in different directions. For example, wood is easy to chop along its grain, not perpendicular to it.

CNTs can be fabricated by vaporizing carbon electrodes in arc-discharge, although more controlled growth can be achieved using CVD method, since other methods produce randomly oriented and arranged tubes on a substrate, which are then difficult to manipulate with. For CVD, hydrocarbon gas is used as a precursor, which decomposes into hydrogen and carbon by the action of a catalyst. The main parameters are thus type of a precursor, type of a catalyst and temperature. For the fabrication of high-quality SWCNTs, methane gas (most stable hydrocarbon at high temperatures, what prevents amorphous carbon formation) and temperatures up to 1000°C are used. For MWCNTs, ethylene or acetylene gas can be used, as well as iron (or other metal) catalyst and temperatures up to 700°C. Good orientation during the CNTs growth can be achieved by CVD on the silicon substrates (silicon pillars) covered by catalyst material. As the nanotube grows, the gas stream keeps them floating. This is still a bottom-up approach. A typical top-down approach is common for a graphene production. It is a simple mechanical cleavage method, peeling of commercially available highly oriented pyrolytic graphite sheets. Another method is chemical oxidation of graphite to graphite oxide and subsequent exfoliation. Chemical treatment is, however, quite invasive, and leads to formation of numerous defects. At the same time, this treatment forms numerous oxygen functionalities on the surface (hydroxyls, epoxides, carbonyls and carboxyls), making the material more soluble in water—an important feature for many biological applications, and making it easier for covalent immobilization for a biorecognition elements. This feature is valid for other carbon nanomaterials, as well [23, 24]. The exfoliation can be assisted by intercalation of small molecules, and most commonly, graphene flakes are produced by sonication. Graphene oxide produced this way can be reduced back to graphene (so-called reduced graphene oxide, rGO) chemically electrochemically or thermally, however, the defects formed previously will be still present. Large-area and high-quality graphene can be produced by thermal CVD on catalytic transition metal surfaces. The formation of a monolayer as well as few-layer graphene is possible using iron, copper, nickel, cobalt, ruthenium, rhenium, rhodium, but also gold or platinum.

Graphene offers many opportunities for researchers and investors, as well. According to Zion™ Market Research, global graphene market will reach almost 194 million USD by 2022. The US National Nanotechnology Initiative (NNI) had a cumulative investment (from 2001 to 2015) of almost 21 billion USD. Using Scopus Document search and "graphene" as a keyword, we'll see that there are more than 80,000 papers published already, with more than 78,000 published since 2010. The nanotechnology and mainly applications of different nanomaterials described above (not just graphene) is still a prominent area for basic and applied research and industry, as well.

1.3.4 Preparation of Novel 2D Nanomaterials—MXene Nanosheets

Even though graphene is no doubt the most popular 2D nanomaterial used so far, there are others such as graphyne, borophene, germanene, silicene, stanene, phosphorene, hexagonal boron nitride (BN, van der Waals bonded structure), metal chalcogenides (MoS_2, WS_2), etc.

Very recently, in 2011, Yury Gogotsi and Michael W. Barsoum with their team published a study about the preparation of 2D nanocrystals from so-called MAX phases (60+ group of layered ternary carbides and nitrides), what opened a door to a family of novel nanosheets. These layered structures were much more compact compared to graphite, where the individual layers stick together by weak van der Waals forces, they exhibited metallic conductivity and were hydrophilic because of the negative charge present on their surface. These so-called MXene nanosheets, synthesized from the MAX phases (where "M" is a transition metal, "A" is an element mostly from group 13 or 14 of the periodic table of elements, "X" is a carbon or nitrogen atom) by hydrofluoric acid (HF) etching, showed also antimicrobial activity in their colloidal form and offered photocatalytic decomposition of some small organic molecules. The name "MXene" denotes the loss of the "A" element from the MAX phase and also structural similarity with graphene. One of the widely used member of this family is $Ti_3C_2T_X$ Mxene, described below. There are other, however, distinguished by different atoms in their structure (Mo or Cr instead of Ti), with three different formulas, namely M_2C, M_3C_2 and M_4C_3 (parental MAX phase has a general formula of $M_{n+1}AC_n$).

Most important for scientific community are, however, its electrical and electrochemical properties. Since it possesses a layered structure (see Fig. 1.7), intercalation of different ions (Li^+, Pb^{2+}, Al^{3+}, but hydrazine N_2H_4 and dimethyl sulfoxide DMSO as well) is possible, as well as adsorption of different gases (NH_3, methanol, ethanol, acetone). The intercalation makes this material extremely useful in the batteries (e.g., Li-ion batteries, where MXene is serving as a capacitor) or gas sensors. Its catalytic activity towards hydrogen peroxide (H_2O_2) has been described, making it a promising platform for different sensing applications [25].

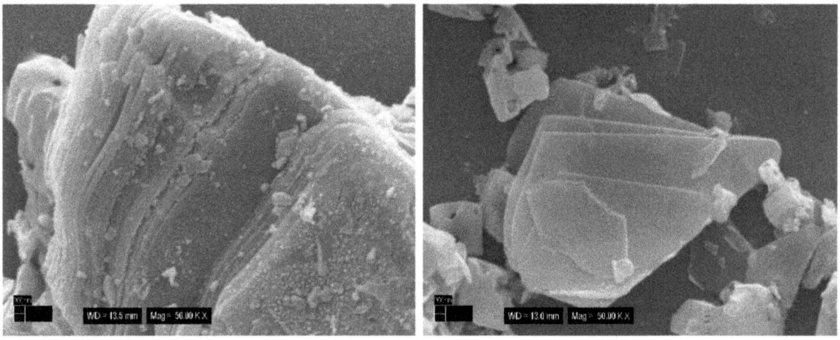

Figure 1.7 SEM images of $Ti_3C_2T_X$ MXene (50,000× magnification) before (left) and after delamination and sonication (right). Layered structure is at first visible, serving as a matrix for the intercalation of different ions, e.g., Li^+—an important feature in batteries research.

Synthesis of the $Ti_3C_2T_X$ Mxene structure (X– = O, –OH, –F) on the picture above is described using following equations:

1. $Ti_3AlC_2 + 3HF = AlF_3 + 3/2H_2 + Ti_3C_2$

2. $Ti_3C_2 + 2H_2O = Ti_3C_2(OH)_2 + H_2$
3. $Ti_3C_2 + 2HF = Ti_3C_2F_2 + H_2$

The first equation is essentially the formation of the MXene nanosheets, followed by the other reactions (Eqs. 2. and 3.). The electrochemical use of MXene (electrochemical sensors etc.) is, however, limited to lower potential windows for now (e.g., for cathodic/reduction processes), since it was shown to undergo a loss of its structure due to TiO_2 nanoparticles formation and subsequent dissolution at the potential value of +430 mV (vs. Ag/AgCl/3M KCl reference electrode) according to the following equations:

1. $Ti_3C_2F_X + 8H_2O = TiO_2 + 2CO + 16H^+ + (16-x)e^- + xF^-$
2. $TiO_2 + 4H^+ + 6F^- = [TiF_6]^{2-} + 2H_2O$

Sensor applications involving oxidation of an analyte molecule is therefore restricted to low anodic potential values. Again, very recently, The structure of MXene was pushed forward to a new application, e.g., Na^+-ion storage with high capacity, by preparation of 3D architectures—hollow spheres using poly(methyl methacrylate)-assisted method.

1.3.5 Doping

To tune a material's properties, not only synthesis, but subsequent modification of a material plays a big role. A common procedure to enhance semiconductor's electronic or optical properties is called doping. Here, a dopant (doping agent—small traces of impurity) is presented to a semiconductor in low concentration levels, commonly taking place of an element in the semiconductor's crystal lattice. Doping shifts the Fermi levels in a material (Fermi level or electrochemical potential for an electron is a thermodynamic work necessary to add one electron to a body; it reflects the energy of the topmost filled level in the ground state of a system containing electrons at absolute zero). In an undoped (intrinsic, or I-type) semiconductor, there is an equal portion of electrons and their corresponding holes (e.g., the ratio of negative "N" and positive "P" charge carriers is 1:1). Once a semiconductor has been doped, it is called extrinsic semiconductor. There are two different types: N-type and P-type semiconductor, based on a type of an element used during the doping procedure. If we used a donor of electrons (i.e., we donated electrons to the system, forming electron energy levels slightly below the bottom of the conduction band), we contributed to electron-type conductivity, thus we prepared the N-type semiconductor (where negative charge carriers are prevalent). P-type is prepared by adding an electron acceptor, the trace elements causing a hole-based conductivity in the material (Fig. 1.8).

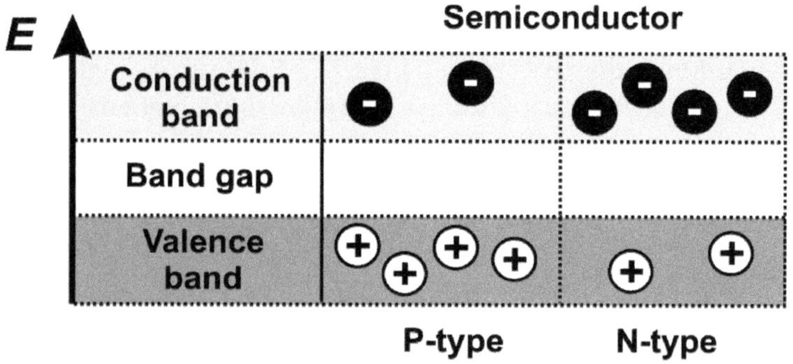

Figure 1.8 Schematic presentation of the valence and conduction bands in a semiconductor material (showing also the band gap between these two bands), with the excess of electrons (black circles with – sign) in case of N-type and with the excess of holes (white circles with + sign) in case of P-type semiconductors.

> **Band gap** generally refers to energy difference between the valence band and conduction band in insulators and semiconductors. It is an energy gap, where no electron states exist.

For example, boron-doped diamond electrodes have been used for the last few years extensively in electrochemical research and sensor preparation. This is due to their unique properties, such as:

- wide solvent window among other electrode materials
- low background and capacitive currents
- reduced fouling compared to other electrode materials
- ability to withstand extreme potentials, corrosive and high temperature/pressure environments

Here, boron atoms have been used as a dopant, to present semi-metallic characteristics to the resulting material, substituting for a carbon atom in the crystal structure and adding a hole in the valence band. In case of diamonds, the impurities also change the diamond color, e.g., boron doping results in blue color, whereas nitrogen doping in yellow.

1.4 Window into the Nanoworld Opened

Even though nanosize objects can't be observed directly with naked eye, or even basic optical (light) microscopes (the main reason will be described in the following section), it is possible to observe their formation/synthesis or size with other methods. The most common approach is to use absorption spectroscopy and Beer's law (Lambert-Beer's or Lambert–Beer–Bouguer's law), describing light intensity passing through a material (this is an indirect method). The exponential attenuation law has a general form:

$$I = I_0 e^{-\alpha l},$$

where I and I_0 is the intensity of light and incident light, respectively, and l is the thickness traversed by the light. The α parameter is called absorption coefficient. For molecular systems, the equation is slightly modified

$$I = I_0 e^{-\sigma[n]l},$$

where n is the concentration of absorbing elements (absorbers) and σ is the absorption cross section. For these molecular systems (for instance QDs can also be considered to be this kind of sample), more often, the following expression is used:

$$A = \varepsilon_{\text{molar}}[c]l$$

In this case, A is the absorbance (optical density, unitless), c is the molar concentration, l is the path length/thickness and $\varepsilon_{\text{molar}}$ is molar extinction coefficient, describing how efficiently a material absorbs light. This is basically a material-dependent parameter.

To estimate the nanostructure dimensions, some other techniques are commonly used. One of them, namely X-ray powder diffraction (XRD) uses scattered electromagnetic radiation. Here, the nanostructures are not visualized directly (just like in previous case), but instead, crystallographic phase and structure of a sample is determined using so-called Bragg's law:

$$n\lambda = 2d\sin\theta,$$

where n is the "order" of reflection (1, 2, 3, ... n). This concept is based on an idea, that atomic planes in crystal lattice diffract incoming X-rays (Fig. 1.9). The above mentioned equation can be thus correlated to Miller indices (hkl), e.g., for cubic lattice

$$d = \frac{a}{\sqrt{h^2 + k^2 + l^2}}$$

with h, k and l being Miller indices mentioned in Section 1.2.1 and a is the lattice constant.

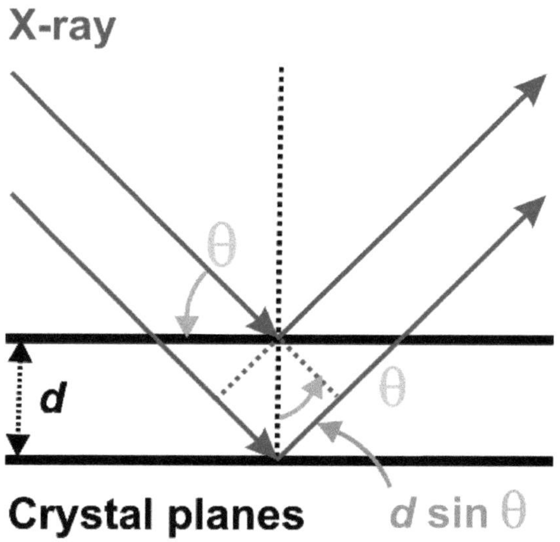

Figure 1.9 Schematic presentation of the incident and diffracted X-rays on the atomic lattices of crystal nanostructure and individual parameters from Bragg's law.

Before moving on to the section about the "real window" to the nanoworld (i.e., advanced microscopy techniques), final scattering technique will be described in this chapter, namely dynamic light scattering (DLS; photon correlation spectroscopy or quasi-elastic light scattering). This is a method used routinely to estimate the size of nanoparticles in their colloidal suspension. In liquid, nanoparticles (or individual molecules) are not static. They move continuously (because of so-called Brownian motion) and as a consequence, scattered light intensity (of an incident light) is changing in time. Via determination of the diffusion coefficient of suspended nanoparticles, their physical dimensions are estimated (their diameter including solvation layers) according to Stokes–Einstein equation:

$$D = \frac{k_B T}{6\pi\eta a},$$

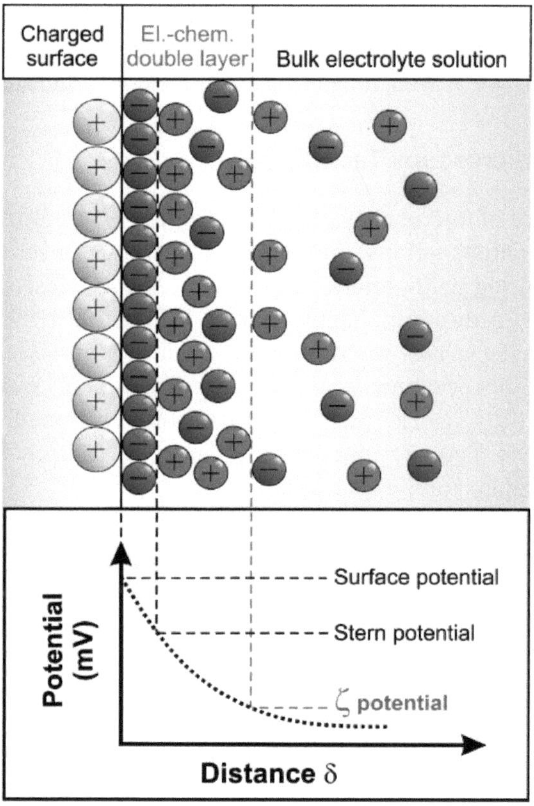

Figure 1.10 Schematic presentation of zeta potential and its relation to surface charge and electrochemical double layer.

where D is the particle's diffusion coefficient, k_B is the Boltzmann constant, T is thermodynamic temperature, η is the solvent viscosity and a is the particle's hydrodynamic radius. To estimate the surface charge and evaluate the stability, ζ (zeta) potential measurements (Fig. 1.10) are popular as a table-top technique. X-ray irradiation of a sample may also be used to determine a material's composition, e.g., in X-ray photoelectron spectroscopy (XPS) or Auger spectroscopy (observing the energy of emitted Auger electrons from a sample). Here, the energy of emitted electrons can be correlated to the identity of emitting atoms.

Even though some of these methods are relatively easy and offer many useful information (also by their combination), nothing beats the possibility to observe the nanostructures with one's own eyes. Even though microscopy is not an answer to everything, this is in fact an area opening a whole new world beyond our own senses to us. Moreover, controlling not simply particle's size, but shape as well, may be crucial for many applications, too.

1.4.1 Microscopy Techniques

The history of microscopy goes back to the 16th century, when two Dutch scientists—Hans Janssen and his son Zacharias—started experimenting with lenses to design a first telescope. Another Dutchman, Anton van Leeuwenhoek (1632–1723), was the first man to make a real microscope. He made a series of important biological observations, he was the first man who observed microbes—his work was later further developed by Robert Hooke, who actually published a first microscopy-based book (Micrographia, 1665; Fig. 1.11).

For objects too small, however, it may be a challenging task to make an observation using common microscopes. Even some viruses (which are still not believe to be truly alive, since they don't have their own metabolism—basically they're just a bunch of biological molecules able to disrupt their host's own metabolism) are in the range of few hundreds of nm (e.g., *Influenza* viruses, family *Orthomyxoviridae*). This is the resolving limit of a light microscope (about a half of the visible light wavelength, 400–700 nm). This prominent physical problem, known as diffraction limit (Abbe's resolution limit or Abbe's principle) basically states that the smallest distance between two lines that can be resolved is proportional to the wavelength and inversely proportional to the angular distribution of the light observed. Mathematically:

$$d_{\min} = \frac{\lambda}{n.\sin\alpha}$$

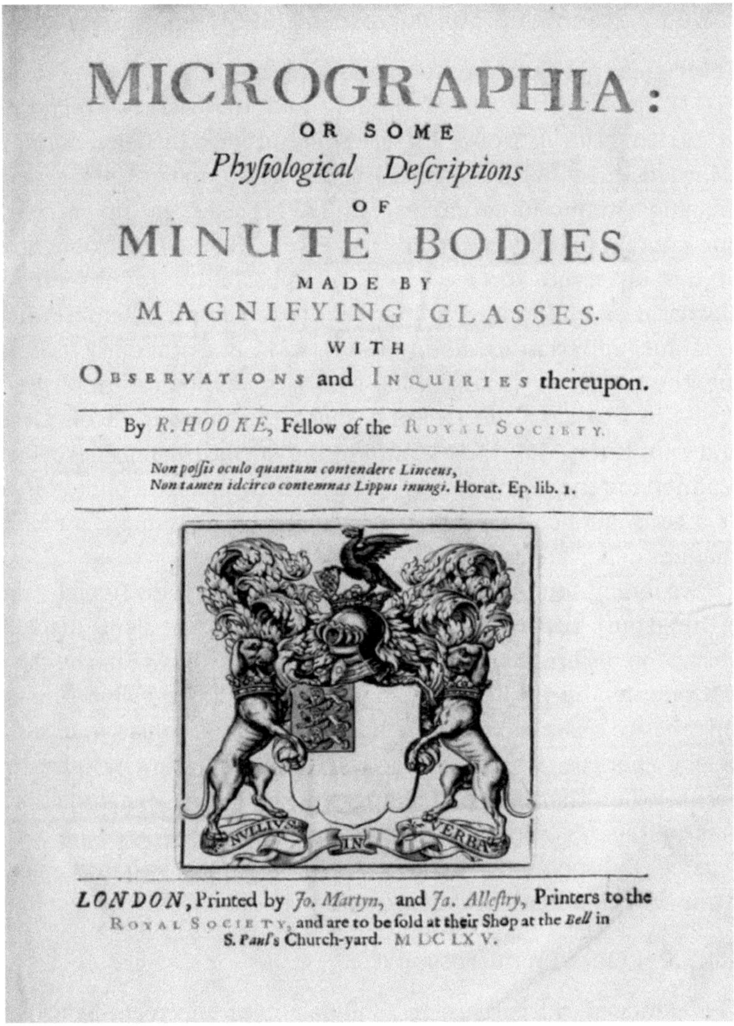

MICROGRAPHIA:
OR SOME
Physiological Descriptions
OF
MINUTE BODIES
MADE BY
MAGNIFYING GLASSES.
WITH
OBSERVATIONS and INQUIRIES thereupon.

By *R. HOOKE*, Fellow of the ROYAL SOCIETY.

Non possis oculo quantum contendere Linceus,
Non tamen idcirco contemnas Lippus inungi. Horat. Ep. lib. 1.

NVLLIUS IN VERBA

LONDON, Printed by *Jo. Martyn*, and *Ja. Allestry*, Printers to the ROYAL SOCIETY, and are to be sold at their Shop at the *Bell* in *S. Paul's* Church-yard. M DC LX V.

Figure 1.11 Title page of Robert Hooke's Micrographia or Some Physiological Descriptions of Minute Bodies Made by Magnifying Glasses. With Observations and Inquiries Thereupon, published in 1665.

To circumvent this problem, other types of microscopy were developed. In nanoscience and nanotechnology, two types are commonly used, namely microscopy using different probes and microscopy using beam of electrons for the visualization of small objects. In each case, the resolution limit is in the range of up to 1 nm.

1.4.1.1 Scanning probe microscopy (SPM)—nano-anatomy

An original probe microscopy was scanning tunneling microscopy (STM), developed by Gerd Binning and Heinrich Rohrer at IBM in Zurich (1981). Nobel Prize in physics for this invention was awarded in 1986. Probe microscopes use a sharp tip for sample imaging. Atomic force microscopy (AFM) uses the tip in contact, non-contact or tapping mode to create a 3D picture (topography) of the observed surface (with the possibility to observe also electrical or mechanical properties of the sample, such as Young's modulus, adhesion or deformation). AFM is a standard technique for the surface visualization, even for biological applications. Probe in this case consists of a sharp tip (silicon/silicon nitride) and cantilever. The tip is scanning the surface using piezoelectric scanner and the cantilever reflects a laser beam into a photodetector as a consequence of a deflection caused by tip-sample forces and changes of surface height.

Scanning tunneling microscope uses a sharp tip (usually made of tungsten) but totally different physical mechanism. Instead of observing different attractive/repulsive forces between the tip and the sample and tip height changes, changes in so-called tunneling current are observed. This method is extremely sensitive to sample height changes, since the electron tunneling (flow of electrons) occurs when the tip is close enough to a conductive sample (when bias voltage is applied), and revolutionized the area of surface science. Subsequently, STM led to some other variants, such as scanning electrochemical microscopy.

1.4.1.2 Electron microscopy

This kind of microscopy techniques uses electron beam as a source of "illumination." The fact electrons can behave like a wave and have smaller wavelength compared to visible light (several tens of thousands times, depending on electron's kinetic energy), much smaller objects can be visualized compared to light microscopy (typically smaller than 1 nm).

Controlling the size (and shape) of prepared and applied nanostructures is a crucial task for every technology. A critical feature for electron microscopy is high degree of spatial resolution. According to Abbe's principle mentioned above, classical diffraction limit of electromagnetic waves is about a half of their wavelength. That's the reason a beam of electrons provides much finer resolution. To provide a mathematical explanation for this phenomenon, for example in transmission electron microscopy (TEM), electrons are accelerated by a voltage (and extracted from a source), so their kinetic energy is

$$E = eV = \frac{1}{2} m_0 v^2$$

and so the we can find an expression relating electron's momentum p ($p = mv$) and microscope's accelerating voltage V:

$$p = \sqrt{2 m_0 e V}$$

Using de Broglie's relationship between wavelength and momentum, we will obtain following equation:

$$\lambda = \frac{h}{\sqrt{2 m_0 e V}} \quad \text{or} \quad \lambda = \frac{1.225}{\sqrt{V}}$$

Meaning that using 100 keV accelerating voltage, the electron's wavelength is $\lambda \approx 3.9$ pm (10^{-12} m). TEM is thus a method analogous to the classical light microscopy, but instead of photons, electron beam is used for thin or diluted samples under high vacuum. A related form of electron microscopy is scanning electron microscopy (SEM), where the electron beam (usually tens of nanometers in diameter or less) is focused and scanned over the sample. Several different events occur here, for instance inelastic electron scattering off the specimen. This can also lead to secondary electrons ejection (from the sample). Sometimes, elemental analyses are used in combination with SEM, namely energy-dispersive X-ray spectroscopy (EDX). Generally speaking, the electrons detected in these microscopies are either transmission (TEM), secondary (SEM) or elastically scattered reflected electrons (REM). For SEM, the sample doesn't require an extensive sample preparation, but usually only dried, conductive samples are visualized. For biological samples (wet and uncoated), environmental SEM (ESEM) is often applied.

In chemistry and particle physics, **inelastic scattering** (as opposed to **elastic scattering**) is a scattering process where the kinetic energy of incident particles is not conserved (some of the energy is either lost or even increased)

1.5 When "Nano" Meets "Bio"

Currently, two different areas of nanotech-oriented research are being extensively developed, namely bionanotechnology and nanobiotechnology. This may be a little confusing, so we need to define both these areas. To put it simply, bionanotechnology is the inspiration by biological systems (e.g., small molecular motors such as F1-ATPase inspired nanotechnologists to create nanomotors), nanobiotechnology is the application of different nanomaterials in biotechnology research (e.g., in drug delivery or diagnostics). These nanomaterials revolutionized the field of biotechnology as well as gene engineering and recombinant technologies several decades ago.

Biotechnology is a fascinating area, represented by different "colors." Since the impact is so broad, it was divided to industrial (white), medical and pharmaceutical (red), environmental (green) or food (yellow) biotechnology.

Biotechnology is the use of living systems (organisms) and their parts (organelles or even individual biomolecules) to develop new technologies or make products that help us improve the quality of our lives.

There are other colors, as well, for instance black (bioterrorism) or blue (marine biotechnology). For nanobiotechnology, bioinformatics and computer related fields, a term "gold biotechnology" was established. Even though some of us are okay with this term (it may look like nanobiotechnology is, at least now, the greatest and purest of all the other fields), it is important to note that very often, there are more than just a single color relevant for a specific problem. Later in this book, you'll read about the application of different nanomaterials and protocols for nanoscale modifications for biosensor research in medical diagnostics, what makes this a red/gold biotechnology. Current problematics of nanobiotechnology involves nanostructured biosensors (what is perhaps the most common example of a biosensor is what we call a pregnancy test for human choriogonadotropin hCG detection, where the red lines are basically gold nanoparticles conjugated to a specific antibody), nanoparticles in therapy (often cancer), nanoengineering of DNA and mechanical nanodevices (for transport of different cargos). The area is exceptionally broad and is still developing [26]. The best source of information on this topic are currently scientific papers published in different journals (based on the application) published periodically on the internet [27].

In the next chapters, the readers will be introduced to different surface modification strategies, bioanalytical methods and glycomics, and subsequently the application of all these fields into diagnostics, enrichment and therapy will be introduced.

References

1. Ketchie, W. C., et al., Influence of Gold Particle Size on the Aqueous-Phase Oxidation of Carbon Monoxide and Glycerol. *Journal of Catalysis*, 2007. **250**(1): pp. 94–101.
2. Edwards, P. P., and J. M. Thomas, Gold in a Metallic Divided State—From Faraday to Present-Day Nanoscience. *Angewandte Chemie International Edition*, 2007. **46**(29): pp. 5480–5486.
3. Sandhu, A., Who Invented Nano? *Nature Nanotechnology*, 2006. **1**: p. 87.
4. Chan, W. W. C., et al., Nanoscience and Nanotechnology Impacting Diverse Fields of Science, Engineering, and Medicine. *ACS Nano*, 2016. **10**(12): pp. 10615–10617.
5. Novoselov, K. S., et al., A Roadmap for Graphene. *Nature*, 2012. **490**: p. 192.
6. Novoselov, K. S., et al., Electric Field Effect in Atomically Thin Carbon Films. *Science*, 2004. **306**(5696): pp. 666–669.

7. Freestone, I., et al., The Lycurgus Cup—A Roman Nanotechnology. *Gold Bulletin*, 2007. **40**(4): pp. 270–277.
8. Kreuter, J., Nanoparticles—a Historical Perspective. *International Journal of Pharmaceutics*, 2007. **331**(1): pp. 1–10.
9. Farokhzad, O. C. and R. Langer, Impact of Nanotechnology on Drug Delivery. ACS Nano, 2009. **3**(1): pp. 16–20.
10. Naguib, M., et al., Two-Dimensional Nanocrystals Produced by Exfoliation of Ti_3AlC_2. *Advanced Materials*, 2011. **23**(37): pp. 4248–4253.
11. Editorial, Plenty of Room, Revisited. *Nature Nanotechnology*, 2009. **4**: p. 781.
12. https://www.nobelprize.org/educational/physics/matter/1.html.
13. Kuno, M., *Introductory Nanoscience, Physical and Chemical Concepts*. 2011, UK: Garland Science. 420.
14. Victor E., and Borisenko, S. O., *What is What in the Nanoworld: A Handbook on Nanoscience and Nanotechnology, 3rd Edition*. 2012, Weinheim, Germany: Wiley-VCH Verlag & Co. 615.
15. https://www.nobelprize.org/nobel_prizes/chemistry/laureates/2011/press.html.
16. Love, J. C., et al., Self-Assembled Monolayers of Thiolates on Metals as a Form of Nanotechnology. *Chemical Reviews*, 2005. **105**(4): pp. 1103–1170.
17. Daniel, M.-C., and D. Astruc, Gold Nanoparticles: Assembly, Supramolecular Chemistry, Quantum-Size-Related Properties, and Applications toward Biology, Catalysis, and Nanotechnology. *Chemical Reviews*, 2004. **104**(1): pp. 293–346.
18. H. R, C., J. D. Schiffman, and R. G. Balakrishna, Quantum Dots as Fluorescent Probes: Synthesis, Surface Chemistry, Energy Transfer Mechanisms, and Applications. *Sensors and Actuators B: Chemical*, 2018. **258**: pp. 1191–1214.
19. Petryayeva, E., W. R. Algar, and I. L. Medintz, Quantum Dots in Bioanalysis: A Review of Applications Across Various Platforms for Fluorescence Spectroscopy and Imaging. *Applied Spectroscopy*, 2013. **67**(3): pp. 215–252.
20. Popov, A. A., S. Yang, and L. Dunsch, Endohedral Fullerenes. *Chemical Reviews*, 2013. **113**(8): pp. 5989–6113.
21. Li, Z., et al., Superstructured Assembly of Nanocarbons: Fullerenes, Nanotubes, and Graphene. *Chemical Reviews*, 2015. **115**(15): pp. 7046–7117.

22. Tasis, D., et al., Chemistry of Carbon Nanotubes. *Chemical Reviews*, 2006. **106**(3): pp. 1105–1136.
23. Karousis, N., N. Tagmatarchis, and D. Tasis, Current Progress on the Chemical Modification of Carbon Nanotubes. *Chemical Reviews*, 2010. **110**(9): pp. 5366–5397.
24. Saito, N., et al., Safe Clinical Use of Carbon Nanotubes as Innovative Biomaterials. *Chemical Reviews*, 2014. **114**(11): pp. 6040–6079.
25. Lorencova, L., et al., Electrochemical Performance of $Ti_3C_2T_x$ MXene in Aqueous Media: Towards Ultrasensitive H_2O_2 Sensing. *Electrochimica Acta*, 2017. **235**: pp. 471–479.
26. Linko, V., A. Ora, and M. A. Kostiainen, DNA Nanostructures as Smart Drug-Delivery Vehicles and Molecular Devices. *Trends in Biotechnology*, 2015. **33**(10): pp. 586–594.
27. Villaverde, A., Nanotechnology, Bionanotechnology and Microbial Cell Factories. *Microbial Cell Factories*, 2010. **9**: pp. 53–53.

Chapter 2

Introduction to Glycomics and Glycan Analysis

Ludmila Belicka

Department of Glycobiotechnology, Institute of Chemistry, Slovak Academy of Sciences, Dubravska cesta 9, Bratislava, 845 38, Slovak Republic

chemluva@savba.sk

Among nucleic acids, proteins and lipids, saccharides are one of the fundamental macromolecules creating all living forms. In biological systems, they are present as free monosaccharides up to extended polysaccharides and intricate saccharidic structures having various "sweetness" and different functions. Moreover, saccharides are the most common (abundant) biomolecules on the earth. This chapter is focused on complex saccharides—glycans, their synthesis, importance, and physiological roles in the mammalian body. Glycans are the key players in human pathophysiology and glycan structure monitoring is necessary for better understanding

Glyconanotechnology: Nanoscale Approach for Novel Glycan Analysis and Their Medical Use
Edited by Tomáš Bertók

ISBN 978-981-4800-16-7 (Hardcover), 978-0-429-44849-2 (eBook)
www.jennystanford.com

of origin and progress of many diseases. Therefore, the major part deals with basic strategies and principles of glycan profiling with the emphasis on mass spectrometry as a gold standard for glycan analysis.

2.1 Glycomics

The term "glycomics" is derived from conventional proteomics or genomics and involves the comprehensive study of all glycan structures of a whole organism. Analogically, the glycome represents the entire glycan repertoire of an organism, cell, tissue or protein.

Glycans are complex saccharides consisting of a number of monosaccharides linked via glycosidic bonds.

Nowadays, glycomics is a rapidly emerging science; however, in comparison to other "omics" sciences, it remains relatively understudied. This is a very disturbing fact, especially when it is well known that glycans are involved in many biological and vital processes. The hard truth is that glycan analysis is much more difficult than it seems. Glycan chains are composed of monosaccharide units differing in ring size and structure, namely nonuloses, hexoses, N-acetyl-hexosamines, deoxyhexoses, hexuronic acids and pentoses (Fig. 2.1). These are linked via different linkage positions forming linear or branched structures.

Besides this, the existence of anomeric forms (α,β) and glycosidic linkages (1,1; 1,2; 1,3; 1,4; 1,6; 2,3; 2,6; 2,8 and 2,9) results in enormous number of possible glycan structures, which are either free oligo- and polysaccharides (hyaluronic acid, milk oligosaccharides) or attached to proteins (N-, O-, C-glycans, glycosaminoglycans, proteoglycans) and lipids (glycolipids, glycosylphosphatidylinositol-anchored proteins). Moreover, the glycan heterogeneity is increased by the additional modification such as phosphorylation, methylation, sulfation or acetylation. It is obvious that the huge diversity of glycans complicates the glycan analysis. Thus, sophisticated analytical approaches and instrumentations are needed.

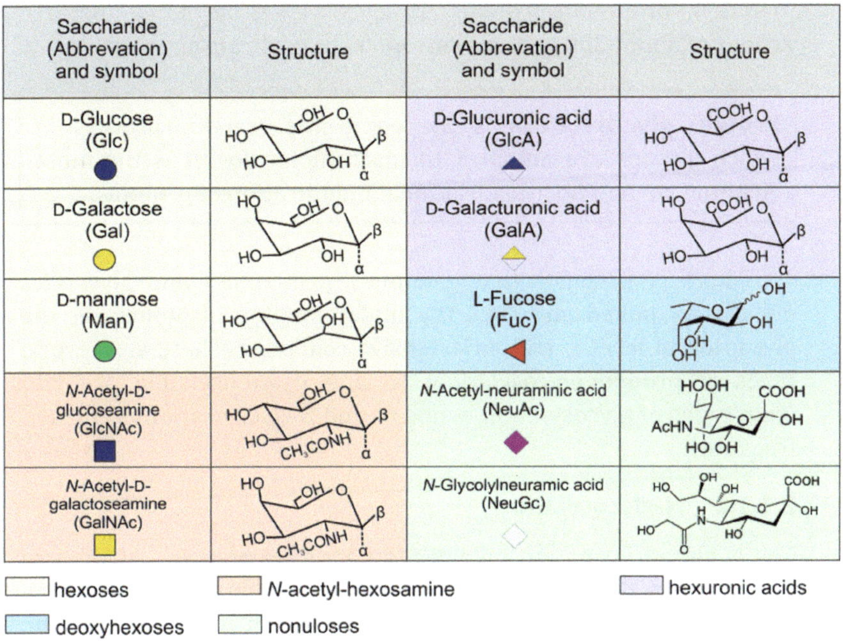

Figure 2.1 Structures of the most common monosaccharide-building units of glycans.

Herein, we target the most common protein glycosylation in eukaryotes (N- and O-glycosylation), the physiological and pathophysiological aspect of glycans and last but not least the glycan/glycoprotein profiling with its potential to use in early stage diagnostic.

2.1.1 Glycosylation

It has been estimated there are between 20,000 to 25,000 proteins encoded in the human genome and the entire proteome contains more than a million proteins. This contradiction is a result of many co- and post-translational modifications of proteins which produce additional classes, of proteins with a huge structural and functional diversity and function. There are more than 300 known protein modifications and glycosylation is one of the most abundant and common posttranslational co- and posttranslational protein modifications. Approximately more than 50% of human proteins are glycosylated by the competitive actions of about 200 genome-encoded glycosyltransferases.

Protein glycosylation is the enzymatic reaction/process in which glycans are attached to macromolecules. It is the most common co- and post-translational modification of proteins.

There is evidence that genomics, proteomics and glycomics are closely linked and thus the understanding of biology at the glycoprotein level is still an extensive challenge. There are several types of protein glycosylation; N- and O-glycosylation are two main types of glycosylation, while C- and S-glycosylation occur less frequently.

2.1.1.1 N-glycosylation

The N-glycan formation involves the addition of glycan structures to side-chain nitrogen ("N") atoms of asparagine. The amino-acid consensus sequence for the N-glycosylation is Asn-X-Ser/Thr, where X is any amino acid except proline. In eukaryotes, N-glycosylation takes place during protein translation and regulates protein folding. The whole process starts in the endoplasmic reticulum (ER) by the synthesis of oligosaccharide precursor (consists of 14 sugar moieties: $Glc_3Man_9GlcNAc_2$) on dolichol-phosphate (Dol-P). This precursor is subsequently transferred to asparagine of the nascent polypeptide chain (core protein) catalyzed by the oligosaccharyltransferase complex. Three glucose residues (involved in quality control) are subsequently removed from the N-linked precursor of the correctly folded protein and after this, the attached N-glycan is modified by trimming and re-building in the Golgi, which results in the formation of high-mannose, hybrid and complex di-, tri- or tetraantennary glycans (Fig. 2.2).

Mature glycans are further modified by elongation, e.g. addition of N-acetylglucosamine (GlcNAc), galactose (Gal), sialic acid (SA) or fucose (F). The addition of β1,4-bound N-acetylglucosamine (GlcNAc) to the beta-linked mannose of the trimannosyl is a specific modification which leads to the formation of "bisecting" N-glycans. The presence or absence of nucleotide sugar donors or glycan modifying enzymes determines N-glycan composition.

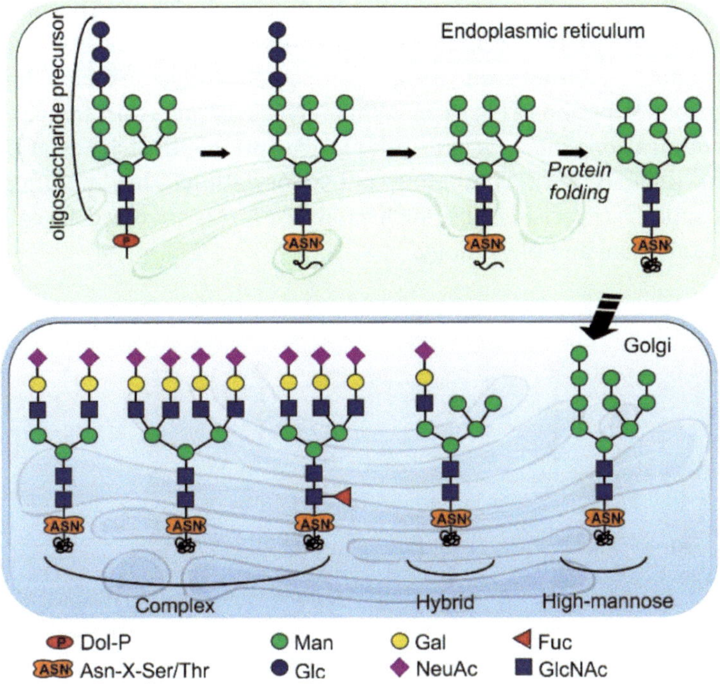

Figure 2.2 Schematic presentation of N-glycan formation in Endoplasmic reticulum and Golgi leading to complex, hybrid and high-mannose N-glycan structures.

2.1.1.2 O-glycosylation

In contrast to N-glycosylation, O-glycosylation takes place in the Golgi (O-mannosylateion is, however, initiated in the endoplasmic reticulum), where monosaccharides are transferred to the sidechain of serine (Ser), threonine (Thr), tyrosine (Tyr), or hydroxylysine residue and linked by O-glycosidic bond. O-glycans do not form such branched structures like N-glycans and also contain fewer mannose residues. The most common O-glycosylation is initiated by the addition of N-acetylgalactosamine (GalNAc) residue to Ser or Thr creating the Tn antigen. The further elongation of Tn antigen leads to the creation of mucin-type O-glycans with a variety of core structures (core 1–4) (Fig. 2.3).

As the O-glycan classification is according to the peptide-proximal glycan structure, besides mucin-type O-glycans, O-fucose, O-xylose, O-mannose, O-N-acetylglucosamine O-galactose and O-glucose have been identified as well. These O-type core structures are then elaborated to high-ordered structures which are present on either cell surface or secreted glycoproteins having various functions (Fig. 2.3). While glycosylation is an irreversible modification, there is one unique exception—O-GlcNAcylation in the nucleus with competing phosphorylation. This modification is rapidly reversible and such modification is involved in cellular metabolism and physiology.

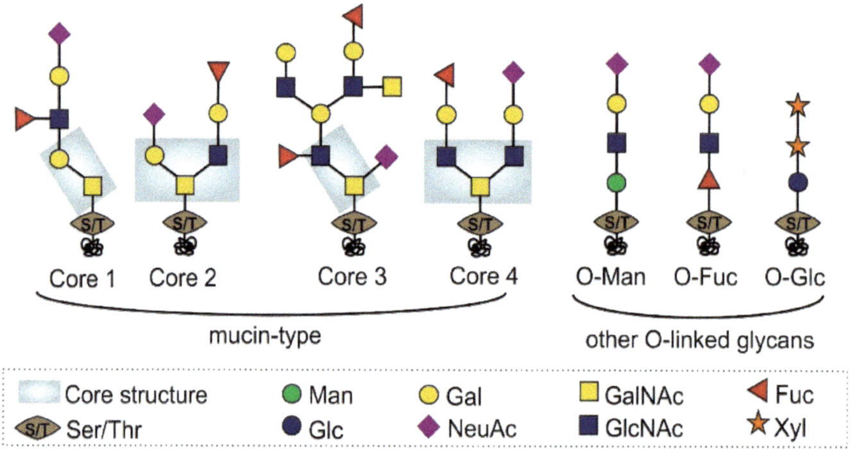

Figure 2.3 Basic structures of mucin-type glycans (core 1–4) and O-fucose, O-mannose, O-galactose in an extended form.

2.1.2 Glycans in Living Systems

Each cell, from bacteria to mammals, is protected by the "sweet glycan coating", which at the same time provides the first information about the cell to the surrounding environment. Because the biosynthesis of glycans depends on the expression of genome-encoded enzymes, the glycan structures on the cell surface reflect the cell type and condition, e.g. the type of organism, type of tissue or development stage.

2.1.2.1 Glycans in physiology

Mammalian cells are decorated by glycan patterns frequently capped with terminal sialic acid residues. Sialic acid is a negatively charged monosaccharide, mostly bound to galactose via α-2,3- and α-2,6-linkage or to other sialic acid through α-2,8- and α-2,9-binding. Even a small change in the configuration or linkage type can have dramatic consequences. For example, while the human influenza virus recognizes α-2,6 linked sialic acid, avian influenza virus recognizes sialic acid bound to galactose via the α-2,3 linkage. Owing to this small difference (α-2,6-versus α-2,3-linked sialic acid), avian influenza cannot infect humans and vice versa. Another example is blood group division. Generally, there are four basic blood groups, namely A, B, AB and 0. Karl Landsteiner was awarded the Nobel Prize in Medicine in 1930 for his discovery of human blood groups. He found that the blood of two people with different blood groups under contact agglutinates, while the blood transfusion between persons with the same blood group did not lead to the destruction of blood cells. This phenomenon is due to the presence of glycan antigens on the blood cell surface. Interestingly, the mentioned antigens differ just in one glycan residue (Fig. 2.4).

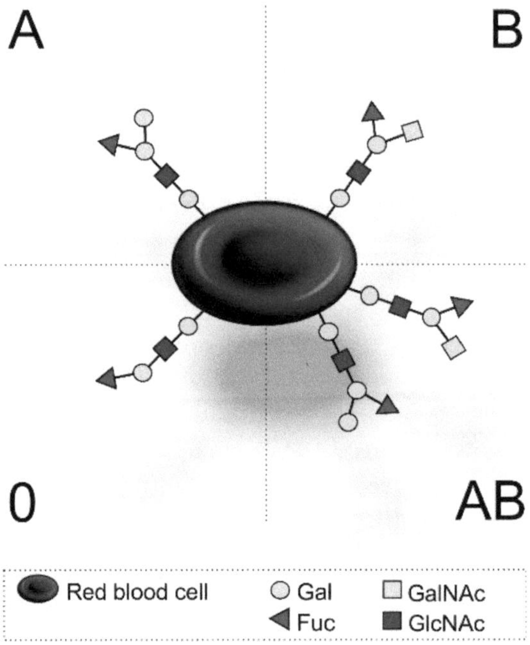

Figure 2.4 Illustration of blood group glycan-based antigen on a red blood cell.

Now we know that glycans play a crucial role in physiological processes. The attachment of glycans to proteins has a significant impact on the charge and volume of glycoproteins, which strongly affects the glycoprotein conformation stability and ligand-binding capability (essential for glycan-binding proteins). Moreover, glycans are involved in cell development, differentiation and apoptosis, cell signalling and adhesion, immune response, sperm-egg binding processes or embryo/cell development and proliferation.

2.1.2.2 Glycans in pathophysiology

As a coin has two sides, glycans are responsible not only for the organism's survival but also for many pathological conditions such as allergic reaction, host pathogen interaction (viral and bacterial invasion) or various diseases including cardiovascular and autoimmune diseases and cancer as well.

Cancer is a second leading cause of death worldwide with estimated mortality over than 8 million per year. The increased occurrence of cancer is a result of the growth and aging of the population, reproductive changes (including lower parity and later age at first birth) and increase of risk factor prevalence (stress, smoking, overweight, lack of physical activity, bad eating habits). The highest mortality rate is reported in patients with stomach, colorectal, liver, lung, pancreas, breast (prevalent in female), ovarian and prostate cancer. With the increasing death rates, it is necessary to know clinical markers-biomarkers.

Biomarker is defined as a characteristic or property that is objectively measured and evaluated as an indicator of normal biological processes, pathological processes or pharmacologic responses to a therapeutic intervention.

It has been demonstrated that almost all types of malignant cells and cells from diseased tissue have different glycan profiles on the surface in comparison to normal (healthy) cells. The aberrant glycosylation is induced by several factors:

- altered glycosyltransferase activity results in under- or overexpression of naturally occurred glycans (incomplete, truncated, branched and novel structures can be also occurred);

- changes in the tertiary conformation of the peptide backbone and that of the nascent glycan chain;
- the availability and abundance of sugar nucleotide donors and cofactors;
- the expression and localization of relevant glycosyltransferases in the Golgi apparatus.

The most widely occurring cancer-associated changes in glycosylation are sialylation (polysialylation, increase in global α-2,6- and α-2,3-linked sialic acid, changed linkage from α-2,6 to α-2,3-linked sialylation, sialylated SLe^a and SLe^x antigens) fucosylation (core and terminal fucosylation, expression of fucosylated SLe^a and SLe^x), occurrence of high mannose glycans, O-glycan truncation (expression of shortened T, Tn, STn antigens) and N-glycan branching (GlcNAc bisecting, polylactosamine-repeats of Galβ1,4GlcNAcβ1,3 capped with sialic acid and fucose; Fig. 2.5). Nowadays, there are at least 100 cancer biomarkers used and most of them are glycoproteins. Representative glycan-based tumour-related biomarkers are shown in Table 2.1.

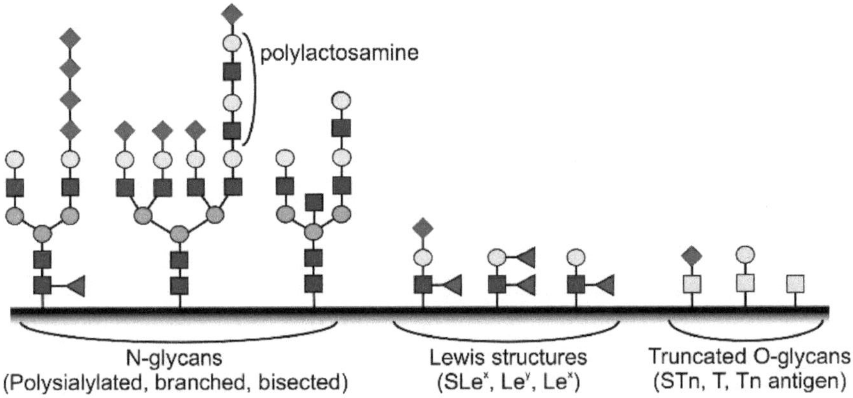

Figure 2.5 The most common cancer-associated changes in glycosylation pattern.

The changes in glycan identity have also a dramatic effect on the progression of various other diseases such as borreliosis, sleeping sickness, Chagas disease (chronic parasitic infection), chronic inflammation, HIV, influenza invasion or autoimmune diseases (e.g. rheumatoid arthritis, systemic sclerosis).

Table 2.1 Selected examples of FDA-approved glycoprotein cancer biomarkers

Biomarker (Abbreviation)	Cancer type	Detection type
Alpha-fetoprotein (AFP)	Liver	Protein concentrations and core fucosylation
B-human chorionic gonadotropin (β-hCG)	Choriocarcinoma and testicular cancer	Protein concentrations
Cancer antigen 125 (CA125/MUC16)	Ovarian	Protein concentrations
Carcinoembryonic antigen (CEA)	Coloteral, breast, lung, gastric	Protein concentrations
Chromogranin A (CgA)	Neuroendocrine tumors	Protein concentrations
Prostate specific antigen (PSA)	Prostate	Protein concentrations
Carbohydrate antigen 19-9 (CA19-9)	Pancreatic, ovarian	SLe^a on mucin glycoproteins
Human epididymis protein 4 (HE4)	Ovarian	Protein concentrations

2.2 Glycan Analysis Using Mass Spectrometry

As we discussed previously, it is well documented that the expression of aberrant glycans is associated with various pathological condition. The monitoring of glycan changes in real samples is still a big challenge due to the huge glycan diversity on the one hand together with enormous similarity on the other. Although glycosylation is prevalent in the human body, the amount of glycans is relatively low (in many cases, it is necessary to analyse trace amounts). Moreover, the complexity of the sample itself (tissue, cells, blood, serum, urine and other body fluids) makes the whole analysis very difficult. Therefore, a sophisticated, sensitive and reliable method is needed. Nowadays, mass spectrometry in tandem with different separation methods is the most popular way for glycan analysis, because it provides highly informative fragments and structural identification of glycans. In general, there are three approaches how to analyse glycan pattern (Fig. 2.6):

- structural analysis of released glycans (the aim of this chapter);
- determination and analysis of glycopeptides;
- glycan monitoring in intact glycoproteins.

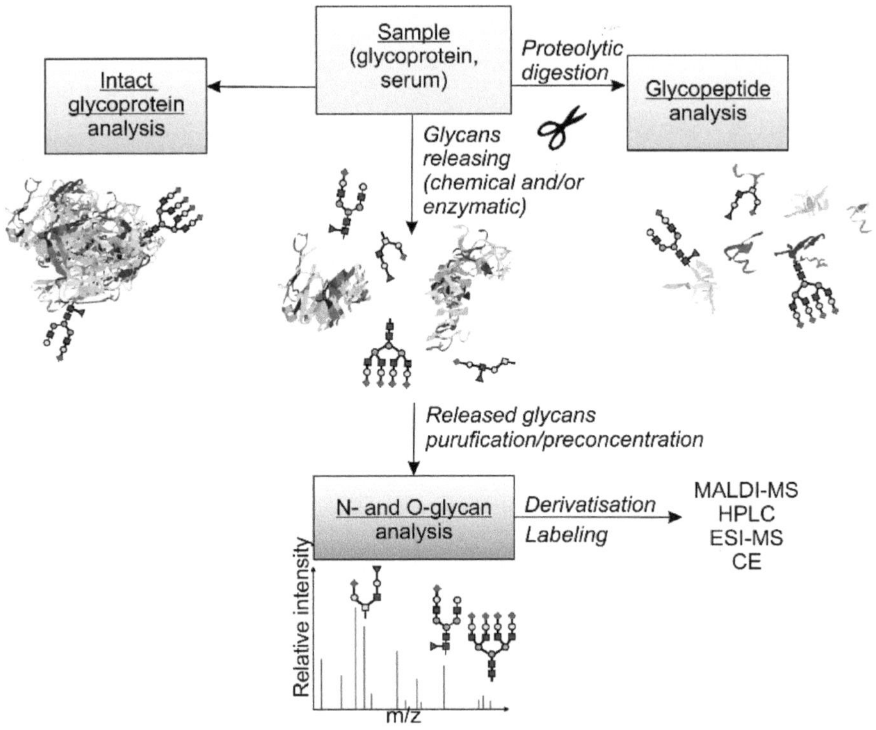

Figure 2.6 Different approaches to glycan analysis; the first involves analysis of whole protein, the second utilises enzymatic digestion (e.g. trypsin digestion) and the last one employes various techniques of glycan release (chemical/enzymatic) with subsequent glycan modification.

From another point of view, investigation of glycan expression can be either targeted (lectin based or antibody based, see Chapter 3) or non-targeted (MS based), which is reviewed in this section with the emphasis on the free-glycan analysis.

2.2.1 Structural Analysis of Released Glycans

Since glycosylation is a non-template driven process, with the potential for significant uncontrolled (not predicted) variability in the glycan distribution, the full glycan structure has to be known for better understanding of pathological states. It is more complicated to determine structural features of glycans while the sugar is still attached to the protein. Thus, for detailed information about the glycan structure at the monosaccharide level, the essential first step is "glycan release".

2.2.1.1 Sample preparation

Sample preparation is a prerequisite of successful experiment and a choice of proper method for glycan releasing depends on the type of glycan connected to protein. There are two ways for glycan isolation: chemical and enzymatic methods. N-glycans containing the same N-glycan core structure (Man3GlcNAc2) and can be easily released via specific enzymes, namely endoglycosidases (very often peptide-N-glycosidases — PNGases) release all or most of the glycan, whereas exoglycosidases remove only a very specific portion of the non-reducing end of the glycan. The most frequently used is PNGase F, which digests virtually all N-glycans except for a core-α(1,3)-fucose, while PNGase A cleaves N-glycans with the core-α(1,3)-fucose, but is ineffective on sialylated N-glycans. Another is endoglycosidase H digesting the chitobiose core of high-mannose oligosaccharides or endoglycosidase F, which cleaves N-glycans between the two N-acetylglucosamine residues. PNGase F is very popular thanks to relative "universality" for N-glycans releasing. In addition, such digest leads to free glycans with intact glycosylamine at the reducing end (that is desirable for labelling or derivatisation), even though the amino group is quickly hydrolyzed and replaced by a hydroxyl group (Fig. 2.7).

In the case of O-glycans, there is no universal O-glycosidase discovered, since O-glycans consist of different core structures as was described previously. Therefore, O-glycans are usually realised by chemical treatment such as β-elimination and hydrasinolysis but there are a few glycosidases for O-glycans liberating. Although the use of hydrazine has been very popular, its use has declined in recent years. In comparison to milder enzymatic methods, hydrazine treatment is nonselective, thus both N-glycans and O-glycans can be released. In addition, this method is capable of preserving the reducing terminus of the glycans. However, anhydrous hydrazine cleaves all amide bonds (including those on N-acetylamino sugars) that results in the protein denaturation (protein doesn't remain for further analysis). More often, the β-elimination for O-glycan releasing is used. This method degrades the reducing end (removing of carbonyl moiety) which is also known as "peeling reaction". To minimize this reaction, a reducing agent (e.g. sodium borohydride) can be added; however, it generates alditols, which cannot be readily modified (derivatized or labelled).

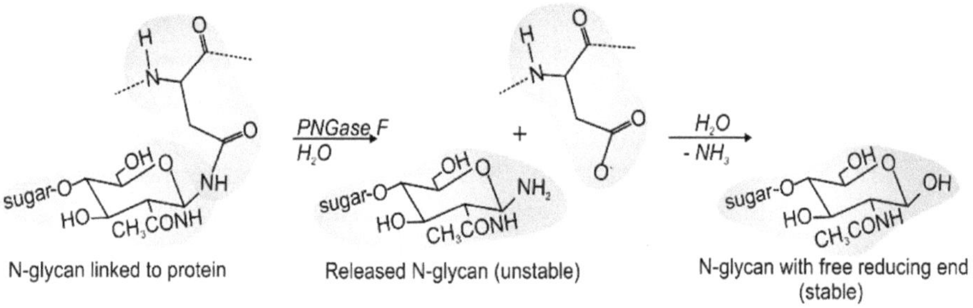

Figure 2.7 The liberation of N-glycan from protein/peptide using PNGase F; released amine-form of the chitobiose (=dimer of β-1,4-linked glucosamine units) core of N-glycan is unstable and in aqueous condition spontaneously converts to stable form–core with free reducing end.

Back to the N-glycan preparation, typically, the first step (before the releasing procedure itself) consists of protein denaturation using various denaturizing agents and conditions (e.g. SDS—increases the accessibility of enzymes to protein glycosylation sites, Cleland's reagent—dithiothreitol reduces disulphide bonds; high temperature). However, standard cleavage protocols are often time consuming; for example, PNGase digestion takes up to 24 h (overnight incubation). An effort to increase the efficiency of whole analysis has stimulated the development of improved strategies such as ultrahigh-pressure or microwave assisted methods with shorted cleavage time (reaction times of less than 1 h). The employment of modified microcolumn or magnetic nanoparticles reduces the reaction times to even 10 min. Recently, a simple, affordable approach has been introduced to prepare both N- and O-glycans together with glycosphingolipid-derived glycans (GSL-glycans) utilising sodium hypochlorite (NaClO). This procedure seems to degrade proteins but leaves N-glycans intact after 1 min of treatment under ambient condition without utilizing any denaturing agents or expensive enzymes.

The releasing process is followed by multiple purification steps, which are essential for removal of (glyco)proteins/(glyco)peptides, lipids, salts and reagents. In order to extract glycans for mass spectrometry analysis, numerous cleansing techniques and glycan enrichment procedures (see Chapter 7) have been developed. This is a crucial step because it is necessary to obtain a small amount of glycan as compared to the non-glycosylated peptides in the sample. The most frequently used is the glycan purification through size exclusion, porous graphitised carbon (PGC) chromatography, HILIC chromatography, HPLC chromatography or tagging-assisted strategies. Very popular are commercially available PGC columns, with the possibility to separate neutral and acidic glycans using different pooling buffers. The other prevalent technique uses the reverse-phase HPLC C18 columns that effectively removed peptidic material and other contaminants such as detergents. Solid-phase extraction (SPE) is a newly emerging approach for glycan analysis utilising the immobilisation of glycoproteins to the solid phase. So-called "Glycoprotein Immobilisation for Glycan Extraction" (GIG) has been successfully demonstrated for the simultaneous analysis of N- and O-glycans on a solid support. It is obvious, there are many possibilities to achieve desired glycan purity that is proper for glycan analysis; however, in many cases further modification (for signal amplification, linkage analysis, quantification, ionization improvement and separation enhancement) is needed.

2.2.1.2 Derivatisation and labelling

The glycan analysis is more difficult than protein or peptide identification, because of their diversity and similarity and relatively low abundance in real samples as explained above. In addition, the ionisation efficiency of glycans is much lower compared to other biopolymers such as peptides. Therefore, subsequent modification of released glycans is often necessary. Generally, in the case of glycan, there are three basic sites allowing glycan modification based on their chemical structure: the reducing end (contains a single reactive carbonyl group), glycan hydroxyl groups and carboxyl group of sialylated glycans.

2.2.1.3 Reducing end modification

Reducing end provides an elegant way to improve ionisation, to attach a chromophore/fluorophore (utilising in HPLC) or add a stable isotope tag (quantitative analysis). Reductive amination is the most often used derivatisation methods working for any glycans containing reducing end. This method is widely used for the introduction of fluorescent tags such as 2-aminobenzoic acid (2-AA), 2-aminobenzamide (2-AB), and 2-aminopyridine (2-AP) (Fig. 2.8), which in addition effectively improve sensitivity and glycan ionisation as well.

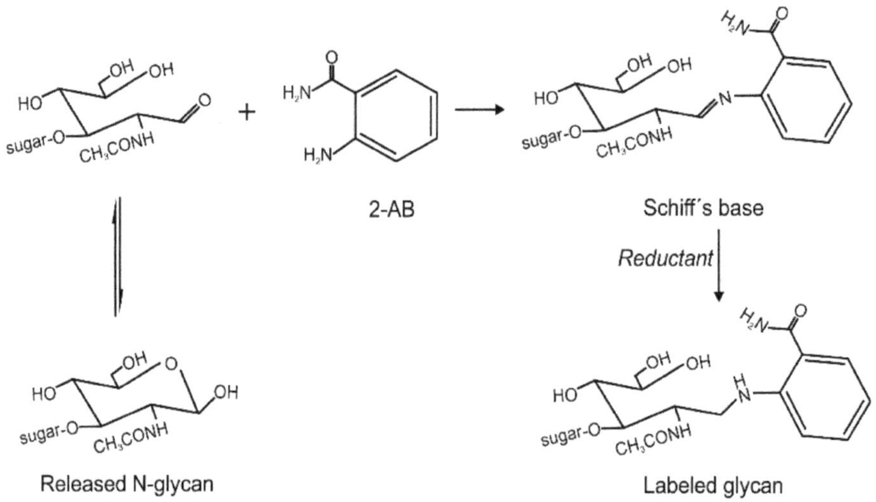

Figure 2.8 Reductive amination of glycan reducing end with the introduction of fluorescent label (2-AB).

Reductive amination is not applicable to O-glycans alditols; however, these can be derivatized and modified with stable isotopes for structural or quantitative analysis too (e.g. β-elimination performed in the presence of 1-phenyl-3-methyl-5-pyrazolone—PMP. A rapid and sensitive approach was introduced using reagent RapiFluor-MS (RFMS) consistin of N-hydroxysuccinimide (NHS), quinolone fluorophore and tertiary amine. It was determined that that RFMS-labelled glycans were detected with superior sensitivity (14 times higher fluorescence and 160 times greater MS signal than 2-AB-labelled glycans). Since many protocols performed the reductive amination in the solution, Rohmer et al. have reported the use of a solid matrix 3-aminoquinoline as a labelling reagent (3-AQ). This procedure does not require additional purification steps (removing of labelling reagents), which involves a risk of sample loss. Moreover, 3-AQ functions as an effective MALDI matrix for carbohydrates in both the positive- and negative-ion modes of detection (for more information about matrix-assisted laser desorption/ionisation—MALDI see Section 2.2.2). Alternatively, hydrazine chemistry has been also used for reducing end modification. This reaction does not require reductive condition; thus it saves one purification step in preparation process. In addition, slightly basic conditions cause minimal loss of sialic acid. The most often applied hydrazine-based derivatisation uses phenylhydrazine. Recently, use of hydrazinonicotinic acid enhanced the signal to noise ratio (S/N) 15-fold in the detection of sialylated glycans compared to the control.

The isotopic labelling strategy is a very promising tool for quantitative glycan analysis. Isotopically coded tags can be incorporated to glycan structure via reductive amination or hydrazine reaction at the reducing end. These tags include aniline, 2-AB, 2-aminopyridine, and 1-phenyl-3-methyl-5-pyrazolone. Additional labels, including ($^{13}C_6$ and $^{12}C_{12}$) 4-phenethyl-benzohydrazide and a unique set of tetraplexed tags, each separated by 4 Da, have been synthesized. An enzymatic ^{18}O-labelling has been also utilised, where the incorporation of ^{18}O is carried out during the N-glycan release from glycoproteins by PNGase F in the presence of ^{18}O-water. In addition, dual isotopic labelling is also possible by employing of PNGase F catalysis and $NaBD_4$ reduction in which both ^{18}O and D atoms were incorporated into glycan structures.

2.2.1.4 Permethylation

The hottest favourite for glycan modification is undoubtedly permethylation (see Fig. 2.9). It was first developed by Hakomori in the early 1960s and employed the reaction of carbohydrate with methyl iodide catalysed by methylsulfinyl carbanion. In 1984,

Ciucanu and Kerek introduced a simpler method utilising sodium hydroxide instead of methylsulfinyl carbanion. This method has been widely adopted and is broadly used today with plenty of variations.

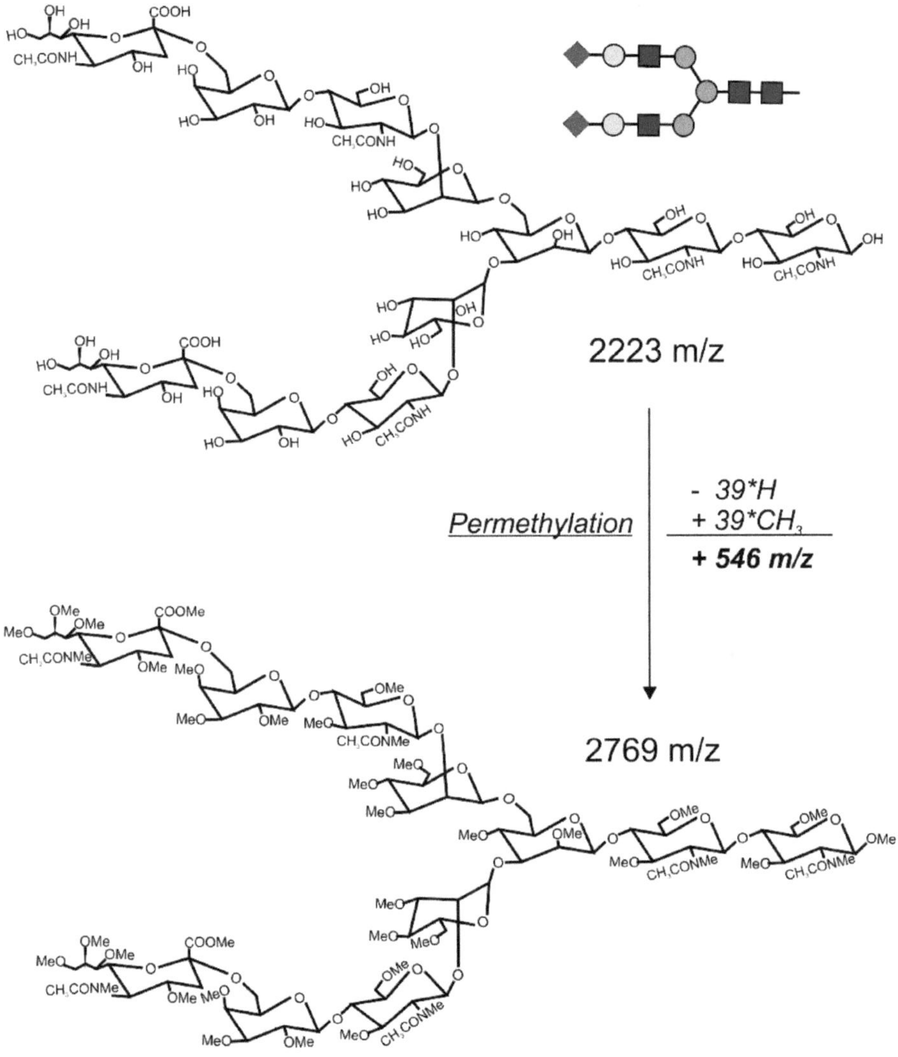

Figure 2.9 Permethylation of most common sialylated N-glycan. This step modifies hydroxyl, carboxyl and amine groups and reducing end as well. In this case, the m/z balance consists of replacing of 39 hydrogens by 39 methyl groups.

Permethylation modifies nearly all functional groups within glycans (hydroxyl groups, carboxyl groups, amine groups and reducing end) in which hydrogen atoms are replaced by methyl groups.

The introduction of hydrophobic methyl groups into glycan structure changes the hydrophilic nature of glycans, enhances their ionisation (decreases the intermolecular hydrogen bonding, thereby increasing the sample volatility) and thus improves the sensitivity and efficiency of MS measurement. In addition, permethylation allows simultaneous analysis of neutral and acidic glycans (oligosaccharides containing acidic groups—sulphate, carboxylate, or phosphate) in the positive ion mode and also leads to predictable fragmentations that give characteristic 'maps' of fragment ions at each amino sugar residue. Moreover, permethylation enhances the formation of diagnostic fragment ions, prevents gas-phase rearrangements (mostly "travelling" fucose) and stabilise sialylated residues. The analysis of permethylated glycans can be also used for interglycosidic linkages analysis that utilises partially methylated alditol acetates, known as PMAAs. During the last several years, the incorporation of heavy and light isotopes (CH_3I or $^{13}CH_3I$ vs. $^{12}CH_3I$) through permethylation has been used to assist in glycans quantification. A variant of isotope labelling is the usage of isobaric labels ($^{13}CH_3I$ and CDH_2) which yields the same nominal mass in low resolution MS but can be differentiated in high-resolution accurate mass. There are also reported a method derived from proteomic technology using isobaric tags referred to as Tandem Mass Tags (TMTs). Aminoxy TMT (aminoxyTMT) reagents are newly commercialized isobaric tags that enable relative quantitation of up to six different glycan samples simultaneously.

2.2.1.5 Sialic acids modification

In the context of diagnostic relevant glycans, sialylated glycans have prominent position as discussed previously. However, sialylated glycans are relatively unstable and the lability of glycosidic bond between sialic acid and glycan residue (α-2,3- and α-2,6-linkage) often leads to their partial loss during the sample preparation (some steps require harsh condition) and MS analysis. In-source and post-source losses of SA often lead to inaccurate representation of the degree of sialylation which may play a crucial role in diagnostics and patient prognosis. To avoid this drawback, modification of carboxylic acid group is necessary. Among permethylation, many other stabilisation strategies such as amidation, methylamidation, perbenzolylation or quinoxaline- and hydrazine-based derivatisation are available. There are some articles describing a facile derivatization in the presence of methylamine and PyAOP ((7-azabenzotriazol-1-yloxy)trispyrroli dinophosphonium hexafluorophosphate) for sialoglycomics. The result shows that both 3′ and 6′-sialyllactose derivatives achieve quantitative conversions regardless of their linkage types. Another interesting approach involves a procedure where methyl ester formation provides information on the sialic acid linkage directly from the MS spectrum using 4-(4,6-dimethoxy-1,3,5-triazin-2-yl)-4-methylmorpholinium chloride (DMT-MM). This method is based on the fact that α-2,3-linked sialic acids spontaneously produced lactones (internal cyclic esters), preventing their modification, whereas α-2,6 linked sialic acids were activated and ultimately esterified. Tkac's group (see Recommended Literature at the end of the chapter) applied the similar approach through the chemical derivatisation of the sialic acid carboxyl groups with 1-ethyl-3-(3-(dimethylamino)propyl) carbodiimide (EDC)/1-hydroxy-benzotriazole (HOBt) reagents with ethanol (Fig. 2.10). They were able to distinguish between α-2,6 and α-2,3-sialylation on PSA (prostate specific antigen) protein which can be used in the early-stage diagnosis of prostate cancer (PCa). While, α-2,6-sialylation is widely present in mammalian healthy tissues, α-2,3-linked sialic acid is rarely found in healthy ones. It has been proven that the increased amount of α-2,3-linked sialic acid in serum-free PSA is closely related with the occurrence of PCa. In this context, more recently, has been introduced a high-throughput approach employing pipette tip (AutoTip) operated by an automated liquid handler, processing up to 96 samples simultaneously. This innovative technique was successfully applied to profile N-glycans in urine samples of patient suffered from PCa. The analysis involves AutoTip preparation, protein immobilisation, sialic acid stabilisation by carbodiimid coupling with subsequent N-glycan releasing. Besides that the methodology results in revealing increased sialylation on urinary N-glycans, it also brings a big promise for high-throughput automated glycan analysis, which can strongly accelerate the glycan analysis and thus improve the glycan-based diagnostics.

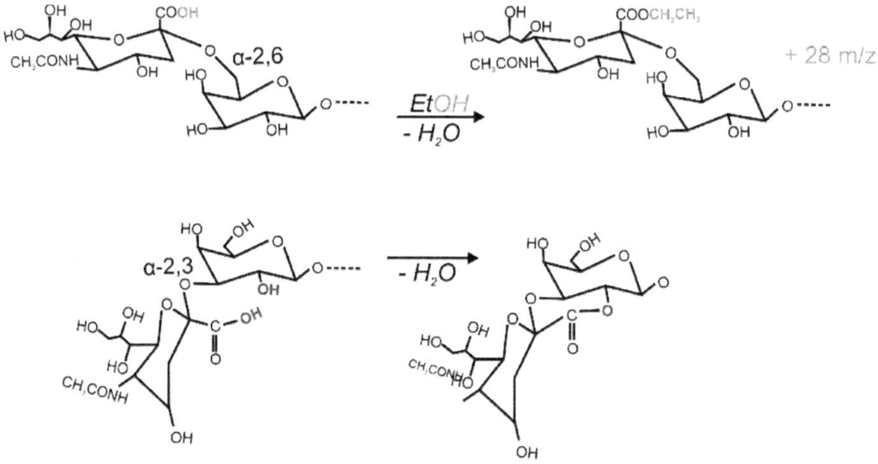

Figure 2.10 Terminal sialic acid modification (ethyl esterification) by 1-ethyl-3-(3-(dimethylamino)propyl) carbodiimide (EDC)/1-hydroxy-benzotriazole (HOBt) reagents with ethanol. While 2,3-linked sialic acids forms lactone (down) that prevent its modification, 2,6-linkages of sialic acid (up) undergo the modification that leads in increase of glycan mass (+28 m/z).

2.2.2 MS Techniques for Glycan Analysis

Mass spectrometry (MS), in its very simple definition, is the production and detection of ions separated according to their mass-to-charge (m/z) ratios. The detection of created ions results in a mass spectrum representing a plot of the relative abundance of the ions as a function of their m/z ratio.

Mass spectrometry is an method for identifying the chemical constitution of a substance by means of the separation of gaseous ions according to their differing mass and charge.

Among numerous ionisation methods, the most widely used ionisation techniques for the analysis of carbohydrates are matrix-assisted laser desorption/ionization (MALDI) and electrospray ionization (ESI). Both methods belong to so-called soft ionisation techniques. It means that the ionization process imparts little excess energy and thus generates few or no fragments for glycans; thus the intact molecular ions can be easily observed in MS spectra. However, prompt loss of labile groups (mainly underivatized sialic acid and fucose) may be observed. ESI generates multiply charged ions (e.g. $[Mr + cation_n]^{n+}$) where the ionization efficiency decreases with an increasing molecular weight, while MALDI yields a series of singly charged molecular ions and the ionization effectiveness for neutral carbohydrates is constant as the size of molecule increases. MALDI, in comparison to ESI, has higher sensitivity for glycans, ionizes well even at higher mass range, and is more tolerant to contaminants.

2.2.2.1 Matrix-assisted laser desorption/Ionization (MALDI) analysis

MALDI was first used for carbohydrates ionization by Karas and Hillenkamp and at the first time was applied to N-glycans by Mock et al. in 1991. Briefly, the term of "matrix-assisted laser desorption and ionisation" means that MALDI employs certain kind of matrix (explained below) adsorbing the UV light from the source (mostly nitrogen laser light, wavelength 337 nm) and converts it to heat energy. This causes the rapid vaporisation of the small part of the matrix together with the sample and generated ions (of various sizes) are accelerated by a high voltage to the detector (Fig. 2.11). Charged molecules then travel in a flight tube at a velocity inversely proportional to the square root of their m/z values (molecules with larger m/z values have longer time of flight—TOF). This technique is called MALDI-TOF and is very popular in glycan analysis thanks to the high-throughput nature of a TOF mass analyser.

Generally, generated ions can be either positive or negative depending on the character of the sample. Neutral carbohydrates are predominantly ionised by forming adducts with alkali metals which yield intense signals in the positive ion mode. Mohr et al. found that the binding affinities of proton and alkali metal ions to carbohydrates followed the order:

$$\mathbf{H \ll Li < Na < K < Cs}$$

Nevertheless, the Na^+ adduct $[Mr + Na]^+$ is the most frequently present and the most prevalent in MS spectra. For the suppression of other ion adducts, an external addition of Na^+ cations is very invited, e.g. addition of NaOH to matrix solution amplifies the signal of Na^+ adduct. In contrast to neutral glycans, acidic glycans are typically detected as deprotonated or anion adducted species in the negative-ion mode, e.g. $[Mr - H]^-$ or $[Mr + Cl]^-$, respectively.

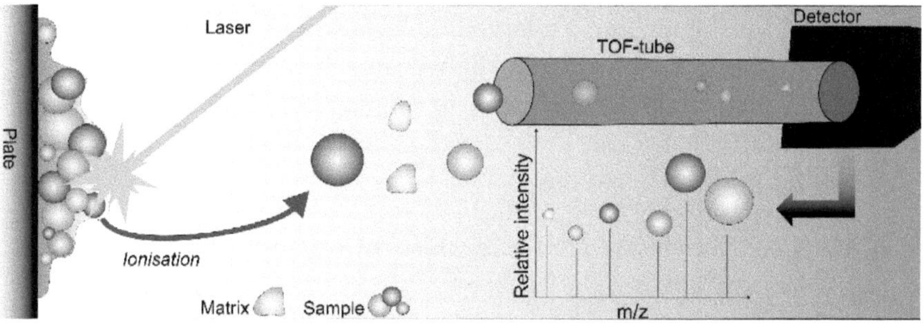

Figure 2.11 Basic principles of MALDI-TOF analysis.

2.2.2.2 Matrices

A suitable choice of the matrix for efficient glycan ionisation has a crucial impact on the MALDI analysis and depends on the character of target glycan. Unlike neutral glycans, sialylated glycans are relatively unstable under MALDI condition due to the migration of the labile acidic proton of the carboxyl group. To prevent the loss of acidic moiety the stabilisation/modification is needed as discussed previously. The utilisation of different matrices can be also helpful to avoid/reduce the sialic acid loss. Very frequently used for sialylated glycan analysis is 2,4,6-trihydroxyacetophenone (THAP), particularly when mixed with ammonium citrate and also 6-Aza-2-thiothymine (ATT), which reduce the sialic acid loss. Besides the classical fragmentation of sialylated glycans, they can by observed in positive mode of operation as well producing a mixture of ions:

$$[M+Na]^+ \text{ and } [M-nH^+(n+1)Na]^+$$

There are many matrices developed, but 2,5-dihydroxybenzoic acid (DHB) as one of the earliest is still the most popular for glycan identification. Its universality allows the analysis of both neutral (in positive mode) and sialylated (in negative/positive mode) glycans without any previous derivatisation. Although the dried droplet method of sample spotting (on MALDI plate) works relative well, DHB often creates large crystals formed around the rim of the MALDI spot pointing toward the centre. This results in analyte/matrix heterogeneity as well as in the decrease in reproducibility. Nowadays, there is an increased interest for the development of novel matrices in order to improve glycan ionisation and enhance standardisation of the analysis. Better performance has been achieved mixing of the DHB with various additives like 2-hydroxy-5-methoxybenzoic acid or 1-aminoisoquinoline. Recrystallization of the already dried spot (sample + matrix) from ethanol is also popular for producing of a more homogeneous surface. The utilisation of ionic liquid matrices (ILMs) consisting of DHB, α-cyano-4-hydroxycinnamic acid (CHCA), or trihydroxyacetophenone (THAP) and some base such as tributylamine or pyridine enabling a relative state of 'liquidity' under vacuum conditions. This specific state provides high homogeneity in analyte/matrix distribution; thus the increased reproducibility. In the context of innovation, there was introduced a direct labelling method with 3-AQ (as mentioned above) using a liquid matrix 3-AQ/CHCA when they achieved high-sensitivity detection. A similar matrix 3-Aminoquinoline/p-Coumaric acid (3-AQ/CA) has been demonstrated as 1000-fold efficient compared to 3-AQ/CHCA or DHB. The successful used of anion-doped liquid matrix (G3CA), p-coumaric acid, and 1,1,3,3-tetramethylguanidine to detect neutral N-glycans in negative-ion mode has been also published. It has been found that prepared ILMs suppressed dissociation of sialic acids; moreover, they are very sensitive (at a low femtomole level) for neutral N-glycans in negative-ion mode that has been proven to be more informative than that of positive-ion fragmentation. Another example is utilisation of a new matrix hydrazinonicotinic acid (HYNIC) forming more uniform crystals allowing the ultrasensitive analysis of oligosaccharides with the detection sensitivity as low as 1 amol loading amount per target well. Besides, HYNIC significantly suppresses ions of other contaminants (e.g. peptides) that minimize the sample preparation steps.

Recently various papers have appeared that describe the novel approach derived from the MALDI technique. Surface-assisted laser desorption/ionisation (SALDI) utilises the specific surface or nanomaterial (e.g. carbon-, metal-, silicon-based)-coated surface that substitutes the chemical matrix. This approach overcomes the crystallisation problem; moreover, it leads to an improvement in sensitivity as well. The use of diamond nanoparticles (DNPs) was successfully applied in the SALDI analysis of underivatized carbohydrates even in the presence of proteins. Owing to the high thermal conductivity of DNPs, low extinction coefficient in the UV-vis spectral range, and stable chemical properties, their proposed technique improved the sensitivity of dextran about 79- and 7-fold compared to the control. Not only DPSs can be applied. The use of an iron oxide nanoparticles core with gluthathione molecules covalently bound to the surface produces a clean mass spectral background in the low m/z region and thus shows also a big potential as far as glycan sequencing.

2.2.2.3 Tandem mass spectrometry

MALDI-MS spectra are highly suitable for mass fingerprinting of glycans and inestimable in the initial glycomic analysis; however, the lack of fragment ions leads to unsatisfied elucidation of glycan structures. The exact glycan structure can be obtained by tandem mass spectrometry (MS/MS or MS^n).

Tandem mass spectrometry (MS^n) is a method involving at least two stages of mass analysis, with some form of fragmentation occuring between the stages.

Tandem mass spectroscopy (see Fig. 2.12) can help reveal fine structural details, e.g. to distinguish closely related isomers such as α-2,6- and α-2,3-linked sialic acid or α-1,6- and α-1,3-attached fucose units, that are relevant in many diseases. A parent ion for MS/MS analysis is selected by a first stage of MS (MS1), then activated and dissociated (MS2) by various techniques, which may be divided into two broad categories. The first ions activation group contains electron-based dissociation methods, when the transfer of an electron to or from a selected multiply charged molecular ion leads to the formation of a radical ion which subsequent undergoes fragmentation. These methods include electron capture dissociation (ECD), electron transfer dissociation (ETD), electron detachment dissociation (EDD), negative electron transfer dissociation (NETD) and negative ion electron capture dissociation (niECD). The second group includes low-energy and high-energy collision induced dissociation (CID, respectively HCD) and infrared multiphoton dissociation (IRMPD). In MALDI, controlled fragmentation of glycans is most usually produced by CID in a collision cell; nevertheless, the fragmentation can occur in the ion source (in-source-decay ISD) or within the flight tube as a post-source decay (PSD). In this type of fragmentation, glycans are bombarded with a buffer gas that increases their bond vibration energies and once the sufficient energy is imparted, the bond cleavage takes place.

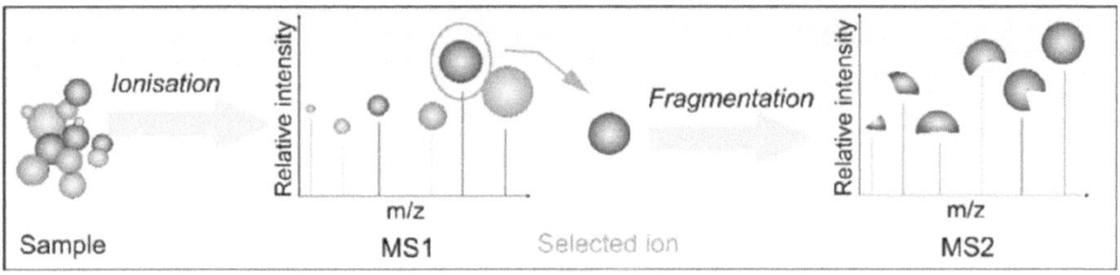

Figure 2.12 Schematic presentation of the tandem MS principle, where MS1 is first stage (MS spectra is obtained and ion (with exact m/z) for fragmentation is selected; MS2 represents the second stage and fragment ions of selected ion (so called parent ion) are observed in MS spectra.

The carbohydrate fragmentation is determined by several factors, such as the type of ion formation, its charge state and the energy deposited into the ion. Generally, the glycans form two major types of ions: glycosidic cleavages (the bond breaks off between the sugar rings accompanied by hydrogen migrations) and cross-ring cleavages (the rupture of two bonds on the same sugar residue). The nomenclature conventionally used for describing these fragment ions has been proposed by Domon and Costello. Ions that retain the charge at the reducing end are named X, Y, and Z, whereas ions with the charge at the non-reducing end are A, B, and C. From the other point of view, A and X correspond to cross-ring cleavages, while B, C, Y and Z correspond to glycosidic cleavages (Fig. 2.13). For positively charged glycans, the most frequently observed ion in MS spectra using low energy CID is B and Y, whereas cross-ring cleavages (ions A and X) are also formed using HCD. For more complex N-linked glycans, the most abundant cross-ring fragments detected with HCD in a MALDI-TOF/TOF mass spectrometer are often the $^{1,5}X_n$ cleavages. In contrast to positive ion mode, the fragmentation spectra of negatively charged ions contain numerous A-type cross-ring fragments, which are needed to distinguish different isomers, and C ions. The exact isomer differentiation can be obtained by employing a sequential MS method. Following an initial CID experiment, a resulting ambiguous ion of interest is isolated and subjected to a subsequent round of fragmentation, and this process is repeated as many times as necessary. Finally, fragments specific to a certain isomer are detected. This approach was successfully used for the discrimination of isomeric structures in ovalbumin, humanized immunoglobulin G or metastatic and nonmetastatic brain cancer cell lines.

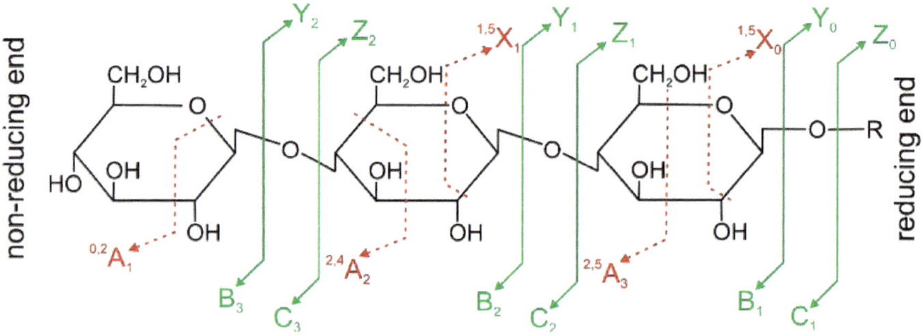

Figure 2.13 Nomenclature for describing the fragmentation of carbohydrates, where ions retaining the charge on the non-reducing terminus are A, B and C; the ions retaining charge on the reducing terminus are X, Y and Z. A and X correspond to cross-ring cleavages; B, C, Y and Z correspond to glycosidic cleavages. Subscript numbers represent the cleavage position, starting at the reducing terminus for the X, Y and Z ions and at the non-reducing terminus for A, B and C. In the case of ring cleavages, superscript numbers are given to show the cleaved bonds.

2.2.2.4 Practical application

MALDI is widely used as a powerful tool to monitor altered glycosylation related to various diseases including cancer. Gomes et al. utilised MALDI-TOF/TOF MS to demonstrate the evidence of circulating proteins displaying abnormal O-glycosylation in patients with precursor lesions of gastric carcinoma and involved also a panel of assumed targets for the non-invasive clinical diagnosis of individuals with gastritis. MALDI-MS was the first MS method employed for global glycan profiling of serum that was reported over 10 years ago for ovarian cancer. Nowadays, such analysis is very often linked with different separation techniques such as electrophoresis (e.g. capillary electrophoresis—CE) or chromatography (e.g. liquid chromatography—LC) yielding to higher sensitivity. Novotny and his group introduced an approach employing microchip electrophoresis combined with MALDI-TOF-MS-based analysis to profiling of sera from patients having colorectal cancer. He observed some unique glycan changes primary associated with fucosyl isomers. In addition, several potential biomarkers were identified to aid in the diagnosis and differentiation of colorectal cancer.

A most remarkable advance has been done via MALDI-MS in combination with ion mobility spectrometry (IMS). This technique has been successfully applied, e.g. to delineation of disease phenotypes associated with oesophageal adenocarcinoma. Currently, a novel method for investigating tissue-specific N-linked glycans employing MALDI-MSI on formalin-fixed paraffin-embedded (FFPE) tissue sections has been introduced. This technique allows can spatially profile N-glycan compositions released from proteins in tissue-specific regions. Such methods were applied in 2D N-Glycan distribution mapping of ovarian cancer and hepatocellular carcinoma tissues. The very similar approach was applied in analysis of formalin-fixed tissues microarray analysis of pancreatic and prostate cancer tissue as well as hepatocellular carcinoma tissue. After the initial antigen retrieval followed by on-tissue PNGase F digestion the MALDI analysis took place where structural composition could be verified directly by on-tissue collision-induced fragmentation. This workflow allows the overlaid of both tissue staining of the protein and glycan profiles thus has a big potential for efficient analysis of the global changes in N-glycosylation associated with cancers.

The defective glycosylation of glycoconjugates leads to development of so-called congenital disorders of glycosylation

(CDG). Conserved oligomeric Golgi (COG)-CDG are genetic diseases characterised by N-glycan and O-glycan processing abnormalities. Sturiale and co-workers applied MALDI analysis to screening of differential serum glycoforms (O-, N-glycans) from patients having COG5-CDG (COG5 subunit deficiency). In the case of O-linked glycans, they observed the decrease of Apolipoprotein C-III_{0a} (apoC-III aglycosylated glycoforms), while apoC-III_1 (mono-sialylated glycoform) decreased significantly. On the other hand, N-glycome of COG5-CDG patients was characterized and showed undersialylation and undergalactosylation of biantennary and triantennary glycans as well as slight increase of high-mannose structures and hybrid glycans. This investigation adds to the current knowledge on glycosylation defects caused by impairment of COG subunits that support and enhance the clinical diagnosis. Also, autoimmune diseases at present are receiving worldwide attention. RA is a long-term autoimmune disorder that primarily affects joints and has a significant influence the life quality. Recently, tandem mass spectroscopy (MALDI-TOF-MS^n) and CE were used for the analysis of immunoglobulin G (IgG) glycosylation in rheumatoid arthritis (RA) patients. The results indicate that CE coupled with MS^n can identify abnormal glycosylation of IgG in RA patients compared with healthy people. They reported a significant decrease of N-glycan galactosylation in RA patients and a slight decrease of sialylated glycans as well. The defect of galactosylation of IgG in RA patients is a useful marker of disease activity and may be used for the assessment of therapy effectiveness.

Diseases are the not only aim of the MALDI-based glycomic investigation. Gestation requires partial suppression of the immune system to ensure maternal-foetal tolerance. Although protein glycosylation (mainly terminal sialic acid linkages) is of prime importance in regulating the pro- and anti-inflammatory immune responses, pregnancy-associated changes of the serum N-glycome have been relatively little explored. Jansen et al. monitored glycan changes during and after the pregnancy focusing on sialic acid linkages. Using high-throughput MALDI-TOF-MS and selective derivatisation allowing the distinction between α-2,3- and α-2,6-linked sialic acids, they followed a total of 77 N-glycans and from individual glycans calculated 42 derived traits including an increase during pregnancy and decrease after delivery observed for both sialylation. In addition, a recovery speed after delivery for sialylated isomers of triantennary glycans was different. The same group also investigated the constant fragment (FC) glycosylation of immunoglobulin G (IgG) isolated from a serum of a pregnant woman. The results showed that Fc glycosylation changes include increase in galactosylation and sialylation and decrease in bisection during pregnancy. Thus, it seems that the in-depth monitoring and determination of such changes during gravidity will be important for unravelling the roles of these glycans in (auto)immunity.

Although the aim of this chapter is free-glycan profiling, the MS technique is very powerful tool for glycopeptides and whole-protein analysis as well. However, it is very difficult to involve all procedures and progress made in this broad field. Thus some excellent review articles available online are listed below.

Recommended Literature

1. Lauc, G., et al., Mechanisms of disease: The human N-glycome. *Biochim. Biophys. Acta*, 2016. 1860(8): pp. 1574–1582.
2. Yamamoto, S., M. Kinoshita, and S. Suzuki, Current landscape of protein glycosylation analysis and recent progress toward a novel paradigm of glycoscience research. *J. Pharm. Biomed. Anal.*, 2016. 130: pp. 273–300.
3. Springer, S. A., and P. Gagneux, Glycomics: Revealing the dynamic ecology and evolution of sugar molecules. *J. Proteomics*, 2016. 135: pp. 90–100.
4. Palaniappan, K. K., and C. R. Bertozzi, Chemical glycoproteomics. *Chem. Rev.*, 2016. 116(23): pp. 14277–14306.
5. Defaus, S., et al., Mammalian protein glycosylation-structure versus function. *Analyst*, 2014. 139(12): pp. 2944–2967.
6. Bennun, S. V., et al., Systems glycobiology: Integrating glycogenomics, glycoproteomics, glycomics, and other 'Omics data sets to characterize cellular glycosylation processes. *J. Mol. Biol.*, 2016. 428(16): pp. 3337–3352.
7. Krasnova, L., and C. H. Wong, Understanding the chemistry and biology of glycosylation with glycan synthesis. *Annu. Rev. Biochem.*, 2016. 85: pp. 599–630.

8. Yang, Y., V. Franc, and A. J. R. Heck, Glycoproteomics: A balance between high-throughput and in-depth analysis. *Trends Biotechnol.*, 2017. 35(7): pp. 598–609.
9. Oliveira-Ferrer, L., K. Legler, and K. Milde-Langosch, Role of protein glycosylation in cancer metastasis. *Semin. Cancer Biol.*, 2017. 44: pp. 141–152.
10. Stowell, S. R., T. Ju, and R. D. Cummings, Protein glycosylation in cancer. *Annu. Rev. Pathol.*, 2015. 10: pp. 473–510.
11. Cummings, Richard D., and J. M. Pierce, The Challenge and Promise of Glycomics. *Chem. Biol.*, 2014. 21(1): pp. 1–15.
12. Jansen, B. C., et al., Pregnancy-associated serum N-glycome changes studied by high-throughput MALDI-TOF-MS. *Sci. Rep.*, 2016. 6: p. 23296.
13. Hakomori, S., A rapid permethylation of glycolipid, and polysaccharide catalyzed by methylsulfinyl carbanion in dimethyl sulfoxide. *J. Biochem.*, 1964. 55: pp. 205–208.
14. Nie, H., Y. Li, and X. L. Sun, Recent Advances in Sialic Acid-Focused Glycomics. *J. Proteomics*, 2012. 75(11): pp. 3098–3112.
15. Ciucanu, I., and F. Kerek, A simple and rapid method for the permethylation of carbohydrates. *Carbohydr. Res.*, 1984. 131(2): pp. 209–217.
16. Pihikova, D., et al., Sweet characterisation of prostate specific antigen using electrochemical lectin-based immunosensor assay and MALDI TOF/TOF analysis: Focus on sialic acid. *Proteomics*, 2016. 16(24): pp. 3085–3095.
17. Landsteiner, K., and P. Levine, On individual differences in human blood. *J. Exp. Med.*, 1928. 47(5): pp. 757–775.
18. Belický, Š., J. Katrlík, and J. Tkáč, Glycan and lectin biosensors. *Essays Biochem.*, 2016. 60(1): pp. 37–47.
19. Karas, M., et al., Matrix-assisted ultraviolet laser desorption of non-volatile compounds. *Int. J. Mass Spectrom. Ion Proc.*, 1987. 78: pp. 53–68.
20. Pan, K.-T., et al., Adapting data-independent acquisition for mass spectrometry-based protein site-specific N-glycosylation analysis. *Anal. Chem.*, 2017. 89(8): pp. 4532–4539.
21. Mock, K. K., M. Davey, and J. S. Cottrell, The analysis of underivatized oligosaccharides by matrix-assisted laser desorption mass spectrometry. *Biochem. Biophys. Res. Commun.*, 1991. 177(2): pp. 644–651.

22. Jansen, B. C., et al., Pregnancy-associated serum N-glycome changes studied by high-throughput MALDI-TOF-MS. *Sci. Rep.*, 2016. 6: p. 23296.

24. Hashim, D., et al., The global decrease in cancer mortality: Trends and disparities. *Ann. Oncol.*, 2016. 27(5): pp. 926–933.

25. Domon, B., and C. E. Costello, A systematic nomenclature for carbohydrate fragmentations in FAB-MS/MS spectra of glycoconjugates. *Glycoconj. J.*, 1988. 5(4): pp. 397–409.

28. Mohr, M. D., K. O. Bornsen, and H. M. Widmer, Matrix-assisted laser desorption/ionization mass spectrometry: Improved matrix for oligosaccharides. *Rapid Commun. Mass. Spectrom.*, 1995. 9(9): pp. 809–814.

29. Snyder, C. M., et al., Complementary glycomic analyses of sera derived from colorectal cancer patients by MALDI-TOF-MS and microchip electrophoresis. *Anal. Chem.*, 2016. 88(19): pp. 9597–9605.

30. Svarovsky, S. A., and L. Joshi, Cancer glycan biomarkers and their detection-past, present and future. *Anal. Methods*, 2014. 6(12): pp. 3918–3936.

32. Palmigiano, A., et al., MALDI-MS profiling of serum O-glycosylation and N-glycosylation in COG5-CDG. *J. Mass. Spectrom.*, 2017. 52(6): pp. 372–377.

33. Banazadeh, A., et al., Recent advances in mass spectrometric analysis of glycoproteins. *Electrophoresis*, 2017. 38(1): pp. 162–189.

34. Jensen, P. H., et al., Structural analysis of N- and O-glycans released from glycoproteins. *Nat. Protoc.*, 2012. 7(7): pp. 1299–1310.

35. Harvey, D. J., Analysis of protein glycosylation by mass spectrometry, in *Analysis of Protein Post-Translational Modifications by Mass Spectrometry*. 2016, John Wiley & Sons, Inc. pp. 89–159.

36. Han, L., and C. E. Costello, Mass spectrometry of glycans. *Biochemistry. Biokhimiia*, 2013. 78(7): pp. 710–720.

37. Gomes, C., et al., Glycoproteomic analysis of serum from patients with gastric precancerous lesions. *J. Proteome Res.*, 2013. 12(3): pp. 1454–1466.

38. Alley, W. R., Jr., and M. V. Novotny, Structural glycomic analyses at high sensitivity: A decade of progress. *Annu. Rev. Anal. Chem.* (Palo Alto Calif), 2013. 6: pp. 237–265.

39. Gaunitz, S., et al., Recent advances in the analysis of complex glycoproteins. *Anal. Chem.*, 2017. 89(1): pp. 389–413.

45. Song, X., et al., Oxidative release of natural glycans for functional glycomics. *Nat. Methods*, 2016. 13(6): pp. 528–534.
46. Lu, H., Y. Zhang, and P. Yang, Advancements in mass spectrometry-based glycoproteomics and glycomics. *Natl. Sci. Rev.*, 2016. 3(3): pp. 345–364.
48. Yang, S., et al., Chemoenzymatic method for glycomics: Isolation, identification, and quantitation. *Proteomics*, 2016. 16(2): pp. 241–256.
49. Nishikaze, T., Sensitive and structure-informative N-glycosylation analysis by MALDI-MS; ionization, fragmentation, and derivatization. *Mass Spectrom (Tokyo)*, 2017. 6(1): p. A0060.

Chapter 3

Bioanalytical Methods

Alena Holazová

Department of Glycobiotechnology, Institute of Chemistry, Slovak Academy of Sciences, Dúbravská cesta 9, Bratislava, SK-845 38, Slovakia

alenasediva15@gmail.com

3.1 Introduction

Bioanalytical methods have traditionally been used in a variety of disciplines, including various diagnostic examinations in medicine, DNA structure research and its modifications in genetics, and more. It is also widely used in the field of food analysis, especially in so-called screening tests.

The use of modern bioanalytical methods within the surveillance or monitoring of food contaminants results in a decrease of not only testing costs but also expenses associated with any lag in storage, processing or marketing of crops and foods.

Fast and cheap bioanalytical methods with sufficient accuracy are essential for the extensive testing of food commodities. The application of biosensors, combining recognition capabilities of

Glyconanotechnology: Nanoscale Approach for Novel Glycan Analysis and Their Medical Use
Edited by Tomáš Bertók

ISBN 978-981-4800-16-7 (Hardcover), 978-0-429-44849-2 (eBook)
www.jennystanford.com

biomolecules with electronic or optical transducers, in the Hazard Analysis and Critical Control Points (HACCP) system can be used either for the real-time risk management or (at least) for the effective Critical Control Point (CCP) verification.

Bioanalytical methods have two types of concepts. The first concept has alternative approaches to the analysis of "biomolecules." The second concept interprets the term as a set of techniques utilizing biological molecules to obtain biospecific interaction that allow qualitative or quantitative determination of the analyte(s).

3.2 Bioanalytical Methods

Bioanalytical methods have two types of concepts. The first concept has alternative approaches to the analysis of "biomolecules." The second concept interprets the term as a set of techniques utilizing biological molecules to obtain biospecific interaction that allow qualitative or quantitative determination of the analyte(s).

3.2.1 Biospecific Interactions in Analytics

In nature, not only living nature dominated the art of creating and digesting covalent bonds, as chemists know, but it was able to use it to realize their intentions and interactions of some molecules. Because one of the interacting molecules is a biopolymer, we call these interactions biospecific.

Biospecific interactions include short-acting forces:

- electrostatic (Coulomb) forces
- hydrophobic interactions
- hydrogen bonds
- van der Waals forces

From the subject "Biochemistry," let's recall some examples of these biospecific interactions:

- pairing of nucleic acid bases (genetic code)
- binding of molecules to receptor proteins
- the transport of certain molecules by blood proteins (e.g., hormone transfer)

Thus, biospecific interactions can recognize the spatial structure of molecules and connect both interaction partners on the principle of known Fischer's theory of "lock and key." In contrast to the formation of a covalent bond, the synthesis of which is to be supplied with energy, and the cleavage of which is often not easy (e.g., hydrolysis of the 6 M HCl peptide chain at 110°C takes 24 h), biospecific interactions form a pair which is often comparable to that of a covalent bond, but can be upset by a small change in conditions (pH, ionic strength of the solution).

From these interactions, the following pairs are used in the bioanalytical processes:

- enzyme
  - substrate (enzyme analysis)
  - cofactor, inhibitor
- antigen
  - antibody (immunochemical analysis)

Modern bioanalytical techniques, in addition to biospecific interactions that ensure high specificity of determination, also use marker, label, tracer marker molecules to allow for easy and sensitive detection of molecules so labeled. An overview of some of the brands used, including the detection limits that can be used in their use, is given in Table 3.1. It is clear that analytical methods using both principles, i.e., biospecific interactions and molecular labeling, are among the techniques that can achieve the lowest detection limits.

Table 3.1 Selected biospecific interactions

Interaction	K_a
(Strept-)Avidin–Biotin	10^{15}
Antibody–Hapten	10^{5}–10^{11}
Antibody–Immunogen	10^{5}–10^{11}
Protein A–Fc Fragment IgG	10^{6}
Lectin–Saccharide:	
Mono- and Oligosaccharides	10^{3}–10^{4}
Polysaccharides	10^{6}–10^{7}
Triazine Colors–Proteins	$\sim 10^{4}$
Enzyme–Substrate	$\sim 10^{1}$–10^{6}
Abzyme–Substrate	to 10^{4}

Table 3.2 Overview of the most commonly used brands

Mark	Approximate detection limit (mol)
^{14}C (β^-)	4×10^{-13}
^{3}H (β^-)	10^{-15}
^{125}I (γ)	10^{-17}
^{32}P (β^-)	3×10^{-18}
^{131}I (β^-, γ)	2×10^{-18}
β-galactosidase	10^{-13}
biotin–streptavidin, galactosidase, fluorescent substrate	10^{-21}*
fluorescein	10^{-14}
luminal	10^{-17}

*In the case of ferritin 600 molecules.

These principles are often combined with other widely known and frequently used techniques throughout the broad field of chemistry, such as chromatographic and electromigration methods. That is why we pay attention to these methodological approaches, especially with regard to their use in bioanalytics.

3.3 Immunochemical Methods

From bioanalytical methods, the front of the technique is based on a specific reaction between antibodies and antigens, and the labeling of one of the reactants can be used to detect the reaction. In the past, radioactive isotopes in the RIA (radioimmunoassay) method have been used in the past to replace the enzyme-linked immunosorbent assay (ELISA) in the vast majority of applications. Some applications based on the specific binding between the antibody and the antigen, such as immunoaffinity chromatography and certain types of biosensors, do not require branding. The reaction between the antigen and the antibody is accomplished by short-range forces (van der Waals forces, hydrogen bonds, Coulomb's and hydrophobic interactions). The strength of these bonds is often comparable to the strength of the covalent bonds, but the energy required for their formation and agitation is much lower. Biomolecules and their conjugates are generally susceptible to small changes in pH, which is used to disrupt the binding between the antigen and the antibody.

First of all, it is advisable to first briefly define the terms monoclonal and polyclonal antibodies and antigen as their properties directly affect the performance characteristics of bioanalytical methods.

Antibodies are proteins produced by white blood cells of vertebrates as a reaction to the entry of a chemical compound or microorganism. Laboratory preparation of antibodies is predominantly for mice, rabbits and for larger volumes then goats, pigs and horses. According to the preparation process, we recognize monoclonal and polyclonal antibodies. When an invasive agent (antigen) is a microorganism or a large molecule, such as a protein, it contains a larger number of specific antigenic epitopes. Therefore, there is a mixture of antibodies against them that are specific to their epitope, but as a mixture they react with a number of antigens and are therefore called polyclonal antibodies. A mixture of antibodies isolated from immunized animals also contains antibodies against all the antigens encountered by the animal during their lifetime, which may induce increases in detection by binding of non-target analytes. Low specificity may also be an advantage if we monitor a whole set of similar analytes and do not require quantification of individual individuals. However, in practice, highly specific monoclonal antibodies that are prepared against a single epitope for white blood cell culture are often preferred.

Antigens are chemical compounds or protein structures on the surface of the microorganism that are capable of inducing antibody formation in an immunized organism.

Haptens are low-molecular-weight substances having only one epitope. They do not in produce antibody production themselves. Antibodies are synthesized against them if they are bound to a protein carrier. Hapten on the surface of the protein carrier is considered to be its next epitope. The antibodies thus produced have the ability to bind the haptens even without the carrier protein. The definition of haptens is met by most chemical contaminants.

The determining factor is non-specific interactions in which neither the antibody nor the antigen, but the various other components of the matrix, react. Non-specific interactions may increase or decrease the analyte determination by partially interfering with the free surface of the solid phase or the reacting molecules. To prevent binding of interferons to the solid phase, so-called blocking proteins are used to immobilize the antigens or antibodies to cover the free surface.

3.3.1 Antigens

The antigen is such a molecular, supramolecular or cellular structure that has the ability to react specifically with the antibody or cellular antigen receptor. In addition, it is appropriate to define the term immunogen. This means any foreign molecule, supermolecular or cellular structure that can elicit a cellular or antibody-type immune response in the recipient. Secondarily, it reacts specifically with the antibodies produced. We can say that each immunogen is also an antigen. However, the opposite statement does not apply. Antigens are divided into (1) complete—they elicit an immune response (they are immunogens) and react with specific antibodies and (2) incomplete—incapable of inducing an immune response, they can only react specifically with antibodies. Incomplete antigens are commonly called haptens. Hapten can only cause an immune response if it is bound to a high-molecular-weight carrier.

From the point of view of origin, the antigens can be divided into (a) natural, (b) artificial (synthetic, whole polymer is synthesized from subunits), (c) conjugated (haptens bound to a high-molecular-weight carrier).

For the preparation of specific antibodies, we must have a specific antigen in sufficient quantity and purity. For isolation of such an antigen from a biological or microbiological material, care must be taken to ensure that the epitope structure is not damaged by the isolation procedure.

3.3.2 Antibodies

The immune system of animals is made up of antibodies that specifically interact with foreign substances (antigens) that have provoked their production. This interaction is part of a more complex defense function of the organism against foreign molecules. Antibodies are globular proteins and are therefore referred to as immunoglobulins (symbol Ig). They form a structurally and functionally heterogeneous population of glycoprotein molecules. All of them have recognition antibody activity, i.e., the ability to specifically bind to the antigen. In addition to this basic function, some Ig are capable of binding to the antigen to activate complement, stimulate certain biochemical processes leading to the differentiation and proliferation of cells of the immune system or to the production or release of mediators from the cells. Immunoglobulins are found primarily in blood serum, as well as in other body fluids, and in the intracellular space and in the cytoplasmic membrane of B-lymphocytes that are antibody producers.

3.3.2.1 Immunoglobulins IgG

In terms of immunochemical methods, the most important and serum of the immunoglobulin class are IgG (about 80% of total Ig). Therefore, the structure (see Fig. 3.1) of the basic immunoglobulin subunit will be described primarily on them.

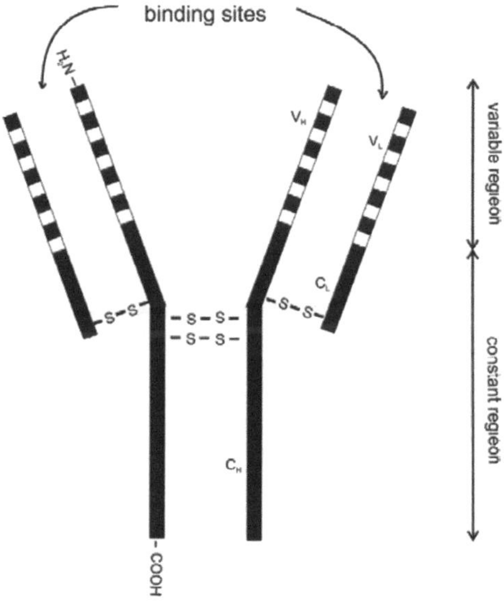

Figure 3.1 Schematic representation of the IgG1 immunoglobulin structure.

Each of the L-chains is connected via a disulfide bridge to one of the H-chains and they are linked by two or more disulfide bonds. In addition to these covalent bonds, non-covalent bonds also contribute to the chain joining. Light chains are attached to the H-strings at the top (forming a doubling of the Y-shaped arms) and the oligosaccharide is attached at the bottom. Only one molecule of oligosaccharide is attached to each chain γ.

The lower portions of both chain types have the same amino acid sequence for all IgGs. These constant segments are indicated by the C symbol. The variable regions are in the upper parts (arm ends) and denoted V. The amino acid sequence here differs in each Ig population. In the case of light chains, the variable part represents the half of the chain. It is VL. The other half is constant and is CL. The VL region consists of 108 to 109 amino acids. The heavy chain variable region (VH) consists of 118 (strands c) to 129 (strands) of amino acids and represents approximately one quarter of the whole chain. The remaining three quarters form a constant region (CH).

In variable regions, amino acids are not freely interchangeable in all positions. This is the case in only three so-called hypervariable stretches (the Complementarity Determining Region [CDR]). These are within the VL and VH intersected by larger, relatively constant stretches (the FR symbol in the Framework Region) that form the framework of variable regions. In each of the variable regions (VL and VH), there are three CDRs comprising a total of 25 amino acids on the L chain and 39 amino acids on the H chain. The position of each of these amino acids can occur in different populations of Ig molecules in the range of 3 to 10 amino acids.

In addition to the inter-chain disulfide bonds that link H- and L-chains, each peptide chain contains extra intra-chain disulfide bonds. These generate peptide chains of loops consisting of 60–67 amino acids. Between these loops there are short straight sections of the string. For L-strings, one loop is identical to the variable (VL) and the second to the constant region (CL) (see Fig. 3.2). Heavy chains consist of a loop identical to the VH region and three loops in the constant region (CH1, CH2, CH3).

This applies to strings of types c, a and d. The H-chains m and e contain another CH4 loop. A high degree of homology was found in the primary sequence of loops of variable and constant Ig regions of different populations of the same class. In many positions, they are always the same amino acids. In immunochemistry, these homologous loops are called domains. All Ig-domains have the same three-dimensional arrangement. The string is spatially packed into a bilayer b-structure of the folded sheet. Between the individual domains, short chain linker peptide segments are in the Ig chain structure. Between the CH1 and CH2 domains, the longer junction section is the hinge region.

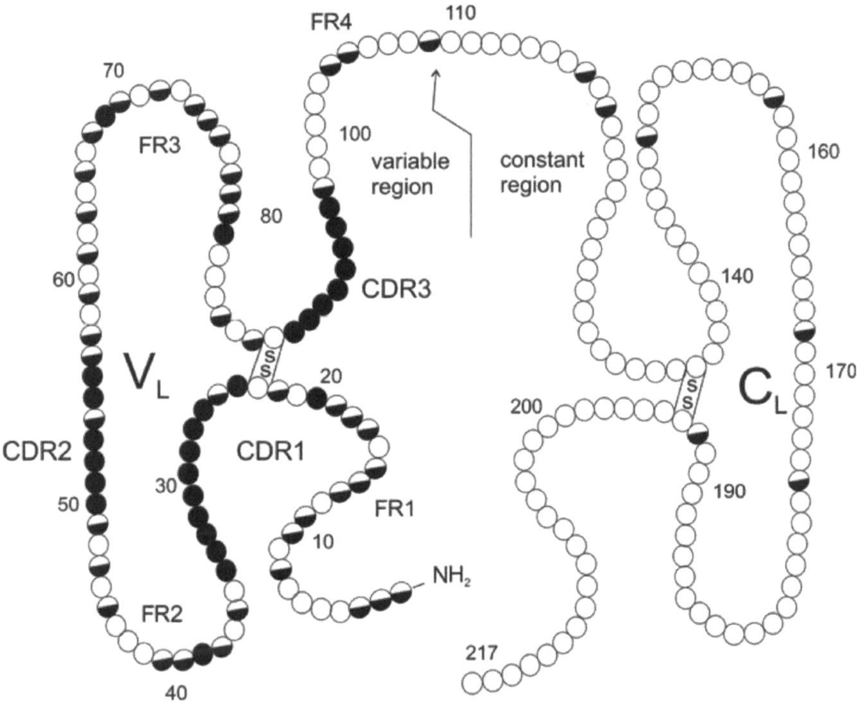

Figure 3.2 Schematic representation of the primary light chain structure in an immunoglobulin molecule with alternating FR and CDRs in the variable region (empty rings = positions with an unmistakable amino acid residue, blackened rings = more than 3 amino acid residues can be rotated in these positions, half of the blackened rings = v these positions alternate with only 2–3 amino acid residues).

Antibody interaction with antigen

Interaction of the antigen (hapten) with the specific antibody results in an immunocomplex. This interaction is non-covalent in nature and involves various types of attractive and repulsive forces as well as other non-covalent interactions between protein molecules. The formation of antibody-antigen binding (PL-AG) supports four types of interactions: electrostatic, disperse, hydrogen and hydrophobic. These forces act only between the molecules that are in close proximity. Therefore, they can only be applied to a pair of PL-AGs whose bonding sites exhibit a high degree of complementarity, thereby largely suppressing repulsive steric forces that otherwise prevent the atoms from being divergent from different molecules.

Primarily acting interactions are Coulomb forces between oppositely charged amino acids. These forces are effective at a distance of up to 100 Å. Van der Waals strength can also play a role here. It consists of the electrodynamic attraction between fluctuating dipoles in one atom and induced dipoles in the other. It acts up to 1000 Å. These primary linkages are established relatively quickly (seconds to minutes), but only a fraction of the binding energy in the complete PL-AG interaction.

Secondarily, hydrogen bonds and hydrophobic bonds are formed in a longer period (hours and days). Hydrogen bonds (hydrogen bridges) are applied between hydrophilic groups (in $-COOH$, $-NH_2$, $-OH$). The formation of hydrogen bonds is an exothermic reaction and therefore the greatest yield is achieved at low temperatures. Hydrophobic interactions play a dominant role in the formation of PL-AG binding. They are formed in close contact with two hydrophobic surfaces where the molecules of water surrounding them surround themselves. This change in the micro-lens of both surfaces reduces the local dielectric constant and increases the strength of electrostatic and ionic bonds. In the structure of proteins, these interactions are mainly leucine, isoleucine, valine and phenylalanine. Dispersing forces arise when the electron clouds interact with two nonpolar groups of atoms.

The presence of nonpolar substances in solution increases the force of Coulombic interactions and reduces the force of hydrophobic interactions. Conversely, increasing ionic strength reduces the force of Coulombic forces and intensifies hydrophobic interaction.

While electrostatic forces are not usually dominant for PL-AG complex formation, but because they act very efficiently over longer distances, they are often the cause of non-specific interactions of antibodies with molecules other than the corresponding antigen.

Immunochemical methods use in vitro antibodies whether in the form of antisera, globulin fractions of the antiserum, IgG fraction, specific IgG isolated by affinity chromatography, IgG molecule fragments, or monoclonal or recombinant antibodies. The basis of all immunochemical methods is the interaction of antibodies with antigen, hapten, complement components, cell receptors. Most abundantly, the immunocomplex of antibodies to the antigen (hapten) is used.

Immunochemical methods can be divided according to different criteria into several groups:

- according to the purpose of the analytical procedure (immunochemical characterization of the molecule, detection, determination)
- according to the test subject (cells, antigens, haptens, anti-exogenous antigens, autoantibodies)
- depending on the nature of the immunocomplex (precipitation, non-precipitation)
- according to the system of immunoassay (with a mark—no brand, competitive—noncompetitive, heterogeneous—homogeneous, with visual—with instrumental evaluation)

For the purposes of further interpretation, it is appropriate to choose a division according to the nature of the immunocomplex formed, namely the following methods:

(a) precipitation: the formation of immunocomplex in the form of a precipitate is used for detection or determination
(b) the non-precipitating: emerging immunocomplex remains soluble, and in order to detect or quantify the resulting immunocomplex; it is usually necessary to mark one of the reactants of the immunochemical reaction in advance with a suitable marker; or a change in the physical parameter of an anchor antibody material in immunosensors is measured, or an immune complex is used in affinity isolation

3.3.3 Precipitation Immunochemical Methods

The simplest immunochemical methods utilize a secondary phenomenon that under certain conditions accompanies the formation of an immunocomplex. It results in the immunocomplex dropping out of the solution as a visible precipitate (immunoprecipitation). While the soluble immunocomplex occurs at any ratio of antigen and antibody concentrations in solution, the immunoprecipitate is generated only within a certain range of the concentration ratios. The area of optimal concentrations is called the equivalence zone (the ratio of the number of antigen and antibody molecules is close to 1:1). Significant excess of one of the immunointegration partners prevents precipitation. The logic of this phenomenon is due to the duality of each Ig molecule and the existence of more determinant sites in the antigen molecule (AG). This allows a chain-to-network structure to be formed where AG and Ig molecules are repeatedly adjacent to each other. While the molecular weight of the conventional AG-Ig-AG immunocomplex may range from 160 to 1500 kDa and allows it to remain soluble, the much larger dimensions of the cross-linked immunocomplex shown in Fig. 3.3 cause its precipitation. With the usual non-skeletal antigen in which each determinant site is unique (not repeated within one molecule), only polyclonal antibodies will provide a precipitation reaction. As can be seen from the properties of monoclonal antibodies, they can only produce a precipitate with such an antigen in which the relevant determinant occurs several times. The monovalent Fab-fragments of immunoglobulins do not, of course, generate a network structure of the precipitate with the antigens. By binding to the antigen, they can inhibit the immunoprecipitation reaction of complete IgG molecules with the antigen. Low molecular haptenes do not produce a precipitate with the antibody. The reason is the small size of their molecule, which prevents the binding of two large antibody molecules.

The precipitation reaction consists of two phases. The former is relatively fast and forms soluble immunocomplexes. The second phase is substantially slower and results in the aggregation of soluble complexes from the first phase into insoluble complexes (precipitates). To maximize precipitation, incubation is required for hours to days. The precipitation rate increases with temperature

(0–56°C), but the maximum amount of precipitate formed increases with increasing temperature. A significant acceleration of the immunoprecipitation reaction in the solution is achieved in the presence of 2% polyethylene glycol. Some ionic detergents (e.g., sodium dodecyl sulfate (SDS), even at low concentrations, greatly inhibit immunoprecipitation. Conversely, non-ionic detergents (Tween, Triton) do not significantly affect this reaction.

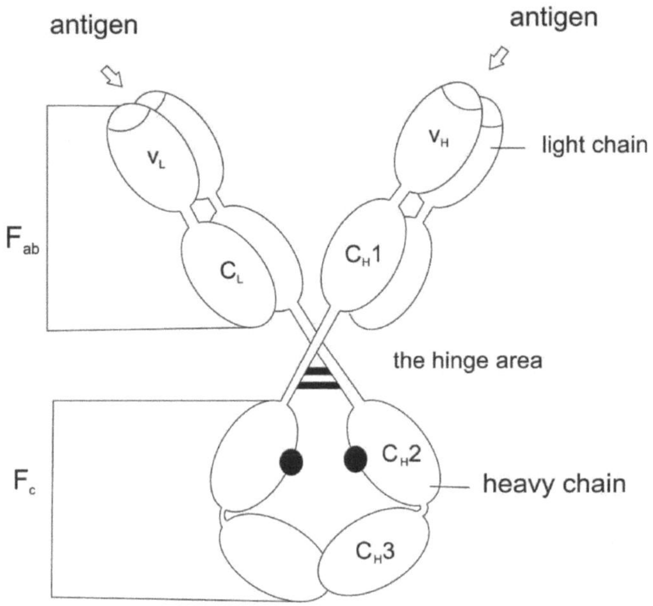

Figure 3.3 Schematic representation of the IgG structure by domains. Black dots on the CH_2 domain show an oligosaccharide unit.

During the second phase of precipitation, in addition to the interaction between the epitope and the paratope, non-specific hydrophobic interactions exist between the Fc region of the antibody molecule and the antigen. This is confirmed by the fact that the F (ab′) 2 fragments, unlike the original whole IgG molecules, do not usually have the ability to precipitate a multivalent antigen.

Immunoprecipitation may take place either in solution or in a gel environment (immuno-diffusion methods).

Immunoprecipitation and agglutination in solution. These techniques represent from the experimental point of view the easiest way to monitor the immunochemical reaction. In the precipitation reaction, the soluble antigen interacts with the antibody to form an insoluble precipitate. The course of production can be monitored by changes in the optical properties of the reaction solution. Thus, the presence of the appropriate antigen or antibody in a sample can be assessed qualitatively only by visual evaluation of whether a visible precipitate has been formed or formed. Quantitative determination is most often performed nephelometrically (or turbidimetric or fluorometric). The intensity of the reflected light depends on the concentration of the precipitate in the solution, and this is, under certain conditions, proportional to the original concentration of the immunoreactant to be monitored. The dependence of the amount of precipitate formed on the antigen concentration is called a precipitation curve. Its characteristic shape is shown in Fig. 3.4. For the purpose of serial determination by this method in clinical biochemistry, automatic analyzers with a laser nephelometer were constructed. To determine a single analyte, 10 μl of the sample is sufficient, and 20 different antigens with a sensitivity of 0.5–2 mg/l can be determined within a few minutes from 0.2 ml of patient serum.

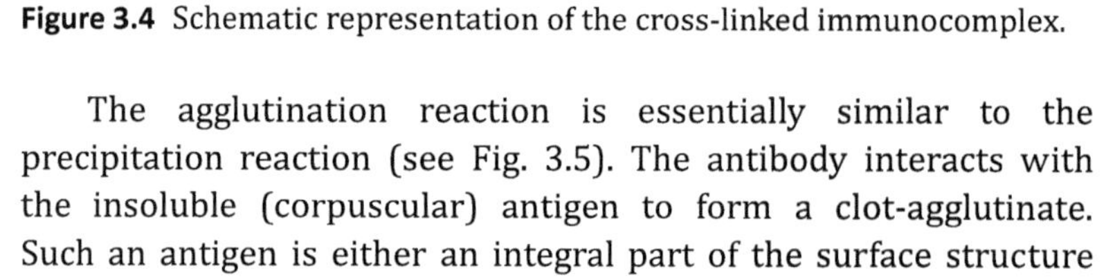

Figure 3.4 Schematic representation of the cross-linked immunocomplex.

The agglutination reaction is essentially similar to the precipitation reaction (see Fig. 3.5). The antibody interacts with the insoluble (corpuscular) antigen to form a clot-agglutinate. Such an antigen is either an integral part of the surface structure of the cells (bacterial cells, animal cells), or it may be purposefully bound to cells or other insoluble particles (passive agglutination). Thus, the low-molecular-weight haptens can be coupled to the particles for the agglutination reaction. Characteristic of this type of antigens (and thus of the relevant determinants) is that they occur repeatedly on one particle many times. This makes it possible to perform agglutination reactions with monoclonal antibodies as well. Agglutination reactions are usually performed in wells of microtiter plates or on slide slides. The result of the reaction is read visually either directly or under a microscope. They are commonly used to identify bacteria and to typify red blood cells.

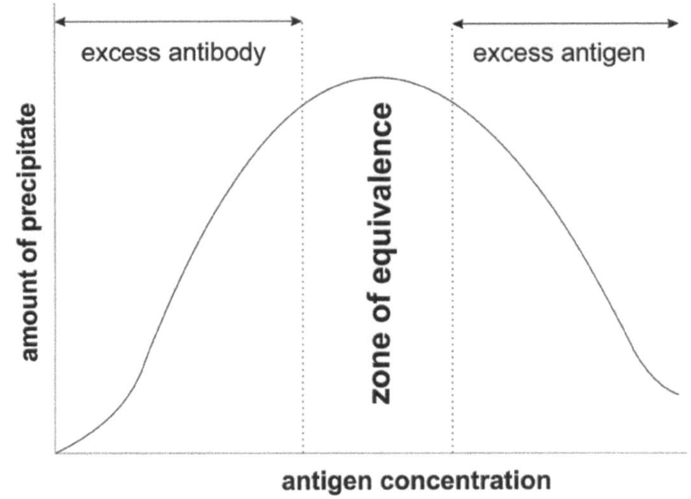

Figure 3.5 Precipitation curve–Dependence of the amount of immunoprecipitation generated on antigen concentration at constant antibody concentration.

Immunoprecipitation in the gel. These methods are based on the diffusion of both or at least one immunochemical reaction partner gel layer and creating a precipitin line in locations where reactant concentration corresponds to the zone of equivalence. The diffusion rate of some techniques increases with DC. The most commonly used format is a 1 mm high layer of agar or agarose gel on a glass plate. When the antibody diffuses into the gel, then we speak of double immunodiffusion. If only one immunoreactant (usually antigen) diffuses and the other is dispersed homogeneously in the gel, then it is a simple immunodiffusion. For both of these variants, electrophoretic distribution of the antigen mixture can be preceded by "precipitation diffusion." In order to improve the visibility of the light precipitation lines, the precipitate can be colored after washing the free immunoreactants from the gel. Depending on the type and purpose of the immunodiffusion, these techniques are divided into two groups:

(A) Double diffusion methods (qualitative or semiquantitative)
- (a) Double radial immunodiffusion (according to Ouchterlony)
- (b) Immunoelectrophoresis (Grabar and Williams)
- (c) Antisense immunoelectrophoresis

(B) Simple diffusion methods (quantitative or semi-quantitative)
- (a) Simple radial immunodiffusion (according to Mancini)
- (b) Electroimmune diffusion (according to Laurella)
- (c) Two-dimensional immunoelectrophoresis

3.3.3.1 Double radial immunodiffusion

The antigen is pipetted into one and the antibody into a second round hole cut into the gel. Both reactants radially diffuse from the wells into the gel, with both gradients being formed in the gel. After 16 to 36 h of diffusion, a characteristic precipitation line is formed between the two wells. Visible precipitates are formed in the gel only in the case of an optimal concentration ratio of the two substances. Therefore, it is advisable to test several immunodiffusion dilutions at the same time in immunocomposing the antigen in the sample against the single antibody solution. This eliminates the possibility of a false negative result where only one dilution of the antigen could achieve the equivalence zone in one of the wells and not in the gel between them (Fig. 3.6).

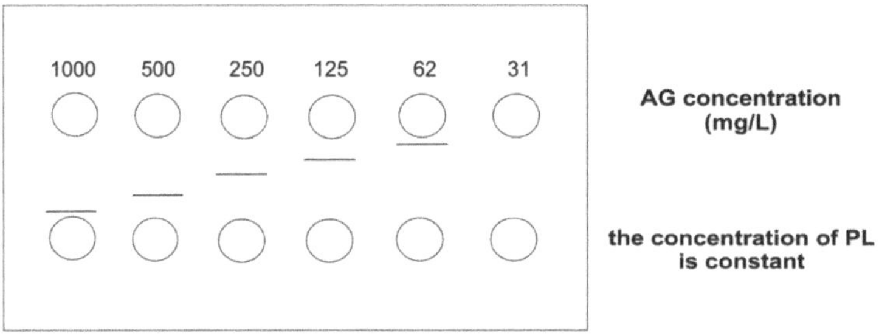

Figure 3.6 One possible arrangement of double radial immunodiffusion; tested is the only antigen at different concentrations.

By evaluating the number of precipitation lines created between wells, we can measure the number of antigens and antibodies in the applied solutions.

If several wells are cut around the polyclonal antibody central well and pipetted to each of them by another test antigen, interacting with the antibody may result in cross-linked or cross-linked precipitation lines. From this, it is possible to find out if the individual antigens contain or not contain identical epitopes, or consider the similarity of these epitopes.

The method is particularly suitable for qualitative assessment. Semiquantitative, the concentration in the samples can be judged by the distance of the precipitation line from the respective well. The implementation of the method is lengthy (2 days).

3.3.3.2 Immunoelectrophoresis

It is a two-stage technique (Fig. 3.7–3.8). In the first step, the mixture of antigens in the gel electrophoretically is divided according to its charge to the individual fractions. In the second step, a polyvalent antiserum (a mixture of antibodies) is applied to the groove cut along the split antigens. Double immunodiffusion is allowed, resulting in curved precipitation lines (curves) that are convex towards the antibody groove. The main advantage of this technique is that it allows the simultaneous detection of multiple antigens. However, the time needed to execute it is 2 days later.

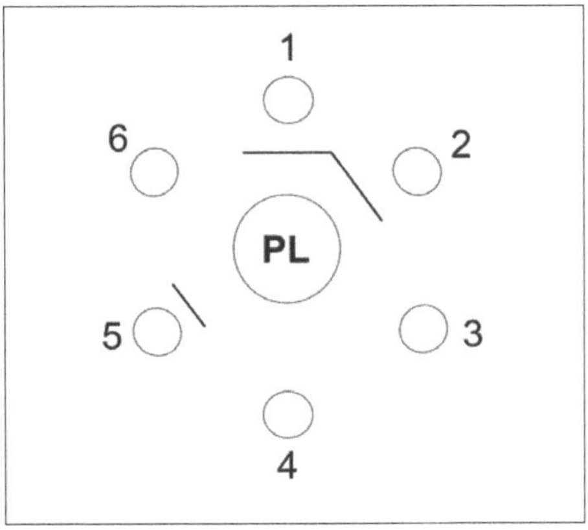

Figure 3.7 Double radial immunodiffusion (about the central well with the antibody, solutions of the test samples are applied to the peripheral wells (1–6) for the presence of the antigen against which the antibody was prepared, the positive reaction is only for the samples in wells 1, 2 and 5).

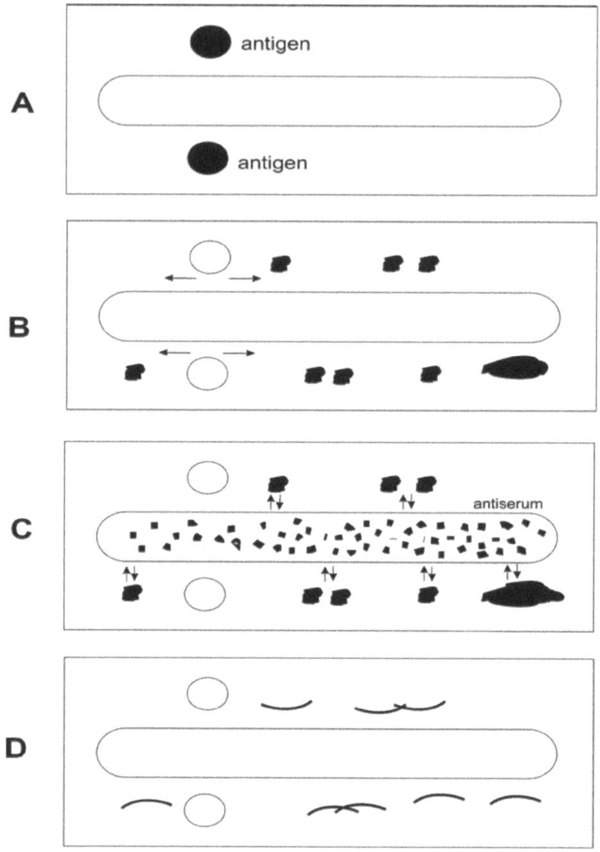

Figure 3.8 Diagram of immunoelectrophoresis: (A) two different antigen mixtures are applied to the wells; (B) the antigens are electrophoretically separated in the gel; (C) an antiserum is applied to the groove; (D), a precipitation line is created after subsequent diffusion of both antigens and antibodies.

3.3.3.3 Antisense immunoelectrophoresis

This is a modification of the double diffusion method. The movement of antigen and antibody molecules is accelerated by direct current, so the precipitation line can be observed after 30 min of diffusion. Because it is working in a pH 8.6 buffer, where the antibodies migrate towards the cathode, the method can only be used to prove antidotes with anodic mobility (Figs. 3.9).

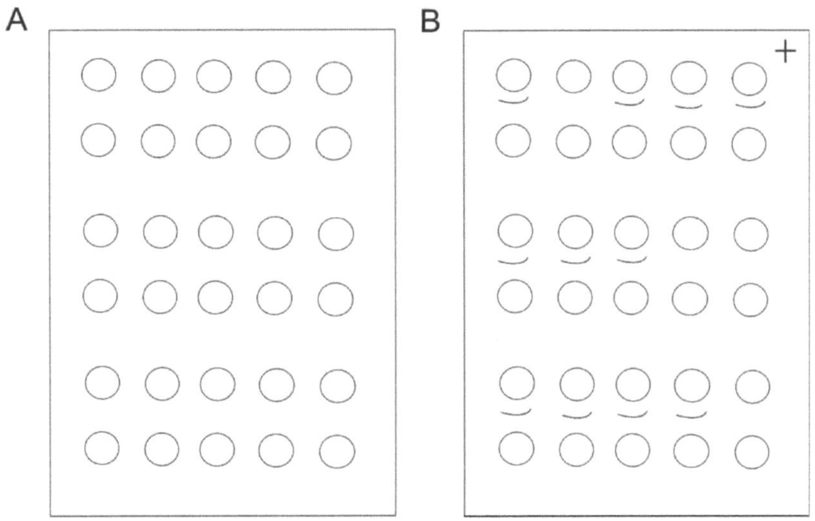

Figure 3.9 Diagram of counter-immunoelectrophoresis: (A) three pairs of rows of wells prior to diffusion; in the upper rows always the antibody, in the lower rows of the sample tested for the presence of a particular antigen; (B) of the precipitation line in positive samples after diffusion.

3.3.3.4 Simple radial immunodiffusion

First, add the antibody to the agarose solution in a 50°C buffer. The solution is poured onto a glass plate and, upon cooling, a layer of homogeneously dispersed antibody gel is prepared. The antigen solution is applied to the cut well. The antigen diffuses radially around the well and forms a precipitation ring in the equivalence zone (see Fig. 3.10). The calibration dependence of the square diameter of the rings on the antigen concentration is linear. The slope and the detection limit of such a calibration line can be affected by the concentration of the antibody in the gel. At a lower antibody concentration, the equivalence zone is formed by the antigen concentration gradient further from the well (the diameter of the precipitation ring is greater).

3.3.3.5 Electroimmune diffusion

It is essentially a prior technique combined with electrophoresis that greatly accelerates antigen diffusion and thus precipitation formation. Most commonly it is called "rocket technology," because

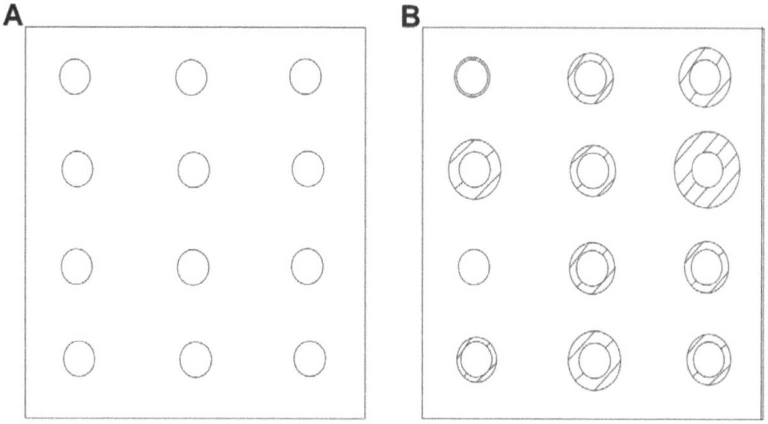

Figure 3.10 Diagram of simple radial diffusion: (A) wells with antigen standards or with test samples at the start of diffusion; (B) a circular precipitation ring around the wells after the diffusion has ended.

the precipitating line is a rocket shape. The principle of this method is the continuous dissolution of the immunoprecipitate with an excess of the traveling antigen and the shifting of the precipitation line in the direction of electromigration until the whole antigen is bound by the antibody. The precipitation line thus receives an elongated pointed shape (Figs. 3.11). The precipitate-defined area is proportional to the antigen concentration. Since the rockets are usually narrow and long, their area can be approximated by the height (i.e., distance of the spike top from the applied antigen well). The concentration of antigen in the samples can be determined from the calibration. In order to successfully implement this method, it is necessary to use an agarose gel that does not exhibit electrodeosmosis. The sensitivity of this method is 10–50 times higher than simple radial immunodiffusion, and it also has the advantage of faster execution (3–12 h). Veronal buffers of slightly alkaline pH are typically used as the electrolyte. Under these conditions, most protein antigens move towards the anode. If migration of antigens of their size (Mh > 100 kDa) is delayed, electrophoresis is performed for a long time, or a mixture of hydroxyethyl cellulose and agarose is used instead of pure agarose. The problem of low antigen mobility due to a small charge (or even cathodic mobility problem) can be solved by carbamylation of the antigen prior to analysis.

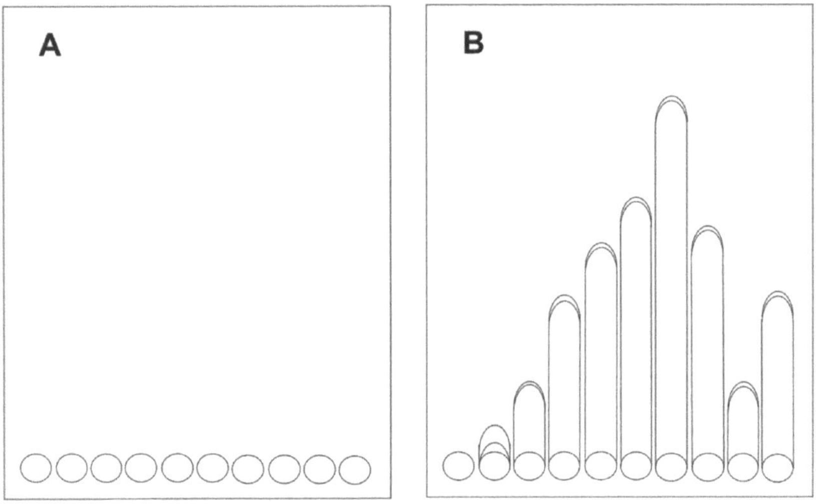

Figure 3.11 Electroimmune diffusion diagram: (A) wells with antigen standards or with test samples prior to diffusion; (B) a rocket-like precipitation line at the end of the diffusion.

3.3.3.6 Two-dimensional immunoelectrophoresis

This method is actually a combination of immunoelectrophoresis (see Section 3.3.3.2) with electro-immunodiffusion technique (see Section 3.3.3.5). The antigen mixture is first electrophoretically separated into the gel. Thereafter, the gel strip is cut off without antigen detection, to contain split antigens, and transferred to the edge of the glass plate, the remaining surface of which contains an agarose gel with a homogeneously dispersed mixture of antibodies (polyvalent antiserum). Further, the second electrophoresis is carried out in the direction perpendicular to the original one. This time, antigens are introduced into the gel containing antibodies and form broad precipitation peaks whose overhangs overlap or overlap (Fig. 3.12). This method (similar to immunoelectrophoresis) is suitable for the evaluation of samples with a larger number of antigens, where in some minor samples it is expected that one of the monitored antigens could fall or be absent. Others may alternatively occur (manifestation of phylogenesis, disease, mutation, technology, etc.). It gives a semi-quantitative image of the concentrations.

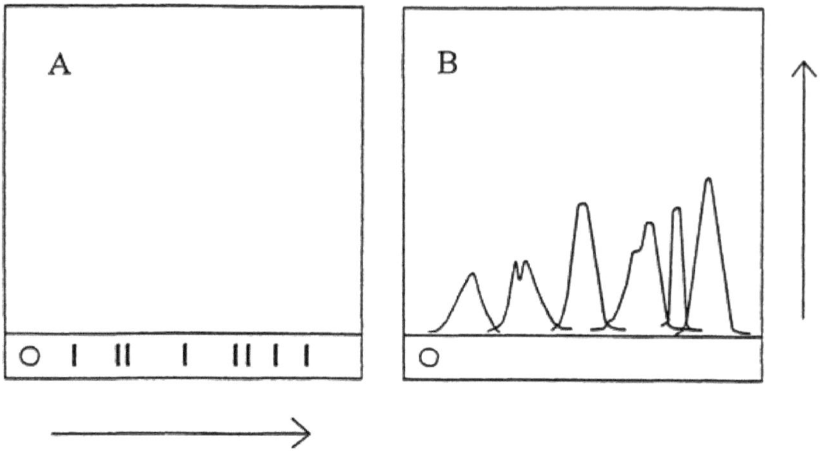

Figure 3.12 Diagram of two-dimensional immunoelectrophoresis: (A) the mixture of antigens is first electrophoretically divided into a gel; (B) subsequently, in the perpendicular direction, the antigens travel to the gel containing the homogeneously dispersed antiserum.

3.3.4 Enzyme Immunoassay

The term enzyme immunoassay (abbreviation EIA) refers to a group of immunoassay methods that use the enzyme reaction in the quantification or detection phase. The enzyme is covalently bound to one of the immunoreactants (antigen, hapten, antibody, second antibody) or in a special arrangement of homogeneous EIA is added to the reaction mixture only after the immunocomplex has been formed. In practice, heterogeneous arrangements are more widely used than homogeneous. The extension of heterogeneous EIA to tertiary laboratories was made possible by the development of an EIA system on a solid phase. Separation of labeled immuno-reactant bound in the immunocomplex from that which remained free in solution is achieved by anchoring the second immunoreactant to the solid support prior to analysis.

3.3.5 Enzyme-Linked Immuno Sorbent Assay

This group of methods is referred to by the abbreviation ELISA of the English Enzyme Linked Immuno Sorbent Assay. (Some authors use the ELISA abbreviation for Enzyme Linked Immuno

Specific Assay.) One of the immunoreactants (PL, AG, hapten) is immobilized on a solid support (membrane, small particles, sticks, tube walls, well walls microtiter plates, etc.). Because there are a number of variants of the ELISA, it is useful to divide them according to whether the immunochemical interaction within the method is competitive or not, as well as the fixed phase immuno-reactant. Noncompetitive ELISAs work with an excess of immunoreactants to capture as much analyte as possible and to obtain the highest signal in enzyme activity measurement. On the other hand, competitive ELISA uses the fact that by decreasing the concentration of immunoreactants the sensitivity of the method increases. This only applies to a certain limit. According to the law on the action of active substance, low concentrations of PL and AG significantly decrease the rate of immunocomplex formation. In addition, this decrease in concentrations is reflected in a deterioration in the accuracy of the assay.

3.3.5.1 Non-competitive ELISA with immobilized antibodies

Immobilization of PL or a solid phase Fab fragment thereof is a widely used procedure for the determination of antigens in the sample. For the application of this method, it is necessary for the antigen to have at least two epitopes. The process consists of several incubations with sequentially added reactants (see Fig. 3.13).

The so-called "second antibody" generates an immunocomplex with the so-called "specific PL" (that interacts with AG). It is prepared by immunization of a different animal species by immunoglobulin fraction of an animal serum—commonly rabbit-mouse, rabbit-pig or mouse-pig.

The most widely used variants for detecting antigen-labeled specific antibodies can be represented as follows for microtiter plates:

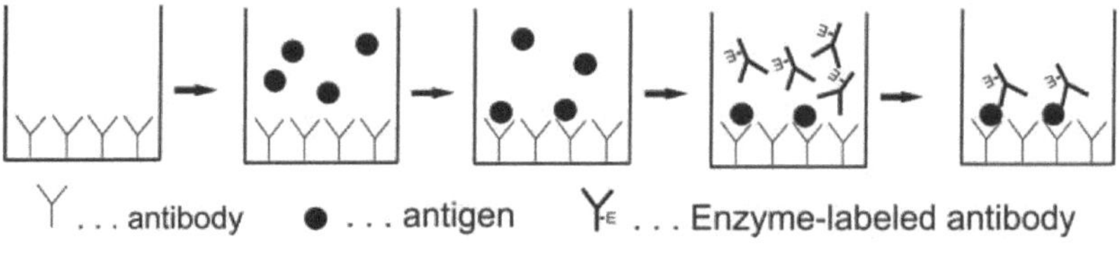

Figure 3.13 Principle non-competitive ELISA with immobilized antibodies.

3.3.5.2 Non-competitive immobilized antigen (hapten) ELISA

Direct method

The immobilized antigen reacts directly with an excess of the antibody-enzyme conjugate.

The method is used to detect proteins after electrophoretic division by immunoblotting. From the gel layer after electrophoresis, the imprint of the separated proteins on the nitrocellulose membrane is obtained by means of a suitable diffusion method, and on that fingerprint is visualized by the respective AG application of an enzyme-specific PL enzyme and subsequently a chromogenic color change in the enzyme reaction. Another application is the detection of AG in tissues and cells by means of an enzyme immuno-histochemical method. These methods are fast but less sensitive, and relatively large amounts of antibodies are consumed in them.

Indirect method

It is an ELISA modification widely used to determine antibodies. In the first stage, the immobilized antigen interacts with a specific antibody and then the second enzyme-labeled antibody.

3.3.5.3 Competitive ELISA with immobilized antibodies

The classic arrangement of the frequently used antigen-locking system (hapten) is to compete antigen-labeled antigen with a test sample of a limited number of antibody binding sites:

The enzyme activity mediated by the solid phase anchor is then inversely proportional to the amount of antigen determined in the reaction. Sometimes this arrangement is called the equilibrium saturation method. Another arrangement of this ELISA variant is called the sequential saturation method (see Fig. 3.14). It does not compete with AG and AG* for PL binding sites at the same time, but they are added to the reaction and incubated sequentially. In a narrow range of concentrations, a higher sensitivity is achieved. The disadvantage of the sequencing method is the need to strictly adhere to the duration of incubation and worse accuracy in the low concentration range.

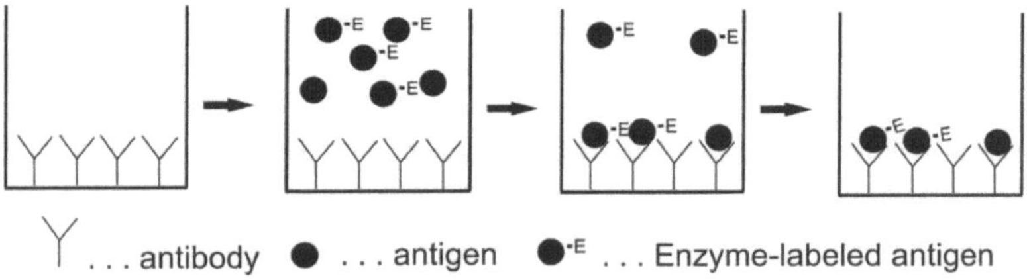

Figure 3.14 Competitive ELISA with immobilized antibodies.

The equilibrium method on microtiter plates can be schematically depicted as shown in Fig. 3.14.

3.3.5.4 Competitive immobilizes antigen (hapten) ELISA

This modification is used to determine antigen (hapten) either in a direct configuration or more often as an indirect ELISA to determine the antigen when a non-enzyme specific PL is used in the first (competitive) phase, and only quantification of immobilized immunocomplexes is performed by a second antibody labeled enzyme (see Fig. 3.15).

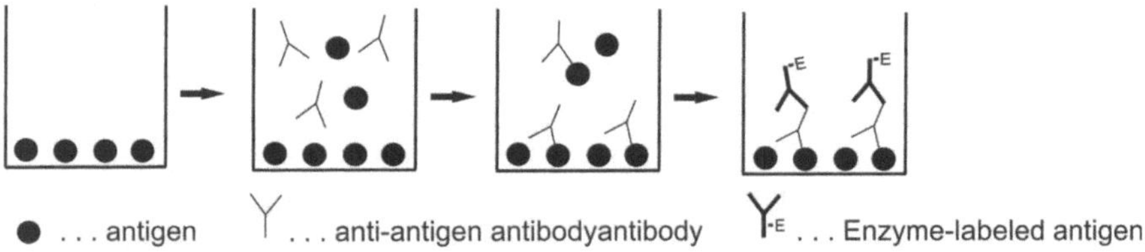

Figure 3.15 Competitive immobilized antigen (hapten) ELISA.

The popularity of this arrangement is due both to achieving high sensitivity and, secondly, to using the second enzyme-labeled antibody universally in indirect ELISA of any other antigen. Such PL2 • E conjugates are additionally commercially available.

Execution of this method on microtiter plates can be schematically shown as shown in Fig. 3.15

3.3.6 Fluorescence Immuno-Assay (FIA)

In FIA, fluorescein isothiocyanate (FITC), tetramethylrhodamine isothiocyanate (TMRITC), umbeliferone, fluorescein, phycoerythrin are most commonly used as fluorophore markers. These fluorophores have the ability to absorb light energy at a certain wavelength and then emit this energy in the form of light with a wavelength of 30–80 nm higher. FIA methods most often comprise competitive homogeneous methods in which the free and immunocomplex-bound labeled antigen (hapten) need not be separated. Their very nature is the fact that binding of the antibody to the labeled antigen will cause certain changes in the fluorescence properties of the label. In these homogeneous FIAs, phenomena such as

- fluorescence polarization;
- excitation transfer of fluorescence;
- graduation of fluorescence;
- quenching fluorescence.

The sensitivity of these methods, however, often limits the interference of ballast substances from biological samples. Therefore, they are used to determine relatively high concentrations (antigens in mg/l, haptens in tens of μg/l).

Between heterogeneous techniques, attention is paid to an arrangement that, instead of traditional fluorophores, uses chelating europium to mark it. This can increase the intensity of fluorescence by up to 5 orders. This increase does not arise directly, but only after the reaction of the primary chelate with the activating solution. At that time, the chelate dissociates the Eu^{3+} ion, which, after irradiation with 340 nm, emits very strong fluorescence with a sharp peak, the peak at 613 nm. This principle used in the arrangement of heterogeneous competitive FIA is known under the abbreviation DELFIA (Dissociation-Enhanced Lanthanide Fluoro-Immuno Assay). A practical example of a heterogeneous noncompetitive FIA is a flow cytofluorometry method that allows for differentiation of cells by membrane antigens. If specific antibodies against certain antigens are previously labeled with a fluorescent substance, cells containing the respective antigens can be quantified using a flow cytofluorimeter in a cell suspension.

3.3.7 Chemiluminescent Immunoassay (CIA)

The most commonly used luminophore is luminol (i.e., 3-aminophthalhydrazide), isoluminol and acridine esters. These substances emit light in the oxidation reaction. In addition to the more manageable heterogeneous CIA, a homogeneous technique based on the transfer of chemiluminescent energy has been proposed. The antigen is labeled with luminoform and the fluorophore antibody. Fluorophore absorbs light emitted by labeled antigen (460 nm) and itself emits light at a higher wavelength (525 nm). So the character of the signal produced by the free labeled antigen differs from the signal that produces the immunocomplex and can be easily distinguished by quantification.

3.3.8 Use of Other Markers in Immunoassay

Dozens of different brands have been used to develop the immunoassay. The biggest application extensions have reached the already mentioned brands-enzymes, radionuclides and fluorophores. Of the others, we include metal compounds (gold, silver, iron, manganese), stable free radicals (so called spin label), bacteriophages, latexes, liposomes. For example, the spin free immunoassay uses the fact that free radical spin (most often dinitrophenyl) produces a magnetic moment that can be measured by a spin resonance spectrometer. The nature of this phenomenon allows us to assemble a homogeneous technique. If labeled hapten is free, we see three intense peaks in the resonance spectrum. The resonance absorption of unpaired electrons varies when the antibody is bound to the labeled hapten. Then we only observe a very weak signal of electron paramagnetic resonance.

3.4 Biosensors

Biosensors are chemical sensors whose recognition is mediated by biochemical mechanisms. This means that the receptor is a biological material that delivers a high degree of selectivity to the biosensor for the measured analyte. At present, enzymes, antibodies, microbial, plant and animal cells, plant tissues, animal tissues, membranes, subcellular particles, and the like are used as receptors. The physico-chemical sensors that cooperate with the bioreceptors can also be of different types: electrochemical, mass, optical, piezoelectric, thermal, etc. The bioreceptor must be suitably captured on the sensor to form an integrated device in which the recognition portion is direct spatial contact with the sensor. Only devices capable of rapid regeneration and repeated calibration can be considered biosensors, so they can be repeatedly used without the addition of reagents.

3.4.1 Electrochemical Biosensors: Enzyme Electrodes

Although optical sensors (laser, luminescence, etc.) are beginning to occupy a significant position among biosensors, the most common biosensors are biosensors on an electrochemical principle. Of these, the most commonly used are those that use different electrodes as a physicochemical sensor. According to the bioreceptor, we distinguish enzyme electrodes, tissue electrodes, cell electrodes, antibody electrodes and others. From these enzymatic (other) electrodes that belong to biosensors, we need to distinguish reactor electrodes that are used to detect enzyme reaction products after passing the analyte (usually substrate) to the immobilized enzyme column. These electrodes are distinctly separated from the enzyme reaction and are applied only as physicochemical sensors.

Enzyme electrodes can measure the concentration of substances that are a substrate, activator or inhibitor of the enzyme that is immobilized on the electrode. The enzyme in the immobilized form is placed on a sensor that measures either the loss of one of the reactants (e.g., oxygen loss in glucose oxidase glucose oxidase) or the resulting product (carbon dioxide in the decarboxylation of fatty acids, amino acids, oxyacids). The substance we want to determine diffuses into the layer containing the enzyme so that the enzyme reaction occurs. The changes occurring in the reaction mixture are measured by the respective electrode and recorded by a connected recorder. The method is inconvenient for both the design and the required chemicals. The electrode can be repeatedly used for as long as the enzyme is active.

A great advantage of enzyme electrodes is their ease of preparation. The electrode must always be commercial, but the enzyme can be immobilized directly in the laboratory where it will be used. If the enzyme is immobilized, for example, on a mesh or on a membrane that only mechanically attaches to the electrode, one electrode can be used for multiple enzymes. It should be noted, however, that enzyme electrodes prepared at workplaces that do not occupy the enzyme immobilization professionally have significantly less stability than enzyme electrodes prepared professionally or commercially.

3.4.2 Immuno-Affinity Biosensors

The principle of affinity cells (electrodes) is the biospecific interaction between the ligand (analyte) and the binding protein, one of the partners of this interaction being in direct contact with the sensitive analytical sensor. The most widespread case of these biosensors are immunoaffinity biosensors utilizing antigen-antibody interactions. One of these pair is covalently immobilized on a metallic electrode. When an antibody electrode is anchored to the thin layer of the hydrophobic polymer (polypropylene, polyvinyl chloride) coating the metallic antibody, an antibody electrode is obtained. In the interfacial space between the surface of the electrode and the solution, there is a charge that depends on the charge of the immobilized antibody. Once the antibody is bound to the antibody, the charge is changed. The charge change can be measured potentiometrically against the reference electrode immersed in the same solution. The potential difference between the two electrodes is proportional to the concentration of antigen in the test solution.

By analogous immobilization of the antigen, we can obtain an antigenic electrode that can be used to quantify the antibodies.

It is possible to determine different haptens and immunogens by antibody electrodes. Detection limits for haptens (e.g., prostaglandins) are in the order of tenths of ng/ml, for immunogens (e.g., enzymes) in the order of ng/ml.

Finally, we would like to point out that the development of biosensors continues very rapidly and new and improved forms are emerging. The revolution in the development of biosensors has meant the use of a lithographic technique that has enabled the so-called chips, i.e., minimal size sensors, to be readily introduced directly into the cell and the measurements performed in vivo.

3.4.3 Optical Enzyme Biosensors

Optical methods are the oldest and still widely used physico-chemical methods in chemical analysis. Their use as sensors in biosensors has enabled the production of high quality and relatively low-cost optical fibers originally intended for communication purposes. Optical fiber biosensors are currently receiving considerable attention because they have a number of advantages from the analytical point of view. The undeniable advantages include the fact that the optical signal is not disturbed by an electric or magnetic field, that the optical sensor does not require a reference signal, and that its biocompatibility and safety can also be used in vivo. They are also very suitable for monitoring in hard-to-reach or dangerous environments, as the detector can be distant up to several hundred meters away from the sensor thanks to the easy signal transmission. The design of the biosensor allows its miniaturization. Optical biosensors also have some disadvantages: the ambient light is influenced by the analysis and the response time is undoubtedly longer than that of the electrochemical biosensors.

Constructive-optical biosensors meet the conditions listed above: the receptor (the most immobilized enzyme) may be in conjunction with an optical detector that may be a monochromator or a fluorescence detector. Laser detectors have been designed, but they have little use for financial demands. During the enzyme reaction, there are changes that are recorded by the respective sensor.

As an example, let us mention the most commonly designed optical biosensors for determining glucose (it has a considerable use in medicine and the food industry). The receptor is usually the enzyme glucose oxidase (GOD), which catalyzes the oxidation of glucose to form hydrogen peroxide and gluconic acid. The changes induced by the enzyme reaction can then be monitored in several different ways:

- Oxygen consumption can be measured by its ability to quench the fluorescence of many organic compounds.
- The formation of hydrogen peroxide can be monitored by using chemiluminescence measurements (with luminol).

- The pH change due to gluconic acid production can be monitored by measuring the changes in acid-base indicators (either changes in absorbance, e.g., using bromocresol green, or fluorescence, e.g., with hydroxypyrenetrisulfonate).
- Changes in the concentration of gluconic acid can also be determined by infrared spectroscopy.

Other enzymes, such as NAD-dependent glucose dehydrogenase, may also be used as the glucose biosensor receptor since, as previously indicated, the NADH that occurs during the enzyme reaction has a different NAD absorbance in the UV region and differs by fluorescence. However, the enzymes we can use for glucose analysis are even more.

The glucose biosensor is given as an example of an optical affinity biosensor. It is based on the ability of A (Con A) to bind in addition to other carbohydrates and glucose. The reaction involves competition for conjugation binding between the glucose present in the sample and the known amount of fluorescently labeled dextran. Changes in fluorescence can then be computed by the presence of glucose present (note the fluorescence decrease is directly proportional to the glucose concentration in the sample).

These examples are only a small example of the use of optical enzyme biosensors that seem to have a great future.

3.4.4 Methods Using Biosensors

In the food industry, biosensors can serve to monitor the presence of certain technologically or nutritionally important substances, but also to detect contaminants such as pesticides, industrial chemicals or natural toxins.

Biosensors reduce the time-consuming nature of the analysis and thus the cost of doing it. Their response is so fast that they are suitable for inclusion in the HACCP system, which is generally considered to be the most effective system for ensuring food safety. Sensitivity of biosensors makes it possible to recognize Escherichia or Salmonella microorganisms, pesticides, herbicides and others in the time horizon of minutes to hours.

While traditional Salmonella cultivation methods may take up to 6 days, and for the detection of low-molecular-weight chemical contaminants, the extraction of conditioned samples followed by chromatography is needed, what is quite time-consuming, the biosensors allow for the detection of these contaminants much faster and in a real time.

The term "sensor" defines a device that reacts to the physical or chemical impulses of the agent and converts it through hardware and software into measurable signals. The biosensor combines two elements: a recognizing biological component (e.g., immobilized antigens, antibodies, enzymes or microbial cells) and a transducer, which may be electronic, optical, piezoelectric, and thermometric, with the greatest development occurring in the field of optical transducers. The biosensors are distributed according to the nature of the transducers or the type of biological activity of the recognition part. According to the biological response, we distinguish two types of biosensors: biocatalytic and bioaffinity (immunosensors).

3.4.5 Piezoelectric Immunosensors

They are based on resonance of a quartz crystal in an alternating electric field. The antibodies are bound on the surface of the crystal. The frequency of oscillation is a function of the crystal mass that changes by binding the antigens to surface-bound antibodies. If a change in the oscillation frequency of the whole crystal is measured, we talk about bulk acoustic (or quartz crystal microbalance [QCM]) devices (frequency decreases with mass increase and is a function of antigen binding). In the case that the acoustic wave only runs on the surface, then it is the so-called surface acoustic wave device.

3.4.5.1 QCM

QCM consists of two circular electrodes between which a cut of crystal is inserted. The alternating current between the electrodes causes the oscillation of the crystal, which affects the binding of the analyte to immobilized antibodies or antigens. Like SPR sensors, they do not even require these labels of antigens or antibodies.

The use of QCM has so far found microbial contaminants such as pathogenic bacteria of the genera Escherichia and Salmonella to detect chemical contaminants not yet common.

3.4.6 Optical Immunosensors

Optical fiber based

On the narrowing of the optical fiber, the biological element, antibody or antigen is immobilized at the lower end. The light enters the wider end and the total inner reflection travels to the narrower end where the radiation is absorbed. Either fluorescence emission of the analytes or, if the analytes themselves do not fluoresce, the fluorescence emission of a suitable reactant is measured. The relationship (direct or indirect) of fluorescence to the analyte concentration depends on the format of the method.

Based on damp waves

Light travels through the waveguide through multiple internal reflections, creating electromagnetic subdued waves that penetrate approximately 200 nm above the surface of the waveguide. Analytes bound to antibodies immobilized on the surface of the waveguide absorb the energy of the wave and then emit it as fluorescence. Fluorescence is a function of analyte concentration.

3.4.6.1 Surface plasmon resonance (SPR)

SPR is an optical phenomenon allowing monitoring of the interaction between the analyte from the sample and the immobilized biomolecule in real time without the need for labeling the reactants. The SPR immunosensor is the most attractive of biosensors for analyzing contaminants because of its sensitivity, selectivity, speed and reliability. The most useful feature is the SPR immunoenzyme's ability to detect small molecules (usually considered in the context of food analysis to be less than 1000 Da) with unusually low LODs in complex matrices such as food. It is most commonly used in flow analysis when a sample passes through the receptor surface where a specific bioaffinity reaction occurs.

Benefits of using SPR biosensors

SPR analysis does not require the labeling of reactants. Labeling is costly and time-consuming to perform, and requires special treatment and equipment when labeling with a radioactive isotope. Marking can even change the reactivity or the specificity of the reaction. Hydrophobic fluorescent labels tend to bind to the background, giving rise to a false positive signal.

The active surface of the sensor can be regenerated with a special solution after use, allowing multiple use of the same chip. One SPR biosensor allows repeated analysis of solutions that do not contain a target analyte because there is no reaction, which is ideal for flow analysis.

Changing the method for other target species by replacing the chip and basically stabilizing it in a buffer solution is an operation lasting in minutes.

3.4.6.2 Atomic force microscopy

Atomic force microscopy (AFM) or scanning force microscopy (SFM) is a very-high-resolution type of scanning probe microscopy (SPM), with demonstrated resolution on the order of fractions of a nanometer, more than 1000 times better than the optical diffraction limit.

The AFM was invented by IBM scientists in 1982. The precursor to the AFM, the scanning tunneling microscope (STM), was developed by Gerd Binnig and Heinrich Rohrer in the early 1980s at IBM Research, Zurich, a development that earned them the 1986 Nobel Prize for Physics. Binnig invented the atomic force microscope and the first experimental implementation was made by Binnig, Quate, and Gerber in 1986.

The first commercially available atomic force microscope was introduced in 1989. The AFM is one of the foremost tools for imaging, measuring, and manipulating matter at the nanoscale.

AFM is a type of scanning probe microscopy (SPM), with demonstrated resolution on the order of fractions of a nanometer, more than 1000 times better than the optical diffraction limit. The information is gathered by "feeling" or "touching" the surface with a mechanical probe: piezoelectric elements that facilitate tiny but accurate and precise movements on (electronic) command enable precise scanning.

The AFM has been applied to problems in a wide range of disciplines of the natural sciences, including solid-state physics, semiconductor science and technology, molecular engineering, polymer chemistry and physics, surface chemistry, molecular biology, cell biology, and medicine.

Applications in the field of solid-state physics include (a) the identification of atoms at a surface, (b) the evaluation of interactions between a specific atom and its neighboring atoms, and (c) the study of changes in physical properties arising from changes in an atomic arrangement through atomic manipulation.

In molecular biology, AFM can be used to study the structure and mechanical properties of protein complexes and assemblies. For example, AFM has been used to image microtubules and measure their stiffness.

In cellular biology, AFM can be used to attempt to distinguish cancer cells and normal cells based on a hardness of cells, and to evaluate interactions between a specific cell and its neighboring cells in a competitive culture system. AFM can also be used to indent cells, to study how they regulate the stiffness or shape of the cell membrane or wall.

In some variations, electric potentials can also be scanned using conducting cantilevers. In more advanced versions, currents can be passed through the tip to probe the electrical conductivity or transport of the underlying surface, but this is a challenging task with few research groups reporting consistent data (as of 2004).

3.4.7 Biological Applications and Other

Force spectroscopy is used in biophysics to measure the mechanical properties of living material (such as tissue or cells). Another application was to measure the interaction forces between from one hand a material stuck on the tip of the cantilever, and from another hand the surface of particles either free or occupied by the same material. From the adhesion force distribution curve, a mean value of the forces has been derived. It allowed to make a cartography of the surface of the particles, covered or not by the material.

3.5 Biological and Microbiological Analytical Methods

Principle and importance of biological and microbiological analytical methods

Compounds that play a significant role in biological systems or which have biological activity can be determined on the basis of this specific activity directly in biological systems. The fundamental difference between these methods and physical and chemical methods is that biological and microbiological methods do not prescribe a chemical individual, but only determine those molecules that actually show biological activity. Therefore, the results of chemical or physico-chemical methods of determining biologically active substances can differ significantly from the results of biological or microbiological methods without any of these results being erroneous. For example, in the case of a chemical determination of a vitamin, we determine the vitamin as a chemical, but we do not take into account that by a small chemical or physical modification this substance could lose its biological activity. By the biological or microbiological method, however, only vitamins that retain their biological activity are determined.

3.5.1 Biological Methods

Biological methods for the determination of biologically active substances are rarely used today and only when there is no other option. The reason is not only the length of these methods, but their high price, the need to use experimental animals, which is now strictly controlled by the animal protectors (and this trend is also suspected in our country) and the impossibility to express results in conventional units.

Biological methods, as we have already mentioned, are carried out on experimental animals for which the test substance is vitally necessary but unable to produce it themselves and must therefore be supplied with food. Such substances are vitamins, essential amino acids, but also some ions. Commonly used experimental animals are young white laboratory rats (strain Wistar). Since each animal is an individual with a number of individual characteristics, animal experiments should always be carried out in at least 10 test group groups and with the same number of control groups.

Biological tests are carried out in two ways, curative or prophylactic.

In the curative method, an animal experiencing a lack of test substance is "treated" by administering a sample with an unknown amount of test substance. The rate at which the symptoms of the observed substance are eliminated is then directly proportional to the content of the substance in the sample. In quantitative terms, the content of the substance can be assessed by comparison with an animal control group which is treated with food with a known content of the test substance after induction of deficiency. Symptoms of lack of test substance may be very specific (e.g., lack of thiamine causes bradycardia, i.e., heart rate acceleration), but in most cases, tests are performed on young animals where any deficiency of vital substances will cause the growth to stop at a first stage. Thus, with exceptions (such as the above-mentioned bradycardia), the curative increments of young animals are followed by the curative method.

The prophylactic method, on the other hand, determines how much of the test sample can still prevent symptoms of lack of test substance. These results are also partially quantified by comparison with the results of the experimental group of animals administered with a known test substance formulation. The method of monitoring the biological effect of the test substance is similar to the curative method.

As mentioned above, obtaining results using biological methods takes several weeks. In the case of curative methods, it is first necessary to induce signs of lack of test substance in the organism (in the case of vitamin testing, the state of deficiency is called hypovitaminosis, with a complete lack of avitaminosis). While it may be possible to induce hypovitaminosis (or lack of other test substances) by administering substances that act against the test substance (for example, vitamin testing may be antivitamins), it may take several weeks. Similarly, the longer time required to show the beneficial effect of the test substance should be taken into account. For the prophylactic method, a longer time is needed to assess the dose, which does not cause any evidence of lack of test substance in the body. All of these factors are the reason why biological methods are used only in the slightest case and today very rarely.

Biological methods are also used to assess the toxicity of drugs or cosmetic products and their effects on the animal organism. When testing drugs, these methods cannot be avoided, but the cosmetics industry has often declared on their products that they have not been tested on animals. However, this type of test cannot be considered as an analytical method.

3.5.2 Microbiological Methods

Microbiological methods are a sort of transition between physical and/chemical and biological methods. Relatively fast (in 1 to 2 days) and with the possibility to express the results in conventional units, the biological effects of the test substances are monitored by the rate of microorganism biomass generation. The lack of vital substances in the culture medium of the microorganism slows or stops the production of biomass. The rate of biomass generation can be set in direct proportions with the content of the test substance in the medium. Biomass growth is monitored either as dry matter of the microorganism or by nephelometric or turbidimetric measurement of turbidity caused by the suspension of microorganisms in the culture medium. The results are compared to a calibration curve constructed by measuring the biomass formation in microorganisms with standard additions of the test substance in the culture medium.

Test microorganisms use microorganisms for which the test substance is vitally necessary but unable to produce it from simpler compounds. The medium in which the micro-organism is cultivated must be of such composition as to exclude the uncontrolled presence of the test substance.

Microbiological methods are of considerable importance in the determination of certain biologically important substances which are very difficult to determine by chemical or physicochemical methods, such as some B vitamins, like folic acid, pantothenic acid, etc. However, it is important to bear in mind that the results obtained, although for the test microorganism expressing the biological effect of the test substance, do not have this informative value for humans because the requirements of the human organism the necessary biologically active substances may be significantly different from the requirements of the test microorganism. The results obtained can then be interpreted analogously to the results of the chemical or physical determination.

References

1. Caponi, L., Migliorini, P. (eds.) (1999). *Antibody Usage in the Lab*, Springer-Verlag (Berlin), p. 154.
2. Fukal, L. (1989). Perspektivy použití imunochemických metod k monitorování některých kontaminantů v životním prostředí, *Chem. Prům.*, pp. 430–433.
3. Fukal, L. (1991). Modern immunoassays in meat-product analysis, *Nehrung*, pp. 431–448.
4. Fukal, L. (1995). Immunoassays—food analysis, *The Encyclopedia of Analytical Science* (London), pp. 2093–2101.
5. Ferenčík, M.(1989). Imunochémia, *Alfa* (Bratislava), p. 591.
6. Kaláb, T. (1995). Imunosensory, *Chem. Listy*, pp. 363–376.
7. Kurtz, D. A., Skerritt, J. H., Stanker, L. (eds.) (1995). *New Frontiers in Agrochemical Immunoassay*, AOAC International (USA), p. 317.
8. Phillips, T. M. (1992). *Analytical Techniques in Immunochemistry*, Marcel Dekker (New York), p. 343.
9. Chard, T. (1990). *An Introduction to Radioimmunoassay and Related Techniques*, Elsevier, p. 290.
10. *Immunology Methods Manual* (1997). Academic Press Ltd.
11. Price, Ch. P., Newman, D. J. (eds.) (1991). *Principle and Practice of Immunoassay*, Stockton Press (New York), p. 650.
12. Hořejší, V., Bartůňková, J. (1998). Základy imunologie, *Triton* (Praha), p. 219.
13. Stryer, L. (1975). *Biochemistry, W. H. Freeman and Co* (New York).
14. Bergmeyer, H. U. (1983). *Methods of Enzymatic Analysis*, Verlag Chemie (Weinheim).
15. Guibault, G. G. (1976). *Handbook of Enzymatic Methods of Analysis*, M. Dekker (New York).
16. Hartmeier, W. (1988). *Immobilized Biocatalysts*, Springer-Verlag (Berlin, Heidelberg, New York).
17. Vrbová, E., Marek, M. (1991). Elektrochemické enzymové elektrody pro stanovení laktátu, *Biochem. Clin. Bohemoslovaca 20*, p. 173.
18. Thévenot, D. R., Toth, K., Durst, R. A., Wilson, G. S. (1999). Electrochemical biosensors: Recommended definitions and classification, *Pure Appl. Chem.*, pp. 2333–2348.

19. Chudobová, I., Vrbová, E., Kodíček, M., Janovcová, J., Káš, J. (1996). Fibre optic biosensors for the determination of D-glucose based on absorption changes of immobilized glucose oxidase, *Anal. Chim. Acta*, p. 103–110.

20. Králová, B., Fukal, L., Rauch, P., Ruml, T. (2007). *Bioanalytické Metody, VŠCHT* (Praha).

21. Mikkelsen, S. R., Cortón, E. (2004). *Bioanalytical Chemistry*, John Wiley & Sons.

22. Sherry, J. (1997). Environmental immunoassays and other bioanalytical methods, *Chemosphere*, pp. 1011–1025.

23. Technical Guide for ELISA.

24. http://www.kpl.com/technical/techdocsearch.cfm?tcat=3.

25. Kodíček, M. (2004). Biochemické pojmy, *Výkladový slovník VŠCHT.*

26. McCarthy, J. (2003). Immunological techniques: ELISA, *Unilever R&D Colworth.*

27. Vass, M., Diblíková, I., Černoch, I., Franěk, M. (2008). ELISA for semicarbazide and its application for screening in food contamination, *Anal. Chim. Acta*, pp. 86–94.

28. Nunes, S. G., Toscana, I. A., Barcelo, D. (1998). Analysis of pesticides in food and environmental samples by enzyme-linked immunosorbent assay, *Trends Anal. Chem.*, pp. 79–87.

29. Mhadhbia, H., Ben-Rejeb, S., Cléroux, Ch., Martel, A., Delahaut, P. (2006), Generation and characterization of polyclonal antibodies against microcystins—Application to immunoassays and immunoaffinity sample preparation prior to analysis by liquid chromatography and UV detection, *Talanta*, pp. 225–235.

30. Ronkainen-Matsuno, N. J., Thomas, J. H., Halsall, H. B., Heineman, W. R. (2002). Electrochemical immunoassay moving into the fast lane, *Trends Anal. Chem.*, pp. 213–225.

31. Gascón, J., Barceló, D. (1994). Rapid magnetic particle-based ELISA assay compared with gas chromatography-nitrogen phosphorus detection for determining atrazine in freeze-dried water samples, *Chromatographia*, pp. 157–162.

32. Kourilov, V., Steinitz, M. (2002). Magnetic-bead enzyme-linked immunosorbent assai verifies adsorption of ligand and epitope accessibility, *Anal. Biochem.*, pp. 166–170.

33. Fujii, S., Ribeiro, R. M. R., Scholz M. B. S., Ono, E. Y. S., Prete, C. E. C., Itano, E. N., Ueno, Y., Kawamura, O., Hirooka, E. Y. (2006). Reliable

indirect competitive ELISA used for a survey of ochratoxin A in green coffee from the North of Paraná State, *Food Addit. Contam.* (Brazil), pp. 902–909.

34. Preston, A., Fodey, T., Elliot, C. (2008). Development of a high-throughput enzyme linked immunosorbent assay for the routine detection of the carcinogen acrylamide in food,via rapid derivatisation pre-analysis, *Anal. Chim. Acta*, pp. 178–185.
35. Mickova, B., Kovalczuk, T., Rauch, P., Moreno, M. J., Abad, A., Montoya, A., Ferri, E., Fini, F., Girotti, S. (2005). Analytical performances of validated chemiluminescent enzyme immunoassays to detect *N*-methylcarbamate pesticides, *Anal. Chim. Acta*, pp. 243–248.
36. AlfaTest: http://vicam.com/products/aflatest.html.
37. AlfaTest: http://vicam.com/products/aflatest-wb.html.
38. AflaCheck: http://vicam.com/products/aflacheck.html.
39. Yau, K. Y. F., Lee, H., Hall, J. H. (2003). Emerging trends in the synthesis and improvement of hapten-specific recombinant antibodies, *Biotechnol. Adv.*, pp. 599–637.
40. Mello, L. D., Kubota L. T. (2002). Review of the use of biosensors as analytical tools in the food and drink industries, *Food Chem.*, pp. 237–256.
41. Patel, P. D. (2002). (Bio)sensors for measurement of analytes implicated in food safety, *Trend Anal. Chem.*, pp. 96–115.
42. Belluzo, S. M., Ribone, M. E., Lagier, C. M. (2008). Assembling amperometric biosensors for clinical diagnostics, *Sensors*, pp. 1366–1399.
43. Piermarini, S., Micheli, L., Ammida, N. H. S., Palleschi, G., Mascone, D. (2007). Electrochemical immunosensor array using a 96-well screen-printed microplate for aflatoxin B1, *Detect. Biosens. Bioelectron.*, pp. 1434–1440.
44. Leonard, P., Hearty, S., Brennan, J., Dunne, L., Quinn, J., Chakraborty, T., O´Kennedy, R. (2003). Advances in biosensors for detection of pathogens in food and water, *Enzyme Microbial Technol.*, pp. 3–13.
45. Shankaran, D. R., Gobi, K. V., Miura, N. (2007). Recent advancements in surface Plasmon resonance immunosensors for detection of small molecules of biomedical, food and environmental interest, *Sens. Actuators*, pp. 158–177.
46. Overview of Surface Plasmon Resonance.
47. http://www.discoversensiq.com/uploads/file/support/spr/Overview_of_SPR.pdf.

48. Surface Plasmon resonance, *Biacore*.
49. http://www.protein.iastate.edu/seminars/BIACore/Technology Notes/TechnologyNote1.pdf.
50. Surface Plasmon resonance, *Biacore*.
51. http://www.biacore.com/food/technology/assay_principles/following_an_interaction/index.html?backurl=%2Ffood%2Ftechnology%2Fassay_principles%2Findex.html.
52. Naimushin, A. N., Soelberg, S. D., Bartolomew, D. U., Elkind, J. L., Furlong, C. E. (2003). A portable surface Plasmon resonance (SPR) sensor system with temperature regulation, *Sens. Actuators*, pp. 253–260.
53. Caldow, M., Stead., S., Day, J., Sharman, M., Situ, C., Elliot, C. (2005). Development and validation of an optical SPR biosensor assay for tylosin residues in honey, *J. Agric. Food Chem.*, pp. 7367–7370.
54. Oflex tylosin product info sheet.
55. http://www.biacore.com/food/service/lit_soft/literature/index.html?backurl=%2Ffood%2Fproducts%25&doc=11890.
56. Cacciatore, G., Petza, M., Rachid, S., Hakenbeck, R., Bergwerff, A. A. (2004). Development of an optical biosensor assay for detection of β-lactam antibiotics in milk using the penicillin-binding protein 2x*, *Anal. Chem. Acta*, pp. 105–115.
57. McGrath, T., Baxter, A., Ferguson, J., Haughey, S., Bjurling, P. (2005) Multi sulfonamide screening in porcine muscle using a surface Plasmon resonance biosensor, *Anal. Chim. Acta*, pp. 123–127.
58. Oflex sulfonamides product info sheet:

    http://www.biacore.com/food/service/lit_soft/literature/index.html?backurl=%2Ffood%2Fproducts%25&doc=11889.
59. Oflex sulfadiazine product:

    http://www.biacore.com/food/service/lit_soft/literature/index.html?backurl=%2Ffood%2Fproducts%25&doc=11883.
60. Oflex sulfamethazine product:

    http://www.biacore.com/food/service/lit_soft/literature/index.html?backurl=%2Ffood%2Fproducts%25&doc=118892.
61. Haasnoot, W., Gercek, H., Cazemier, G., Nielen, M. W. F. (2007). Biosensor immunoassay for flumequine in broiler serum and muscle, *Anal. Chim. Acta*, pp. 312–318.
62. Traynor, I. M., Crook, S. R. H., Bowers, J., Elliott, C. T. (2003). Detection of multi-beta-agonist resuidues in liver matrix by use of a surface plasma resonance biosensor, *Anal. Chim. Acta,* pp. 187–191.

63. Oflex beta agonist product info sheet:

http://www.biacore.com/food/service/lit_soft/literature/index.html?backurl=%2Ffood%2Fproducts%25&doc=11886.

64. McCarney, B., Traynor, I. M., Fodey, T. L., Crooks, S. R. H., Elliot, C. T. (2003). Surface Plasmon resonance biosensor screening of poultry liver and eggs for nicarbazin residues, *Anal. Chim. Acta,* pp. 165–169.

65. Patent US4724318—Atomic force microscope and method for imaging surfaces with atomic resolution.

66. Binnig, G., Quate, C. F., Gerber, C. (1986). Atomic force microscope, *Phys. Rev. Lett.*, pp. 930–933.

67. Lang, K. M., Hite, D. A., Simmonds, R. W., McDermott, R., Pappas, D. P., Martinis, J. M. (2004). Conducting atomic force microscopy for nanoscale tunnel barrier characterization, *Rev. Sci. Instrum.*, pp. 2726–2731.

68. Murugesapillai, D., et al. (2014). DNA bridging and looping by HMO1 provides a mechanism for stabilizing nucleosome-free chromatin, *Nucleis Acid Res.,* pp. 8996–9004.

69. Murugesapillai, D., et al. (2016). Single-molecule studies of high-mobility group B architectural DNA bending proteins, *Biophys. Rev.*, pp. 17–40.

70. Takenaka, M., Miyachi, Y., Ischii, J., Ogino, C., Kondo, A. (2015). The mapping of yeast's G-protein coupled receptor with an atomic force microscope, *Nanoscale*, pp. 4956–4963

71. Radmacher, M. (1997). Measuring the elastic properties of biological samples with the AFM, *IEEE Eng. Med. Biol. Mag.*, pp. 47–57.

72. Perkins, T. (2016). Atomic force microscopy measures properties of proteins and protein folding, *SPIE Newsroom Retrieved.*

73. Thomas, G., Ouabbas, Y., Grosseau, P., Baron, M., Chamayou, A., Galet, L. (2009). Modeling the mean interaction forces between power particles. Application to silica gel-magnesium stearate mixtures, *Appl. Surf. Sci.*, pp. 7500–7507.

Chapter 4

Functional Nanomaterials in Sensing and Biosensing Applications

Lenka Lorencová

Institute of Chemistry, Slovak Academy of Sciences, Dúbravská cesta 9 Bratislava, 845 38, Slovak Republic

chemlelo@savba.sk

This chapter deals with the description of electrochemical sensors and biosensors aiming at the identification and quantification of various chemical species. The introduction of conceptually new biosensors or an improvement of the existing ones is benefiting from the integration of functional nanomaterials. In recent years, great progress has been made in applying nanomaterials to a design of novel chemical sensors and biosensors. Nanomaterials, when used in (bio)detection systems, have shown to be extremely valuable tools to improve the analytical performance of laboratory methods and advance biosensing technology. The development

Glyconanotechnology: Nanoscale Approach for Novel Glycan Analysis and Their Medical Use
Edited by Tomáš Bertók

ISBN 978-981-4800-16-7 (Hardcover), 978-0-429-44849-2 (eBook)
www.jennystanford.com

of new techniques capable of performing very sensitive detection and quantifying certain parameters plays also an important role in medical diagnosis applications. These parameters can be correlated with the presence of specific molecules and their quantity. Unfortunately, these conventional techniques are demanding, expensive, and often complicated. Currently, most of the enthusiasm comes from new possibilities and significant properties promised by nanomaterials. (Bio)sensors combined with nanomaterials are serving as a bridge between advanced detection/diagnostics and daily/routine tests.

4.1 Introduction

Nowadays, nanomaterials are considered a pivotal tool for numerous applications in part due to their high surface area, compared to their respective bulk forms. These nanometer-size structures, which have outstanding physicochemical properties and a size of less than 100 nm or smaller, are extremely useful in many areas of interest (Fig. 4.1). During the past years, there has been an improvement in the synthesis and characterization of different nanomaterials, such as carbon-based nanomaterials, hydrogels, magnetic nanoparticles, polymer nanoparticles, and two-dimensional materials.

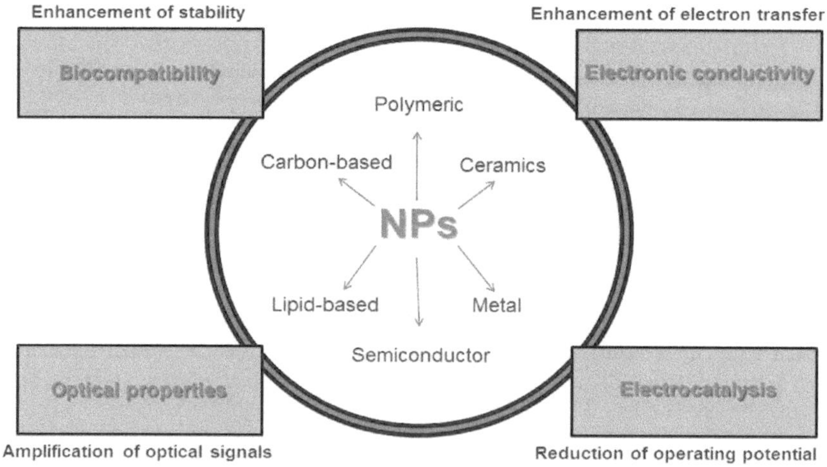

Figure 4.1 Properties and functions of nanomaterials in biosensor applications.

Various electroanalysis-based devices, including ion-selective electrodes, biosensors, lab-on-a-chip devices and other electrical/electrochemical platforms, are developed with the aim on the analysis of DNA, proteins, cells and other clinically important analytes.

> *Analyte*: The target molecular species. This is the molecule we wish to identify and quantify.

In recent years, one of the leading areas of state-of-the-art nanoscience and nanotechnology has been the development and application of novel (bio)sensing systems. The application of nanomaterials to (bio)sensing platforms offers exceptional electronic, magnetic, mechanical, and optical properties (Table 4.1). Nanomaterials can increase the surface of the transducing area of the sensors, which in turn provides enhanced catalytic activity. Nanometer-size structures have large surface-to-volume ratio, controlled morphology, and structure that would scale down the characteristic size, a clear advantage when the sample volume is critical.

Table 4.1 Commercial application of nanoparticles

Nanoparticle component	Application	Indication
Carbon nanotubes	In vitro diagnostics	Respiratory function monitoring
Dendrimers	Therapeutics	HIV, cancer, ophthalmology, inflammation
Gold nanoparticles	In vitro diagnostics	HIV
	In vitro diagnostics, imaging	Labeling reagents (PCR, RNA, Western blotting), angiography and kidney
Liposomes	Drug delivery	Cancer
	Drug delivery	Vaccines: influenza, hepatitis A
	Drug delivery	Fungal infection

(*Continued*)

Table 4.1 (*Continued*)

Nanoparticle component	Application	Indication
Magnetic nanoparticles	In vitro diagnostics	Cancer
	Imaging, therapeutics	Liver tumors, cardiovascular disease, anemia
	Therapeutics	Cancer
Quantum dots	In vitro diagnostics, imaging	Labeling reagents: Western blotting, flow cytometry, biodetection

The emerging development of novel multifunctional nanosystems (Fig. 4.2), in which the combination of different functions into a single nanoparticle provides biocompatibility, biostability, and biodistribution, provides new potential for therapeutic applications that will undoubtedly revolutionize the medical landscape.

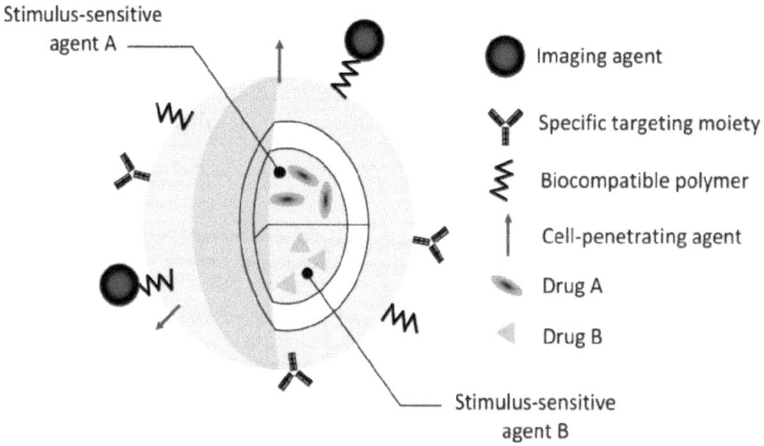

Figure 4.2 Multifunctional nanoparticle for drug delivery.

The design of nanoparticles that combine several properties is an elaborate process that requires different steps, such as the depositing of metal layers onto a supporting nanoparticle core, modifying the biocompatible polymer used to stabilize the nanoparticle and the use of different linkers.

However, advances related to nanomaterials biofunctionalization are crucial to achieve a higher specificity in biosensing. To that end, nanomaterials can be "decorated" with different (bio)receptors offering specific recognition for biosensing. Currently the biosensing field represents an emerging technology with high capabilities and versatility to detect a large number of analytes in different matrices, for instance pesticides in food or water sample, toxic chemicals in food and clinically relevant (bio)molecules in urine or blood (e.g., the pregnancy test and glucose sensors). Given the importance of this application field, the research and development of biosensing technology has continuously blossomed during the last decade. This has been supported by the rise of nanomaterials, building blocks for design of innovative biosensors or improvement of the existing ones. One of the challenges of bringing biosensing devices to the end-user relies on the modification of current technologies so as to make them run in everyday life, rather than in specialized laboratories.

Recent advances in nanomaterials and nanotechnology research allow the development of highly improved existing biosensing systems or novel ones fulfilling the requirements on cost-effective and user-friendly devices.

4.2 The Origins of Bioelectrochemistry

In ancient times the known electrical phenomena were lightning, electrical attraction and the discharges delivered by the torpedo fish. Probably the most intriguing natural phenomenon was the production of electricity by a living organism such as a fish (not only the torpedo fish but also the gymnotus, or eel of Surinam, and the electric catfish living in the Nile waters). It is generally accepted that the science of electricity began in 1600 with the treatise *De Magnete* by William Gilbert. At that time, electricity was considered a "fluid" that flowed within living organisms like water in pipes. The earliest observations on the chemical effects produced by electricity were reported by Beccaria (ca. 1750), who observed gas evolution on passing electric sparks through water placed in tube. The first experiments on the effect of electric currents on living organisms were perhaps described by Johann Georg Sulzer at the same time. Sulzer found that when a plate of lead and another of silver laid on the tongue touched one another, a "vitriolic taste" was perceived by the experimenter. Unfortunately, Sulzer did not connect the observed phenomenon with electricity. The possibility of using Leyden jars opened new ways for investigating the effects of electricity on living organisms. In 1752, Leopoldo Caldani, an Italian physiologist from Bologna, concluded from experiments on the crural nerves of various animals that "electric matter is the most effective of stimulating agents" in living organisms. From the rapidity of the muscular responses to repetitive stimuli via the nervous system, it became generally accepted that a "nerval fluid" or "animal spirit" must exist in living organisms, rapidly identified as "animal electricity." In October 1786, Luigi Galvani, then professor of anatomy in Bologna, gave an account about his famous experiments on the muscular contractions that convulsed frog's legs when touched with a metallic arc, or better a composite (ion + copper or silver) arc. Galvani supposed that the observed phenomena was due to the "animal electricity," the interior or exterior muscles of the frog's legs forming a kind of small Leyden jar, and the nerve behaving as a conductor for the jar (Fig. 4.3). Volta, then professor of physics at the University of Pavia, first shared Galvani's views.

Figure 4.3 The prepared frog and the electric machine on the left allude to the spark experiment.

From 1790 to 1792, he carried out a series of careful experiments dealing with the effect of the electric discharges on the convulsion of frog's legs, varying the nature and the combination of metals by using rather sophisticated heterogeneous metallic arcs. In 1794, he took a different stance, concluding that the "galvanism is nothing but artificial electricity set in motion by the contact of heterogeneous conductors." In conclusion, he claimed that there was only one type of electricity. The construction of the "voltaic pile" (1799) derives, in fact, from Volta's experiments on contact between metals. This discovery can be regarded as the real foundation of the science of electrochemistry and, indirectly, of bioelectrochemistry, considered as an "affiliate" of the electrochemistry. Later, Johann Wilhelm Ritter (ca. 1800) claimed that the contact potential difference between unlike metals could not be a source of energy required to contract the muscles of the frog's legs and it should originate from chemical reactions providing the necessary energy. Volta's breakthrough stimulated a high level of research activity on the effect of the electric current. The development of electrochemistry progressed throughout the nineteenth and twentieth centuries, thus allowing several fundamental as well as applied aspects to be investigated, leading to highly valuable discoveries and theories. Three main avenues were opened, including studies on new power supplies (development of cells and batteries), the problem of the transmission of charge through solutions, and transfer of charge at metal-solution interfaces (Faraday's laws). During this time, bioelectrochemistry developed more slowly, without striking discoveries compared to those that have marked the evolution of "mother" electrochemistry. It has benefited, however, from increasing knowledge in different areas including biochemistry, membrane phenomena, electrophysiology, medical diagnosis, and so on. Progress in the understanding of biological and physiological processes using electrochemistry can be examined on the basis of two distinct sets of phenomena related to (1) the existence of potential differences at phase boundaries and (2) the electrical polarization and the electron-exchange process.

4.2.1 Electrokinetic and Membrane Phenomena

At the end of the nineteenth century, research was undertaken on electrocapillary and electrokinetic phenomena that arise from matter in motion. Three related effects were considered, namely, electroosmosis, streaming potentials, and electrophoresis. The electroosmosis is one of the earliest known electrochemical effects. It was established by Blinks in the middle of the last century that true electroosmosis is negligible compared to osmotic pumping in some large plant cells of fresh water and marine algae.

Streaming potentials are produced by a flow of liquid forced through a capillary system. In 1943, Miller and Dent presented experimental evidence that streaming potentials are the cause of wave portions of an electrocardiogram. Later, several experimenters attempted to correlate electrocardiogram profiles with different physiological events, such as the effect of the pulsatile flow of saline electrolytes that can generate electrokinetic potentials remarkably similar to in vivo cardiograms (e.g., Findl and Kurtz). Electrophoresis is concerned with the migration of particles under the influence of an external field. The method was developed and refined by Tiselius (1937, Nobel Prize in 1947). Since then, there have been a large number of applications in biology and medicine, in particular in the analysis of sera and the separation of amino acids and proteins. Recently, the technique was improved by coupling capillary electrophoresis to an electrochemical detector (CEEC) that can be implanted in vivo for studying microdialysis. The first ion/selective electrodes were constructed in 1936 and then largely developed in the 1960s, when Eisnman established the relationship giving the electrical potential difference between two aqueous solutions separated by an ion-exchange membrane. Of particular interest for measurements in biological media was the construction of ion-selective electrodes capable of detecting calcium ions. Afterwards, other membrane electrodes were constructed (e.g., the urea electrode, by Guilbault and Montalvo). In 1958, Hill calculated that high electric fields could bring about separation of two molecule chains of nucleotides in DNA. It has been suggested that the electric fields and their variations at biological interfaces might act as the trigger for division of genetic material in the cell prior to self/duplication. On the basis of the analogy between a cell surface/biological fluid interface and an electrode-solution interface, investigations were carried out around the 1970s using differential capacitance measurements (a.c. polarographic method) in conjugation with ellipsometry measurements. The dependence on applied potential of the adsorption of several biomolecules on mercury electrodes was investigated, thus giving insights into the different orientations and bindings.

Membranes are organized structures specifically designed to perform several specific functions. They act as a barrier in living organisms to separate the regions, and they must be able to control the transport of matter. Moreover, alternation in transmembrane potentials can have a profound effect on key physiological processes such as muscle contraction and neuronal activity. In 1875, Gibbs stated thermodynamic relations that form the basis of membrane equilibrium. The theory of ionic membrane equilibrium was developed later by Donnan (1911). From the theoretical considerations, Donnan obtained an expression for the electric potential difference, commonly known as the membrane potential between two phases.

4.2.2 Electron Transfer Reactions in Biological Compounds

Curiously, a new field of research was opened up toward the study of electron transfer reactions through the discovery of polarography by Heyrovsky in 1927 (Nobel Prize in 1959). A large series of compounds of biological interest have been investigated through the second and third quarter of the last century using polarographic methods, for example purines, pyrimidines, vitamin B12 and related cobalamines, nucleic acids, pteridines, flavins and flavin nucleotides, porphyrins, cytochromes, and so on. The tests for cysteine and proteins, important in clinical analysis, were introduced by Brdlička (1993). Such a test has been shown to be efficient for detecting hepatitis cases and possible cancers in more than 90% of proved cases. During the second half of twentieth century, electroanalysis has enjoyed a renaissance because of the development of several new techniques (e.g., single and cyclic voltammetry, pulse voltammetry, etc.). Bioelectrochemistry has benefited from these new techniques that allowed other (e.g., solid) macro- and microelectrodes to be used. Microelectrodes were pioneered in the 1940s to measure oxygen concentrations inside biological tissues. Cyclic voltammetry at the platinum electrode has been shown to be suitable for analyses in blood serum by Koryta. The measurement of the time dependence of ascorbic acid concentration in the cortex of an isolated kidney gave information about circulation within the organ (1973). Adams succeeded in probing the concentration of neurotransmitters by directly implanting microelectrodes inside the living brain of rat. Thereafter, microelectrodes were miniaturized to ultramicroelectrodes to be used to probe chemical reactions inside even single biological cells. Another major success for bioelectrochemistry was achieved in the field of redox protein electrochemistry. The first work using this approach was reported in 1965 by Griggio and Pinamonti, concerning cytochrome c. In the 1980s, reports on different families of metalloproteins were published concomitantly by several groups, thus demonstrating that the direct electrochemistry of redox proteins could be observed provided that well-defined experimental conditions were fulfilled. A new route was thus opened towards the understanding of electron transfer processes through electron-carrier chains using the electrochemical model. Thus, the new realm encompassed study of electron transfer processes in living organisms, the analysis of biological fluids, the control of the composition of the intracellular medium (as established by Neher and Sakmann, Nobel Prize winners in 1991), the electrochemical detection of immunological reactions, the construction of biosensors (which constitute a significant portion of the total effort).

4.3 Biosensors

Chemical sensors are miniaturized analytical devices that can deliver real time and online information on the presence of specific compounds or ions in complex samples. A chemical sensor is a device that transforms chemical information, ranging from the concentration of a specific sample component to the analysis of the total composition of a sample, into an analytically useful signal.

Like many fields in science, chemical sensors have benefited from the growing power of computers, integrated electronics, new materials, novel designs, and processing tools. Chemical sensors use chemically formed materials (nanomaterials, molecularly imprinted polymers, etc.) and have advantages such as the capacity to resist harsh conditions such as strong pH, extreme ionic strength, and a wide variety of organic solvents. An analytical device with no recognition function is not a chemical sensor but a concentration transducer. The signal from a sensor is typically electronic in nature, being a current, voltage, or impedance/conductance change caused by the change in analyte composition or quality. While chemical sensors contain a physical transducer and a chemically sensitive layer or recognition layer, the micro-instrument or spectrometer sends out an energy signal, which can be thermal, electrical, or optical, and reads the change caused by the chemical of interest. Chemical sensors are divided into two groups characterized by different functionality: affinity-based sensors and catalytic sensors. Catalytic sensors are modified with different molecules that show catalytic properties. Affinity-based sensors, which are molecularly imprinted polymer-based sensors, have a specific recognition patter mechanism for the target analyte.

The setting for sensor use plays a critical role; it makes a huge difference whether the device is to be operated by a layperson, possibly the patient herself, or by a healthcare professional. Aside from personal use, the most obvious application is in critical care monitoring. A rather optimistic wish list is summarized below:

- An accurate and stable sensor for an essential variable; it must be accurate and stable indefinitely; physicochemical detection must be reversible to minimize sensor recovery and response times
- Non- or minimally invasive
- Continuous monitoring function with the ability to display trends
- Easy to use and a display that is easily understood
- Small size and weight
- Ruggedness and transportability

The term "biosensor" is now often applied to any sensor measuring a chemical concentration in a biological system.

"Detect-to-protect" biosensors are compact devices and analytical tools and a type of chemical sensors converting the biochemical reaction into analytical and measureable signal. Due to their high specificity, which is directly dependent on the receptor (biomolecules or synthetic compounds) that is used, their sensitivity, compact size and user-friendly properties, biosensors are the main choice in detection of chemical and biological components. Principally, biosensors are formed by two components named transducer (where the signal of the biosensor is obtained and transformed into a measurable signal) and recognition part (consisting of a biological or synthetic receptor that utilizes a specific biochemical or chemical reaction mechanism. The biosensors may utilize enzymes, antibodies, nucleic acids, organelles, plant and animal tissue, whole organism, or organs. Biosensors that contain biological catalysts (enzymes) are called catalytic biosensors. These types of biosensors are the most abundant, and they have their largest application area in medicine, ecology, and environmental monitoring. There are two major problem areas in developing biosensors:

(1) the incorporation/immobilization of (bio)receptors in suitable matrix;
(2) monitoring/quantifying the interactions between the analytes and these receptors.

Matrix: everything else present in the sample apart from the analyte. Matrix interference, where sensor response is inadvertently elicited by so-called spectator species present in the sample, is a major problem in bioanalysis due to the complexity of biological systems. Matrix interference can usefully be divided into two categories:

(a) signal caused by non-target molecules, e.g., ascorbic acid or paracetamol, both of which are oxidized at potentials used for the measurement of glucose;

(b) sensor inactivation or "poisoning" due to adsorption of proteins or other surface-active material. For example, albumin is typically present at around 4% (w/v) in plasma. Adsorption of albumin can occlude or scatter light transmission from fiber optic sensors or prevent electrocatalysis on amperometric or voltammetric sensors (vide infra).

Biosensors are defined as analytical devices in which the recognition system is based on biochemical or biological mechanisms. Thanks to the high specificity of the biological recognition systems, biosensors are considered a high selectivity tool for detecting the analytes.

The parts comprising a typical biosensor (Fig. 4.4):

(a) bioreceptors that specifically bind to the analyte
(b) an interface architecture where a specific biological event takes place and gives rise to a signal picked up by
(c) the transducer element; the transducer signal (which could be anything from the in-coupling angle of a laser beam to the current produced at an electrode) is converted to an electronic signal and amplified by a detector circuit using the appropriate reference and sent for processing by, e.g.,
(d) computer software to be converted to a meaningful physical parameter describing the process being investigated; finally, the resulting quantity has to be presented through
(e) an interface to the human operator

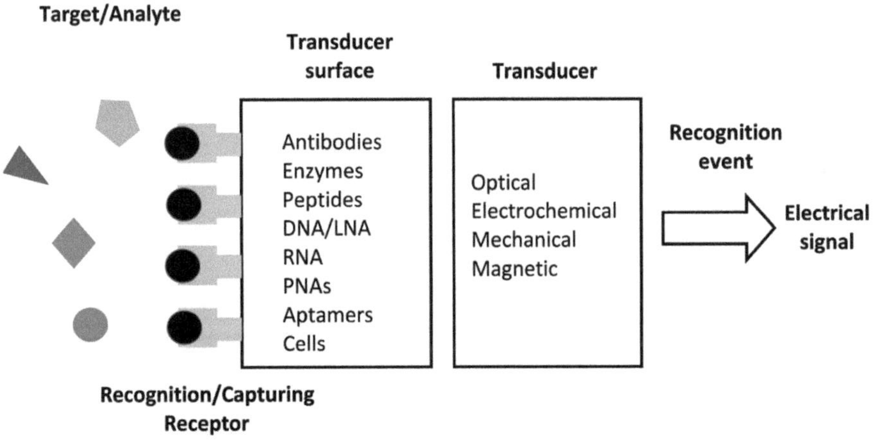

Figure 4.4 General scheme of a biosensor.

In order to design a successful biosensor for the non-specialist market a number of conditions must be met:

1. The biocatalyst must be highly specific for the purpose of the analysis, they must be stable under normal storage conditions and show a low variation between assays.
2. The reaction should be independent of physical parameters such as stirring, pH and temperature. This will allow analysis of samples with minimal pre-treatment. If the reaction involves cofactors or coenzymes these should, preferably, also be co-immobilized with the enzyme.
3. The response should be accurate, precise, reproducible and linear over the concentration range of interest, without dilution or concentration. It should also be free from electrical or other transducer induced noise.
4. If the biosensor is to be used for invasive monitoring in clinical situations, the probe must be tiny and biocompatible, having no toxic or antigenic effects. Furthermore, the biosensor should not be prone to inactivation or proteolysis.
5. For rapid measurements of analytes from human samples it is desirable that the biosensor can provide real-time analysis.
6. The complete biosensor should be cheap, small, portable and capable of being used by semi-skilled operators.

In biosensing technology, the signal of interest is generated by correlating a transducing signal with the capture of the analyte using a (bio)receptor. Then, the transducer transforms this signal into a quantifiable parameter (piezoelectric, optical, electrochemical, etc.), which is the output signal. The key part of a biosensor is the transducer which makes use of a physical change accompanying the reaction. This may be

- the heat output (or the heat absorbed) by the reaction (*calorimetric biosensors*);
- changes in the distribution of charges causing an electric potential (*potentiometric biosensors*);
- movement of electrons produced in a redox reaction (*amperometric biosensors*);
- light output during the reaction or a light absorbance difference between the reactants and products (*optical biosensors*); or
- effects due to the mass of the reactants or products (*piezo-electric biosensors*).

When the recognition event takes places in the biosensor, a significant change occurs in the properties of the system so the transducer can read it and transform the perturbation into an electric signal. In some cases, signal labels are necessary in the system to achieve this readable signal. This type of sensors are referred to as label-based biosensors. Usually labels are organic molecules, nanoparticles, or proteins (proteins or enzymes) which can generate a fluorescent or electrochemical signal. The other alternative is the so-called label-free biosensors, in which the interaction between the target and the recognition element generates a change, physical or chemical, that can be directly red by the transducer and hence used to monitor the interactions in the biosensor. As a general rule, label-free biosensors are the preferred option owing to the disadvantages associated with label-based sensors: the costs of both the labels themselves and labeling the recognition elements, the additional steps in the detection procedure, the alteration of the binding properties of the recognition element due to the attachment of the label, and the nonspecific signal issues associated with the labeling. Unfortunately, the main limitation of label-free assays is the lack of significant changes in recognizing the event of the target. The whole area of biosensors started with the introduction of the first generation glucose oxidase (GOx) biosensor in 1962. The GOx sensor is still the most widely used, although many improvements (generations) have been added since the 1960s. As exemplified by the glucose sensor, electrochemical biosensors do not suffer the drawback of complexity and cost. This is due to their close link to developments in low-cost production of microelectronic circuits and their easy interfacing with standard electronic read-out and processing. Other inherent advantages of electrochemical biosensors are their robustness, easy miniaturization, excellent detection limits, also with small analyte volumes, and ability to be used in turbid biofluids with optically absorbing and fluorescing compounds. The combination of knowledge in bio- and electrochemistry, solid-state and surface physics, bioengineering, integrated circuit silicon technology and data processing offers the possibility of a new generation of highly specific, sensitive, selective and reliable micro (bio-) chemical sensors and sensor arrays addressing these remaining issues.

4.3.1 Mating Nanomaterials and Biological Receptors for Biosensing Applications

In the past years, there has been an improvement in the synthesis and characterization of different nanomaterials, such as magnetic nanoparticles, hydrogels, polymeric nanoparticles, carbon-based nanomaterials and two-dimensional materials. However, advances related to nanomaterials biofunctionalization are crucial to achieve even higher specificity in biosensing.

4.3.1.1 Surface architecture

Sensor surface needs to be conductive and, depending on the detection technique, also transparent. A surface coating is to provide inertness and/or functional groups for tethering of the recognition element, which is often an antibody. The recognition elements (antibodies) are surface immobilized to specifically catch biologically relevant molecules (antigens for an antibody-based sensor) that need to be detected. Many biosensors with some notable exceptions employ labeling to quantify the binding, e.g., the binding of an additional antibody with a fluorescent label in the case of ELISA (enzyme-linked immunosorbent assay).

Antibodies: are proteins with Y shape, with the ability to recognize a specific molecule; that is the antigen, via the antigen-binding (Fab) fragment variable region. This region is highly specific and can be tailored, thus each antibody can be produced with a completely different specificity. Antibodies have two light chains; exactly the same, and two heavy chain, also identical, each component chain contains one NH_2-terminal and one or more COOH-terminal, both chains are linked by disulphide bridges.

In electrochemical biosensors the label is usually an enzyme that catalyzes certain reactions in cases where the bound molecule in itself does not significantly alter the charge transfer process across the electrode interface. Finally, the electrons generated during the recognition event or usually the enzymatic label reactions with a substrate also need to be detected. If the reaction takes place away from the electrode, interface mediators can be used to shuttle the electrons between the reaction site and the surface.

Antibodies represent a very common bioreceptor for analyses based on immunoassays, which exploit different techniques (Chapter 3). On the other hand, it is important to remember that the conjugation of nanoparticles with antibodies combines the outstanding properties of nanomaterials with the highly specific and selective recognition ability of the antibodies targeting a number of antigens.

Specific changes in the heavy chain produce different immunoglobulins (Fig. 4.5). Such immunoglobulins are glycoproteins synthesized by the plasma cells as immune response to neutralize pathogens such as bacteria and viruses.

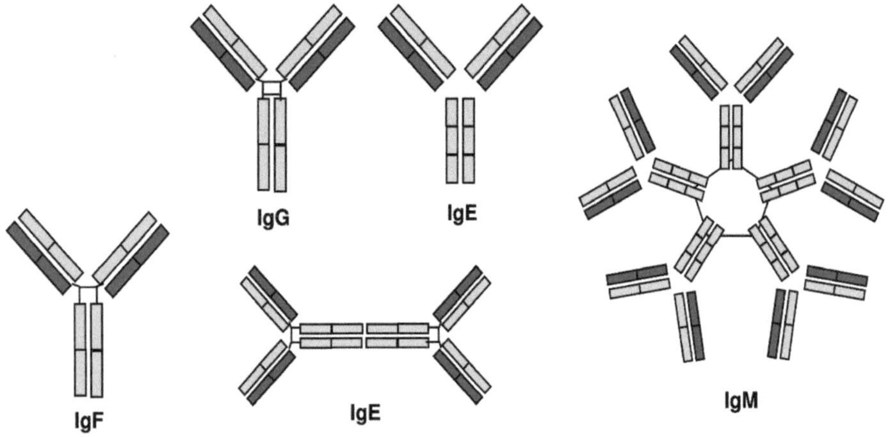

Figure 4.5 Types of immunoglobulin.

There are five immunoglobulin classes found in serum: IgG, IgM, IgA, IgE and IgD. The difference between them is the type of heavy chain they contain. IgG have γ-chains; IgMs μ-chains; IgAs α-chains; IgEs ε-chains; and IgDs δ-chains. The variations in heavy chain polypeptides allow each immunoglobulin class to work as a different type of immune response or during a different stage of the body's defense.

Table 4.2 Monoclonal vs. polyclonal antibodies

Antibodies	
Monoclonal antibodies	**Polyclonal antibodies**
Expensive to produce	Inexpensive to produce
Generate large amounts of non-specific antibodies	Generate large amounts of specific antibodies
Relatively quickly Produced	Hybridomas take a relatively long time to produce
Recognize multiple epitopes on any one antigen	Recognize only one epitope on an antigen
Basic skills required for production	Training is required for the technology used
Can have batch-to-batch Variability	Once a hybridoma is made, it is a constant and renewable source

Due to their high specific surface area nanomaterials provide a better efficiency of the immobilization of host materials and thus enhance a number of (bio)receptor units per area. Covalent binding is one of the most common strategies to immobilize (bio)receptors onto nanomaterials. Depending on the surface chemistry of the involved materials, this type of linking can be carried out by a classical amide reaction. The main drawback is that the anchoring of the biomolecule is uncontrolled and the sensing surface can be negatively affected hindering the biorecognition phenomenon targeted by the biosensing system. Another strategy can be the immobilization of biomolecules by supramolecular or coordinative interactions. One of the most utilized is the biotin/streptavidin system, whose simple—albeit effective—process is also used in the field of biosensors. Physical non-covalent adsorption is widely used in immobilization of biomolecules on surfaces due to its simplicity and cost effectiveness. However, the formed sensing surfaces are often not stable, as the weakly bound receptors may be exchanged for molecules with a higher affinity. Therefore, it is desirable to find a simple method to form a stable sensing surface with high density coverage by antibodies. The biotin-streptavidin system is characterized by a very strong non-covalent biological interaction with a dissociation constant of 10^{15} M. This high affinity ensures high coupling efficiency between the biotin/biotin-derivatives and streptavidin. These biotin-streptavidin complexes, once immobilized, are not disturbed by extreme changes in pH, temperature, organic solvents and other denaturing agents. Therefore, surface immobilized-streptavidin offers a suitable platform for building optimized immunosensors.

Different types of assays (Fig. 4.6) are applied:

- Direct assay: Antigen is directly absorbed on the surface where the biosensing is performed, a detection antibody against the antigen is used.
- Indirect assay: Antigen absorbed on the biosensing platform substrate and a primary antibody is used as intermediate between the analyte and labeled antibody.
- Capture assay "sandwich": A capturing antibody is preimmobilized on the substrate surface, a sandwich with the analyte is formed between the capturing antibody and the detecting antibody.

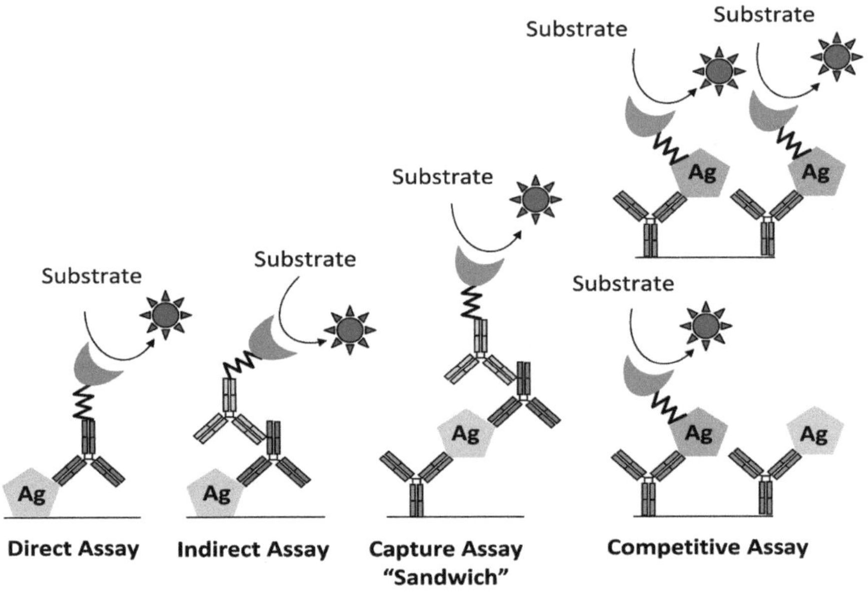

Figure 4.6 Examples of types of assay in biosensor using antibodies as selective molecules.

A labeled antibody against the detection antibody is used to detect the analyte.

- Competitive assay: Artificial analyte or analogues to the analyte is externally labeled, analyte from the sample competes with these labeled pseudoanalytes to capture the antibody. Native analyte in the sample displaces the labeled analyte from the capturing antibodies, washing away the labeled antigen; in the example displayed, the amount of analyte in the sample is related to the decrease in signal of the assay.

4.3.1.2 Lateral flow immunoassay

Lateral flow immunoassays (LFA) can be catalogued within paper-based devices. Their features such as low-cost, robustness, specificity and low limit of detection make them interesting for biosensing applications. Lateral flow is mostly used for semi-quantitative and qualitative assays, including pathogens detection, drugs, hormones and metabolites in biomedical, phytosanitary, veterinary, feed/food and environmental monitoring. However, LFA has some drawbacks such as the qualitative information obtained by an unaided eye, which could be eventually turned into a semi-quantitative signal by a reading device. Moreover, the sample has to remain in a liquid state to avoid clogging of the pores in the lateral flow membrane, which in some cases can be a problem due to unspecific absorptions on the membrane. Nowadays there are different lateral flow assays, the most common and widely used being based upon gold nanoparticles for colorimetric detection. Gold nanoparticles are employed in the colorimetric detection because of their easy synthesis and manipulation, their stability and biocompatibility. Moreover, their intense red color contrasting the typical white background on to the nitro-cellulose strip makes them perfect for visually detectable assays, such as LFA (Fig. 4.7). Currently LFA are reaching new goals by integrating more complex nanomaterials such as magnetic nanogold microspheres, with nano-Fe_2O_3 as a core and gold nanoparticles, carbon nanotubes, quantum dots (CdSe@ZnS QDs, QD and GO).

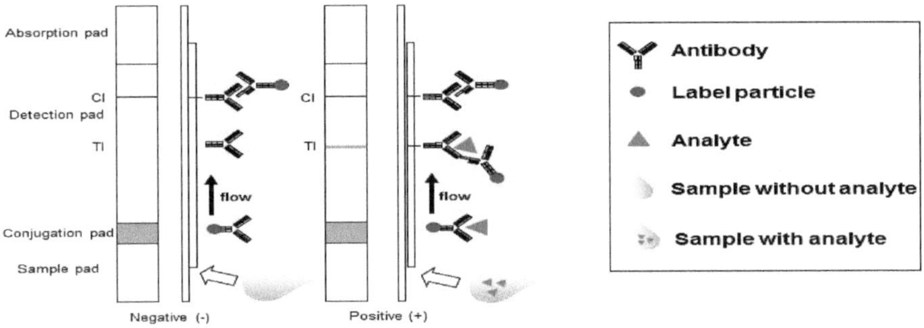

Figure 4.7 Schematic representation of the different parts of lateral flow strip.

4.3.1.3 Molecular imprinting technology and nanomaterials

Molecular imprinting technology is implemented as either bulk polymerization (3-D) or surface imprinting polymerization (2-D). Bulk polymerization produces bulk materials, hence additional preparation steps are required to attain the recognition element (Fig. 4.8) Surface imprinting is more advantageous due to controlled nature of the process and its convenience for sensor technology.

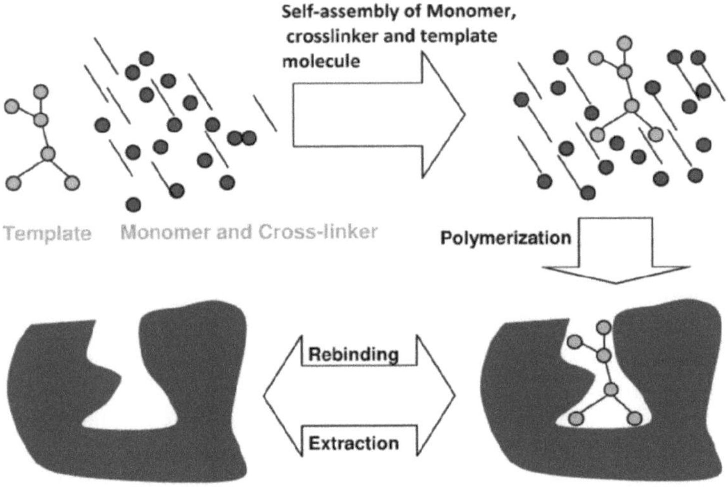

Figure 4.8 MIP preparation steps.

Molecular imprinting is defined as the formation of artificial receptors for a specific target molecule on a polymer or on self-assembled materials. Natural receptors are widely used for sensor technology to target the analyte, leading to electrochemical, optical, mass or magnetic changes on transducers.

Molecularly imprinted polymers (MIP's) are synthetic materials obtained by polymerization of a monomer and a cross-linker, which are located around the target molecule. This assembly of a monomer around a target molecule is encouraged by covalent and non-covalent interactions. It is easier to remove the target molecule from a non-covalently formed MIP-target molecule complex than removing from a covalently formed MIP-target molecule complex. MIPs are synthetic polymers, which can only be used as plastic antibodies for now.

Therefore currently, MIPs on sensors are only used as affinity sensors and not as a catalytic biochemical enzyme mimicking sensors. MIPs are highly stable and resistant to physical and chemical treatment, including heating, acids, bases and organic acids—these polymers are physicochemically stable at extreme pH and temperatures levels. MIPs are also easily synthesized and handled, and they have a low fabrication cost, high specificity, and excellent reproducibility. The application of MIPs may face major limitations in certain fields, including heterogeneous binding, small binding capacity with the analyte, lower selectivity, and poor site accessibility.

Molecularly imprinted polymers are a group of compounds in which the functional and crosslinking monomer are copolymerized in the presence of a target analyte. During the final step the analyte is removed and the cavities (acting as non-biological receptor) are ready to be used. These cavities are a kind of mold complimentary in terms of size and shape regarding the imprinted analyte. MIPs represent an effective alternative to conventional biosensors as they are cheaper and easier to manipulate in comparison with enzymes or antibodies and depending on their size, MIPs can bear thousands or millions of binding sites, whereas biological receptor show a few binding sites.

For the sensor/analytical device preparation, there are two methods to design MIPs; first is the in situ technique with the imprinted polymer prepared on the transducer, whereas the second is an ex situ technique with the imprinted polymer prepared separately from transducers. The MIPs are then immobilized on transducers and the result is a molecularly imprinted sensor. The integration of MIP with the transducer can be achieved in different ways: e. g., polymerization, spin coating, electropolymerization, open sandwich MIP (using antibodies combined with MIP technology), and other procedures, with excellent results. The most frequently used MIP-based biosensing techniques are fluorimetry, conductometry, chromatography, or a number of electrochemical techniques. Molecular imprinted nanomaterials are excellent candidates to fabricate sensing layers. There are two great advantages to consider, high surface area and small size. In fact, precipitation, emulsion, free radical polymerization and core–shell approaches are the most common techniques for synthesizing MIPs at the nanoscale level. Nanomaterials can be combined with MIPs by the surface imprinting technique, which allows to print the complex onto surfaces such as screen printed electrodes, plastics, or other biological/inorganic particles. Currently, the fabrication of imprinted nanospheres can be achieved via precipitation polymerization. A combination of molecularly imprinted polymers and transducers form a synergistic device. MIPs are used as recognition elements in the design of sensors due to their higher thermal stability than biological receptors, reusability, and selectivity. These polymeric materials bind to the target molecules causing variations in physical parameters, such as mass, absorbance, or refractive index depending upon the shape, charge, and functionality of the target molecule. The design of these synthetic materials, which are able to mimic the recognition processes found in nature, has become an important and active area of research making molecular imprinting one of the strategies followed to create materials with recognition ability comparable to the natural systems in recent years. Therefore, their usage in sensor technology is very beneficial. This technology is also quite suitable and advantageous for non-electroactive molecule detection. Non-electroactive species are molecules that cannot be transformed by electrochemical reactions such as pesticides, drugs, etc. Therefore, they can be measured by affinity techniques, or catalytic secondary molecule usage. Secondary molecule usage has disadvantages such as secondary molecule and target molecule interaction, solvent problems, where template and secondary molecule may not be solved in the same solvent or harsh conditions can affect the reaction of target molecule, hence the measurement. Affinity measurement is used to detect molecules depending on the affinity between target molecule and an auxiliary molecule. In biosensor technology DNA, antibody, protein, and receptor-based systems are designed which could be collectively called affinity-based systems. However, these biocompounds are expensive, hard to immobilize onto transducers, and challenging to study at their optimum conditions.

Since then, MIP-based sensors have been constructed as electrochemically, optically, and piezoelectrically active.

Electrochemical techniques of molecularly imprinted sensors (Fig. 4.9) mostly measure surface properties of the transducers, binding kinetics and polymer rearrangements.

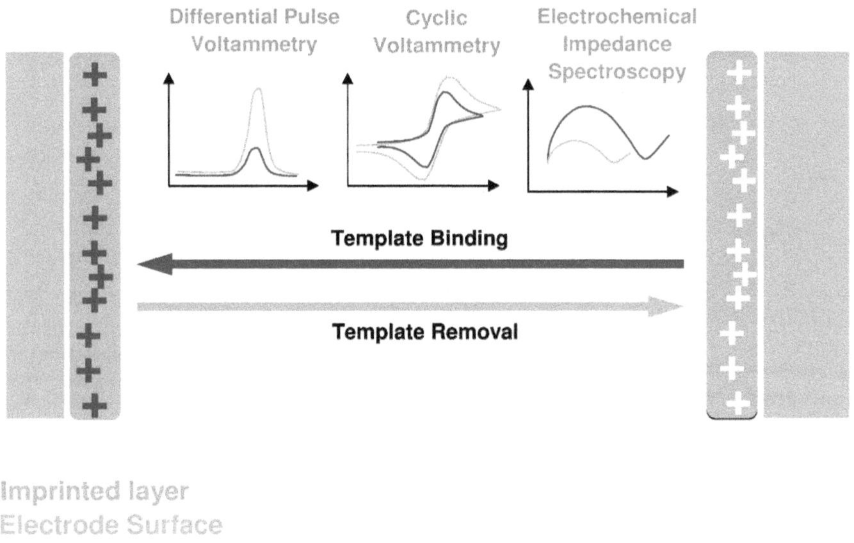

Figure 4.9 A representative molecularly imprinted electrochemical sensor system.

4.4 Electrochemical Sensors and Biosensors

One of the main challenges facing the analytical chemist is the development of methods that respond to the growing need to perform rapid in situ analyses. These methods must be sensitive and accurate, and able to determine various substances with different properties in "real-life" samples. In recent years, many of the methods developed with this end in sight have been based on the use of electrochemical techniques due to their high sensitivity, selectivity and low cost.

> *Sensitivity*: the change in sensor output per unit change in analyte concentration. This could be measured in μA mol^{-1} dm^{3} for an amperometric biosensor or absorbance units per mole for a spectrometric device.
>
> *Selectivity*: the ability of the analytical method to respond only to the target analyte. The expected error can be quantified for each expected interferent as follows:
>
> Maximum error = (Effect of interferent/Effect of analyte) × 100%, (4.1)
>
> where the interferent concentration is maximum expected in the sample and the analyte concentration is the minimum expected. Where there are more than one expected interferents, these can be summed (assuming no synergistic effects).

Electroactive properties of nanoparticles towards certain reactions have been widely exploited in biosensing. In electrochemical sensors, NPs can be detected by direct or indirect electrochemical detection. In the former case, NPs are detected by their own redox properties; therefore, they can be detected without any preliminary step, meaning that the metal forming the NPs is electrically detected in a direct mode. As a strong point, direct detection offers rapid responses with reasonable limits of detection (LOD), short analysis time, and the possibility to perform a compact/integrated assay.

> *Limit of Detection* (LOD): the level of analyte that leads to a sensor signal which is statistically significantly different from the background signal obtained in the absence of analyte. A frequently used definition of LOD is a concentration that gives a signal greater than three times the standard deviation of a blank sample consisting entirely of matrix.

On the other hand, indirect electrochemical detection exploits, e.g., the electrocatalytic properties of the AuNPs.

As electrocatalyst, nanoparticles take part in the catalysis and contribute to the electron transfer between the electrode and the reagents involved. However, an alternative procedure to perform indirect electrochemical detection is the preliminary oxidative dissolution of the AuNPs in acidic medium followed by the detection of the metal ions released from the particle. Electrochemical signal from NPs can be easily monitored by electrochemical means such as chronoamperometry, differential pulse voltammetry, square wave voltammetry, electrochemical impedance spectroscopy. Low LODs, in order of pM have been achieved.

4.4.1 Electrochemical Sensing Platforms

In electroanalysis, various conventional electrodes are applied for preparation of sensors and biosensors. Conventional electrodes include solid stationary phases made from different metallic and carbon compounds.

4.4.1.1 Classical solid electrodes

Widely used electrode materials include gold, microporous gold, graphite, glass carbon and indium tin oxide (ITO). Noble metals such as platinum, gold, sometimes copper are used as solid electrode materials. Less noble metals are not so suitable because they participate in side reactions with the solution components. Besides mechanical polishing, metal electrodes must be electrochemically cleaned and activated as well (especially platinum). The working electrode is allowed to polarize at certain potentials in a suitable solution (e.g., in dilute mineral acid). The most common is the cyclic change of the electrode potential between such positive and negative potential values that allow for the evolution of oxygen and hydrogen on the electrode. The advantage of electrochemical cleaning is that it can be continuously applied during the measurement so that the electrode is polarized by altering between measurement and purifying pulses.

It is relatively simple to clean carbon electrodes of different types, the surface of which can be polished with fine metallographic or abrasive paper, diamond paste and Al_2O_3 slurry. Carbon electrodes, however, cause difficulties due to their porous nature, resulting in the retention of the solution, leading to high residual current and poor reproducibility of the measured signal.

Porosity is reduced by impregnation of the electrode, e.g., by cerine, or low porous pyrolytic graphite, or most commonly by glassy carbon.

Glassy carbon (GC) is a very popular electrode material for its excellent mechanical and electrical properties, a wide potential range and relatively well reproducible results. However, the glassy carbon is sensitive to large currents and rapidly ages (recrystallizing to graphite), thereby increasing the noise and non-reprocessing residual current.

Glassy carbon is prepared from pre-modulated polymer phenol-formaldehyde resins by careful (controlled) heating under an inert atmosphere. The carbonation process runs slowly in the range 300 to 1200°C, removing O, N and H from the resin, leaving only a carbon skeleton in which the individual graphite-like layers are cross-linked with thin, entangled "ribbons." Due to its high density and low pore size, glassy carbon requires no impregnation. Only the surface is polished to a "mirror" gloss gradually by a more gentle suspension of Al_2O_3.

The rotating disk electrode (Fig. 4.10) represents the most studied and best defined hydrodynamic electrode. It was introduced and described by Levič on the basis of hydrodynamic ideas. A metal or carbon disk is inserted into the inert material (teflon). The electrode rotates about the axis and also sets the solution in motion.

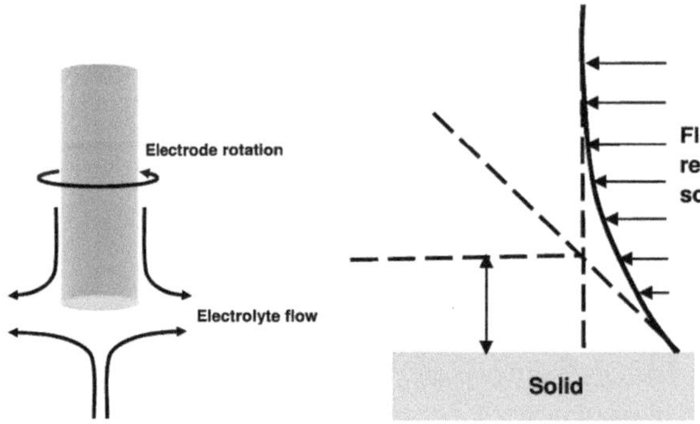

Figure 4.10 Rotating disc electrode (left). Gradient of fluid flow rate on RDE surface (right).

4.4.1.2 Screen-printed electrodes (SPEs)

Since the 1990s, screen-printing technology, adapted from the microelectronics industry, has offered high-volume production of very affordable, and yet highly reproducible and reliable single-use sensors; a technique which holds great promise for on-site monitoring. Therefore, the use of screen-printing technology in the mass production of disposable low-cost electrodes for the electrochemical determination of a wide range of substances is currently undergoing widespread growth. The main advantage associated with down-sizing of the electrochemical sensors is the reduction of sample volume required, to as low as a few microliters, which in turn helps in reducing the overall size of the diagnostic system into which the device will be integrated.

Screen-printed technology consists of layer-by-layer depositions of ink upon a solid substrate, through the use of a screen or mesh, defining the geometry of the sensor. This technology has advantages of design flexibility, process automation, good reproducibility, a wide choice of materials. SPEs usually include a three electrode (Fig. 4.11) configuration (working, counter and reference electrodes).

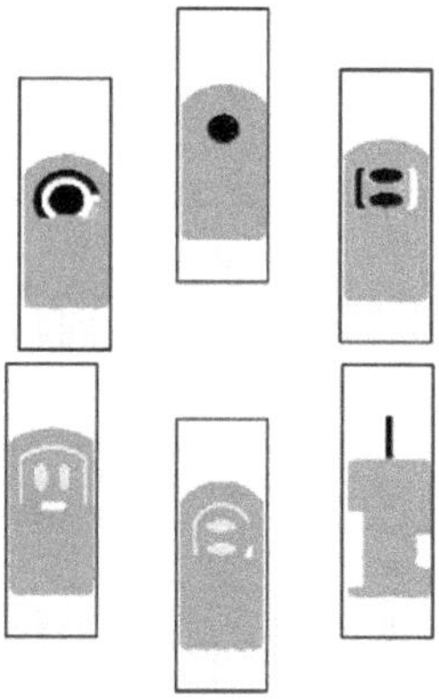

Figure 4.11 Various electrodes prepared by screen-printing technology.

SPEs are devices that are produced by printing different inks on various types of plastic or ceramic substrates. Polyester screens are generally used for printing with patterns designed by the analyst in accordance with the analytical purpose in mind. The composition of the various inks used for printing on the electrodes determines the selectivity and sensitivity required for each analysis. Alternatively, a wide variety of devices of this type are commercially available. The great versatility presented by the SPEs lies in the wide range of ways in which the electrodes may be modified.

Table 4.3 Applications of SPEs

Working SPE	Analyte
Au, Bi, Ni-coated SPCE	Metals
Hg-film-modified SPCE	Metals
Metal-based SPEs	Metals
SPCE	H_2O_2, metals
Enzyme-modified SPE	H_2O_2, metals
Metallic NPs-modified SPE	H_2O_2, metals
SPCE	aurothiomalate, creatinine, cysteine and tyrosine, dopamine,
	chlorophyll, phloroglucinol derivates, procaine, vitamin B_2, uric acid
Enzyme-modified SPE	ethanol, glucose, herbicides, cholesterol, phenolic compounds, pesticides
SPE immunosensor	allergy antibody (IgE), amphetamine, DNA, food pathogens,
	hormones, human cytomegalovirus, *Listeria monocytogenes*,
	mycotoxins, Mycobacerium tuberculosis, salmonella, seafood toxins, pneumococcal pneumonia, myocardial and tumor markers, *Vibrio cholera*

The composition of the printing inks may be altered by the addition of different substances such as metals, enzymes, polymers, complex agents, etc. On the other hand, the manufactured electrodes can be modified by means of depositing various substances on the surface of the electrodes such as metal films, polymers, enzymes, etc. This versatility, its compact size, and the portability allow for highly specific on-site determination of target analytes (Table 4.3).

Electrodeposition along with the screen-printing process is beginning to allow mass production of electrochemical sensors with various catalytic activities (Fig. 4.12). The sensor strips fabricated by this process are promising tools with higher detection rates. Drop casting, the addition of nanomaterial in the ink during the SPE printing, the electrodeposition, the Langmuir-Blodgett film approach, and the deposition by means of the electrospray technique are among most popular modification methods.

Figure 4.12 SEM images of electrodeposited silver (left) and nickel (right) NPs onto SPCEs.

The application of various nanoparticles and nanocomposites also helps to increase the stability during a cyclic recharging in supercapacitors. In addition the modification of prepared materials with NPs would improve catalytic properties towards hydrogen evolution reaction (HER), hydrogen oxidation reaction (HOR) and oxygen reduction reaction (ORR). Pt nanoparticles as significant ORR catalysts are commonly used in proton exchange membrane fuel and direct methanol fuel cells. The overall HER and its reverse, HOR, $H^+ + e^- \leftrightarrow (1/2)H_2$, could proceed via the Volmer–Heyrovsky or the Volmer–Tafel mechanisms.

$$H^+ + e^- \rightarrow H^* \quad \text{(Volmer)}$$

$$H^* + H^+ \rightarrow H_2 \quad \text{(Tafel)}$$

$$H^* + H^+ + e^- \rightarrow H_2 \quad \text{(Heyrovsky)}$$

Carbon-based nanomaterials are often materials of choice because of their unique properties, such as a high surface-to-volume ratio, high electrical conductivity, chemical stability, biocompatibility, and robust mechanical strength. Carbon-based NPs also increase the electroactive surface area, enhance electron transfer, and promote adsorption of molecules.

Graphene-based nanomaterials offer excellent electron conductivity, a large edge plane/basal plane ratio, high surface area, superior hydrophobicity, rapid electrode kinetics and enhanced mechanical strength. Graphene-based electrodes have been commonly used as chemical sensors, such as gas sensors, mechanical sensors, resonant sensors, humidity sensors, biofuel cells, environment sensors and optical sensors. Graphene-based biosensors are suitable for sensing of ascorbic acid, dopamine, glucose, hydrogen peroxide, cholesterol, metal ions, pesticides, uric acid, dinucleotide (NADH) and reduced b-nicotinamide adenine molecules.

The noble metal NPs on MXene showed a highly sensitive surface-enhanced Raman Spectroscopy detection of several common dyes with calculated enhancement factors on the order of 10^6. These findings open a pathway for SERS applications of novel 2D hybrid materials in sensors, catalysis, and biomedical applications. Applying the $Ag@Ti_3C_2T_x$ nanocomposites as nanocarriers, an acetylcholinesterase biosensor was fabricated through drop-casting method for electrochemical detection of organophosphate pesticides (OPs), via enzymatic inhibition pathway.

In the construction of biosensors, enzymes (Table 4.4) are often applied as useful biological elements due to their high extraction, isolation and purification costs.

They rapidly and cleanly form selective bonds with the substrate. Enzymes were the first biocatalysts used in biosensors and remain by far the most commonly employed. Enzymes demonstrate a significantly greater level of substrate specificity as a key property, primarily because of the constraints placed on the substrate molecule by the active site environment. This fact involves factors such as molecular size, polarity, stereochemistry, functional groups and relative bond energies. Disposable biosensors based on enzyme immobilization on SPEs have been widely used for the analysis of several analytes.

Table 4.4 Disposable enzyme-based biosensors

Enzyme-modified SPEs		
Determination	**Enzyme**	**Reaction**
pesticides and herbicides	acetylcholinesterase (AChE)	Acetylthiocholine + H_2O + AChE → thiocholine (TCh)+acetuc acid 2TCh (red) → TCh (ox) + $2H^+$ + $2e^-$
heavy metals	urease + glutamate dehydrogenase (GLDH)	Urea + H_2O + urease → CO_2 + $2NH_3$ Urea + H_2O + urease → CO_2 + $2NH_3$ NH_3 + α-ketoglutarate + NADH + H + GLDH → l-glutamate + NAD^+
cholesterol	cholesterol oxidase (ChOX) + peroxidase (POD)	Cholesterol + O_2 + H_2O + ChOX → cholest-4-en-3-one+H_2O_2 H_2O_2 + H^+ + $2K_4Fe(CN)_6$ + POD → $2K_3Fe(CN)_6$ + $2K^+$ + $2H_2O$ $2K^+$ + $2K_3Fe(CN)_6$ + $2e^-$ → $2K_4Fe(CN)_6$
glucose	glucose oxidase (GOx), glutaraldehyde (GA) and tetrathiafulvalene (TTF) /glucose oxidase (GOx)	Glucose + $2TTF^+$ → gluconic acid + 2TTF + $2H^+$ 2TTF + $2TTF^+$ + $2e^-$ Glucose + O_2 GOx → gluconic acid + H_2O_2
ethanol	alcohol oxidase (AOX) alcohol dehydrogenase (ADH)	RCH_2OH + O_2 + AOX → RCHO + H_2O_2 RCH_2OH + NAD + ADH → ADHRCHO + NADH + H^+
phenolic compounds	tyrosinase (TYR)	Phenol + O_2 + TYR → catechol Catechol + O_2 + TYR → o-quinone + H_2O o-Quinone + $2H^+$ + $2e^-$ → catechol Catechol → o-Quinone + $2H^+$ + $2e^-$
	cellobiose dehydrogenase (CDH)	Cellobiose + CDH → cellobionolactone + $2e^-$ + $2H^+$ o-Quinone + $2H^+$ → $2e^-$ + catechol Catechol → o-Quinone + $2H^+$ + $2e^-$
Oxidizable H_2O_2 or NADH is easily generated as a result of these enzymes:		Substrate + O_2 oxidase → Product + H_2O_2 Substrate + NAD dehydrogenase → Product + NADH

4.4.2 Electrochemical Detection Techniques

In biosensing the measurement of electric properties to extract information from biological systems is usually of electrochemical nature. A bioelectrochemical component serves as the main transduction element. Although biosensing devices employ a variety of recognition elements, electrochemical detection techniques use predominantly enzymes. This is mostly due to their specific binding capabilities and biocatalytic activity. Other biorecognition elements are, e.g., antibodies, nucleic acids, cells and micro-organisms. An immunosensor uses antibodies, antibody fragments or antigens to monitor binding events in bioelectrochemical reactions. Typically in (bio-)electrochemistry, the reaction under investigation would either generate a measurable current (*amperometric*), a measurable potential or charge accumulation (potentiometric) or measurably alter the conductive properties of a medium (*conductometric*) between electrodes. References are also made to other types of electrochemical detection techniques, such as *impedimetric*, which measures impedance (both resistance and reactance), and *field-effect*, which uses transistor technology to measure current as a result of a potentiometric effect at a gate electrode.

4.4.2.1 Cyclic voltammetry (CV)

Voltammetry belongs to a category of electro-analytical methods, through which information about an analyte is obtained by varying a potential and then measuring the resulting current. It is, therefore, an amperometric technique.

Since there are many ways to vary a potential, there are also many forms of voltammetry, such as polarography (DC voltage), linear sweep, differential staircase, normal pulse, reverse pulse, differential pulse and more. Cyclic voltammetry is one of the most widely used forms and it is useful for obtaining information about the redox potential and electrochemical reaction rates (e.g., the chemical rate constant) of analyte solutions.

In this case, the voltage is swept between two values at a fixed rate, however, when the voltage reaches E_λ the scan is reversed and the voltage is swept back to E_i. The scan rate,

$(E_\lambda - E_i) = (t_2 - t_1)$, is a critical factor, since the duration of a scan must provide sufficient time to allow for a meaningful chemical reaction to occur (Fig. 4.13). Varying the scan rate, therefore, yields correspondingly varying results. The voltage is measured between the reference electrode and the working electrode, while the current is measured between the working electrode and the counter electrode. The obtained measurements are plotted as current vs. voltage, also known as a voltammogram. As the voltage is increased toward the electrochemical reduction potential of the analyte, the current will also increase. With increasing voltage towards E_λ past this reduction potential, the current decreases, having formed a peak as the analyte concentration near the electrode surface diminishes, since the oxidation potential has been exceeded.

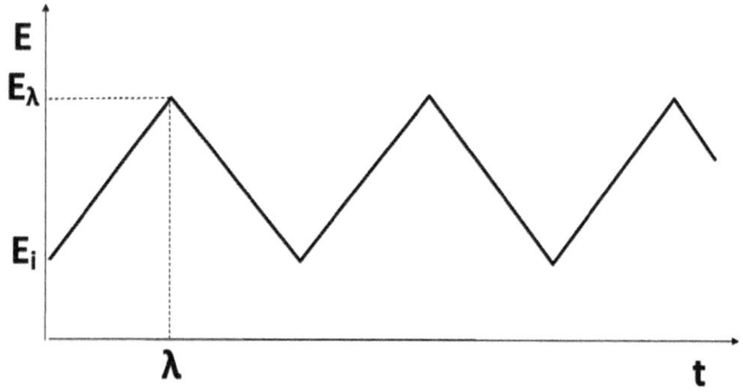

Figure 4.13 The potential ramp at cyclic DC.

As the voltage is reversed to complete the scan towards E_i, the reaction will begin to reoxidize the product from the initial reaction. This produces an increase in current of opposite polarity as compared to the forward scan, but a current decreases again, having formed a second peak as the voltage scan continues towards E_i. The reverse scan also provides information about the reversibility of a reaction at a given scan rate.

The important parameters in a cyclic voltammogram (Fig. 4.14) are the peak potentials (E_{pc}, E_{pa}) and peak currents (i_{pc}, i_{pa}) of the cathodic and anodic peaks, respectively. If the electron transfer process is fast compared with other processes (such as diffusion), the reaction is said to be electrochemically reversible, and the peak separation is

$$\Delta E_p = |E_{pa} - E_{pc}| = 2.303RT/nF \tag{4.2}$$

Thus, for a reversible redox reaction at 25°C with n electrons ΔE_p should be 0.0592/n V or about 60 mV for one electron. In practice this value is difficult to attain because of such factors as cell resistance. Irreversibility due to a slow electron transfer rate results in ΔE_p > 0.0592/n V, greater, say, than 70 mV for a one-electron reaction. The formal reduction potential (E^0) for a reversible couple is given by

$$E^0 = \frac{E_{pa} + E_{pc}}{2}. \tag{4.3}$$

For a reversible reaction, the concentration is related to peak current by the Randles–Sevcik expression (at 25°C):

$$i_p = 2.686 \times 10^5\, n^{3/2} A c^0 D^{1/2} v^{1/2}, \tag{4.4}$$

where i_p is the peak current in amps, A is the electrode area (cm^2), D is the diffusion coefficient (cm^2 s^{-1}), c^0 is the concentration in mol cm^{-3}, and v is the scan rate in V s^{-1}.

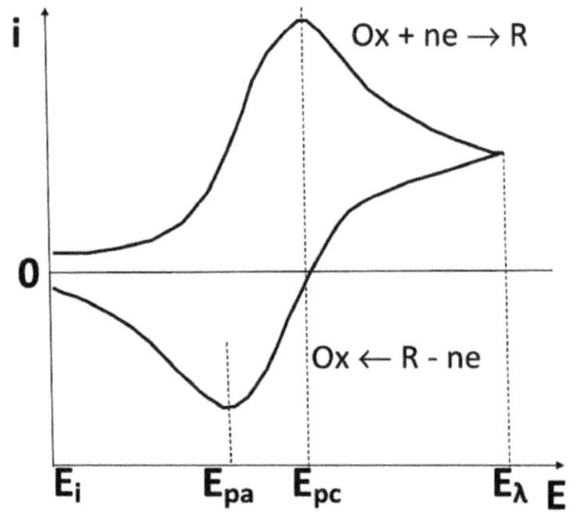

Figure 4.14 Cyclic voltamperogram.

The shape of the voltammogram for a given compound depends not only on the scan rate and the electrode surface, which is different after each adsorption step, but also on the catalyst concentration (Fig. 4.15). For example, increasing the concentration of reaction specific enzymes at a given scan rate will result in a higher current compared to the non-catalyzed reaction.

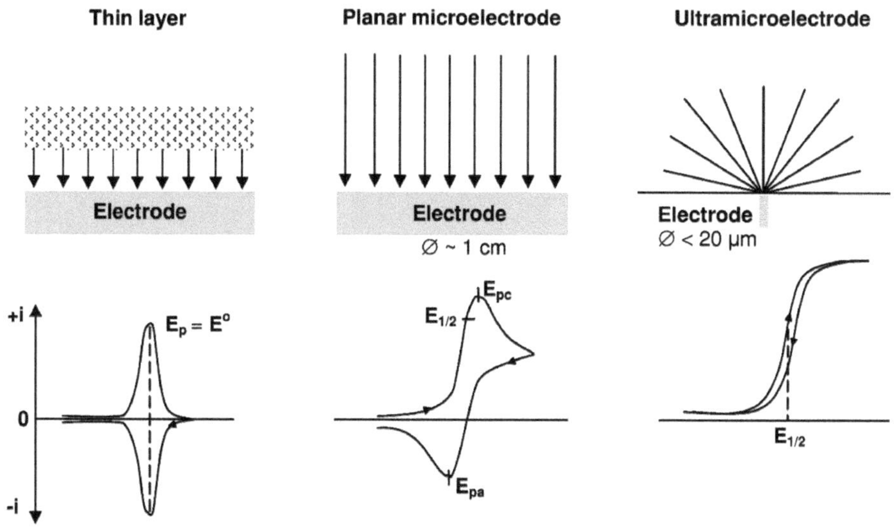

Figure 4.15 Diffusion of the electroactive species to the indicator electrode and the shape of the voltamperogram at different working electrodes.

4.4.2.2 Pulse voltammetric techniques

Standard pulse voltammetry (NPV) uses a series of potential pulses of increasing amplitude. The current measurement is made near the end of each pulse, which allows time for the charging current to decay. It is usually carried out in an unstirred solution at either DME (called normal pulse polarography) or solid electrodes. The potential is pulsed from an initial potential E_i. The duration of the pulse, t, is usually 1 to 100 ms and the interval between pulses typically 0.1 to 5 s. The resulting voltammogram displays the sampled current on the vertical axis and the potential to which the pulse is stepped on the horizontal axis.

Differential pulse voltammetry (DPV) is comparable to normal pulse voltammetry in that the potential is also scanned with a series of pulses. However, it differs from NPV because each potential pulse is fixed, of small amplitude (10 to 100 mV), and is superimposed on a slowly changing base potential. Current is measured at two points for each pulse, the first point (1) just before the application of the pulse and the second (2) at the end of the pulse. These sampling points are selected to allow for the decay of the non-Faradaic (charging) current. The difference between current measurements at these points for each pulse is determined and plotted against the base potential.

Differential normal pulse voltammetry (DNPV) is mainly used in in vivo measurements (on experimental animals) because the shape of the applied voltage minimizes the time that the electrolytic process takes place, saving the surrounding tissue from damage (Fig. 4.16). The registered current signal is similar to the DPV method (peak).

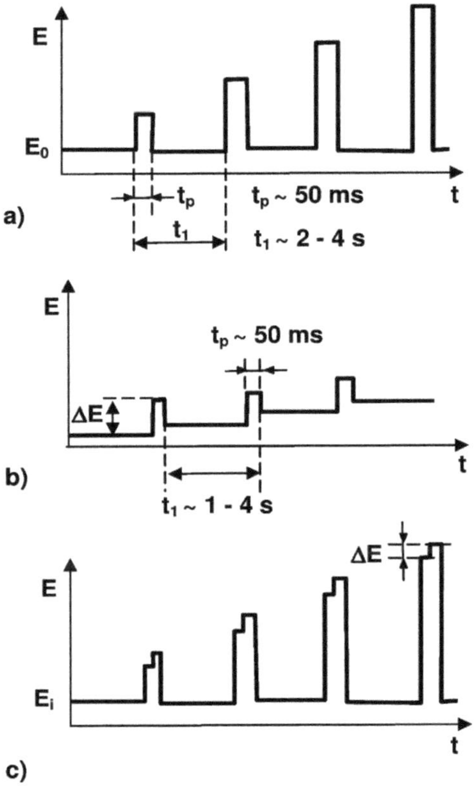

Figure 4.16 Time course of polarization potential at (a) NPV, (b) DPV and (c) DNPV.

The excitation signal in square-wave voltammetry (SWV) consists of a symmetric square-wave pulse of amplitude E_{sw} superimposed on a staircase waveform of step height ΔE, where the forward pulse of the square wave coincides with the staircase step. The net current, i_{net}, is obtained by taking the difference between the forward and reverse currents ($i_{for} - i_{rev}$) and is centered on the redox potential. The peak amplitude is directly proportional to the concentration of the electroactive species and direct detection limits as low as 10^{-8} M are possible. Square-wave voltammetry has several advantages. Among these are its excellent sensitivity and the rejection of background currents. Another is the speed (for example, its ability to scan the voltage range over one drop during polarography with the DME). This speed, coupled with computer control and signal averaging, allows for experiments to be performed repetitively and increases the signal-to-noise ratio. Applications of square-wave voltammetry include the study of electrode kinetics with regard to preceding, following, or catalytic homogeneous chemical reactions, determination of some species at trace levels, and its use with electrochemical detection in HPLC (high-performance liquid chromatography).

4.4.2.3 Chronopotentiometry and chronoamperometry

Chronopotentiometry is an electroanalytical technique in which an unstirred solution of an electroactive species, in presence of excess of a supporting electrolyte, is electrolyzed at a programmed current density between an indicator electrode and a counter electrode and the potential-variation of the former vs. a suitable reference is recorded as a function of time (Fig. 4.17).

Current functions which have been used are constant current, current varying linearly with time or as a square, square root or an exponential function of time, the current is stopped or reversed after a pre-determined time, an alternating current (a.c.) of a small amplitude is superimposed over the electrolysis current etc. The technique enables study of the kinetics of electrode processes with or without chemical kinetics and of the adsorption at the electrode surface.

Another amperometric technique is known as chronoamperometry, where a square-wave potential is applied to the working electrode and a steady state current is measured as a function of time. Alterations in the current arise from the expansion or reduction of the diffusion layer at the electrode. The concept of a diffusion layer was introduced by Nernst and states that there is a stationary thin layer of solution in contact with the electrode surface. The local analyte concentration drops to zero at the electrode surface and diffusion controls the transfer of analyte from the bulk solution of higher concentration to the electrode. This results in a concentration gradient away from the electrode surface. In the bulk solution the concentration of analyte is maintained at a value of c_0 by convective transfer.

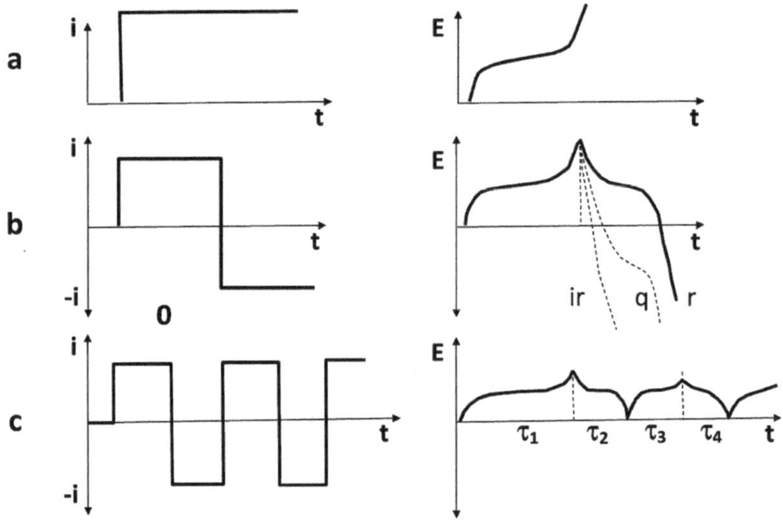

Figure 4.17 Different types of galvanostatic techniques. a: chronopotentiometry with constant current, b: current reversal potentiometry (r: reversible process, q: quasireversible process, ir: irreversible process), c: cyclic chronopotentiometry.

The technique of chronoamperometry is, therefore, closely related to the Cottrell equation, which is shown in Eq. 4.5.

It defines the current-time dependence for linear diffusion control at a planar electrode. In the Cottrell equation, the current i is dependent upon F Faraday's constant, n the number of transferred electrons per molecule, A the electrode area, c_0 the analyte concentration, D the diffusion coefficient and time t.

$$i = nFAc_0\sqrt{\frac{D}{\pi t}} \tag{4.5}$$

According to the Cottrell equation, the current depends on the rate at which the analyte diffuses to the electrode.

4.4.2.4 Electrochemical impedance spectroscopy (EIS)

Impedimetric biosensors have been arranged by immobilizing bioreceptors such as antibodies, bacteria, cells, enzymes, lectins and nucleic acids at the surface of the electrode. After binding target molecule to the electrode surface; a shift in impedance, a change in capacitance or admittance at the bulk of the electrode interface owing to the insulating properties are observed (see Fig. 4.18).

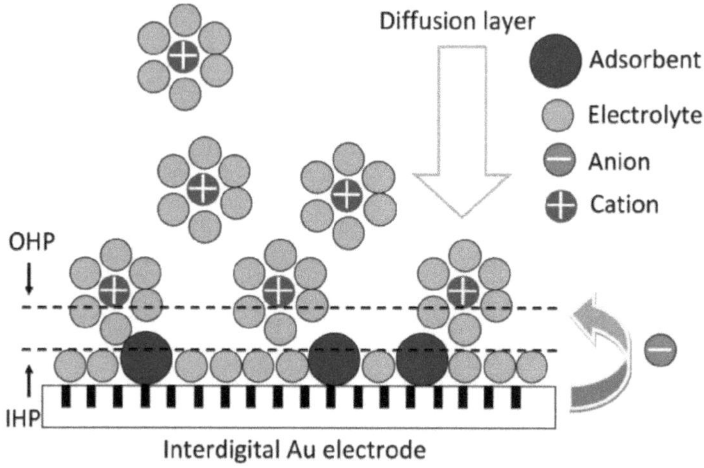

Figure 4.18 Kinetic processes taking place at electrode–electrolyte interface (Randle's cell model).

Lectins (L. *legere*, to select or choose) are sugar-binding proteins from many sources regardless of their ability to agglutinate cells. They also precipitate polysaccharides and glycoproteins. The agglutinating and precipitating activities of lectins are very similar to those of antibodies. They can likewise be specifically inhibited by low molecular weight compounds (haptens) which, in the case of lectins, are monosaccharides or oligosaccharides. Lectins have been found in plants, viruses, microorganisms, and animals but despite their ubiquity, in many cases their biological function is unclear.

The most important step in developing biosensors is to immobilize bioreceptors on electrode surface appropriately. The covalent and affinity immobilization, physical adsorption, entrapping bioreceptors in films or gels, layer-by-layer deposition, cross-linking are used for the immobilization of receptors. Lectins have been found in plants, viruses, microorganisms, and animals but despite their ubiquity, in many cases their biological function is unclear.

Nanomaterials have recently become of interest for the design of electrochemical impedimetric biosensors thanks to their high sensitivity, stability, and conductivity. Electrochemical impedance spectroscopy is a sensitive label free electrochemical technique for monitoring biorecognition events at the electrode surface and as a result it provides low detection limits. Determination is rapid and requires short detection time. Impedimetric biosensors provide reproducible results, when the biorecognition elements are immobilized on the electrode surface by using strong chemical bonds such as SAM immobilization method.

Many studies on impedimetric biosensors are focused on aptasensors and immunosensors. The aptamer-based biosensors offer a number of advantages, such as high specificity of binding affinity, better stability, and larger shelf life. Cell-based impedimetric biosensors are useful in external in situ analysis. Bacterial, procaryotic, and mammalian cells have been used to monitor the amount and activity of microorganisms. Enzyme-based impedimetric biosensors are based on detection of substrate or product of an enzyme reaction.

The first publication of electrochemical impedance spectroscopy dates back to 1975.

The basic Ohm's law in Eq. 4.6 defines the resistance R in terms of potential, U, and current I as:

$$R = \frac{U}{I}, \tag{4.6}$$

whereas Ohm's law for alternating current defines impedance Z in terms of time dependent alternating potential E_t, and current I_t as

An aptamer-based sensor is a particular class of biosensor where the biological recognition element is a DNA or RNA aptamer. In an aptasensor, the aptamer recognizes the molecular target against which it was previously in vitro selected. The aptamer-target reaction is independent of both the type of detection system and the type of transducer employed. Aptasensors can be easily multiplexed to detect a variety of aptamer-target reactions simultaneously.

An immunosensor is an integrated device consisting of an immunochemical recognition element in direct spatial contact with a transducer element. Electrochemical immunosensors employ either antibodies or their complementary binding partners, i.e., antigens or haptens as biological recognition elements in combination with electrodes or field-effect transistors.

$$Z = \frac{E_t}{I_t}. \tag{4.7}$$

The impedance, Z, now can be expressed in term of a magnitude of Z_0 and a phase shift, θ. It can also be expressed in term of Euler's relationship given by

$$e^{i\theta} = \cos\theta + j\sin\theta. \tag{4.8}$$

The impedance, Z, can be expressed in term of potential, E, and current response, I, given by

$$E_t = E_0 e^{j\omega t}$$

$$I_t = I_0 e^{j(\omega t - \theta)}. \tag{4.9}$$

Therefore, the impedance, Z:

$$Z(\omega) = \frac{E_t}{I_t} = \frac{E_0 e^{j\omega t}}{I_0 e^{j(\omega t - \theta)}} = Z_0 e^{j\theta}$$

$$Z(\omega) = Z_0(\cos\theta + j\sin\theta) \tag{4.10}$$

The impedance now is in the form of real part (Z_0 $\cos\theta$) and imaginary part (Z_0 $\sin\theta$) represented as follows:

$$ReZ = Z' = \text{Zreal} = Z_0\cos\theta$$

$$Z'' = \text{Zimag}' = Z_0\sin\theta \tag{4.11}$$

Through the application of a small sinusoidally varying potential U, one measures the resulting current response I (see Fig. 4.19). By varying the excitation frequency f of the applied potential over a range of frequencies, one can calculate the complex impedance, sum of the real and imaginary impedance components, of the system as a function of the frequency (i.e., angular frequency w). Therefore, EIS combines the analysis of both real and imaginary components of impedance, namely the electrical resistance and reactance, as shown in Eq. 4.12.

$$Z(j\omega) = \frac{U(j\omega)}{I(j\omega)} = Z_r(\omega) + jZ_i(\omega); \qquad \omega = 2\pi f \tag{4.12}$$

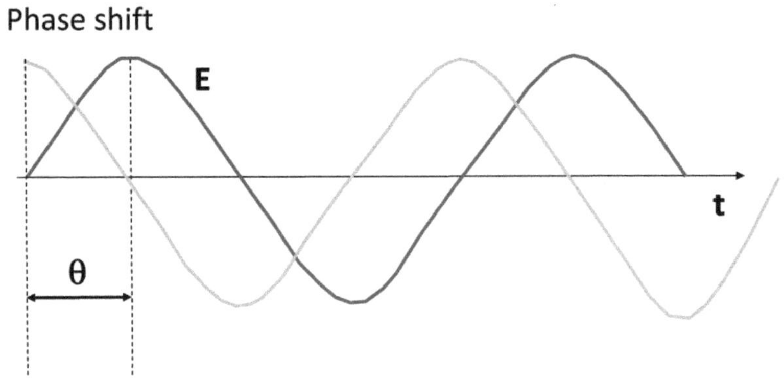

Figure 4.19 Phase shift in I_t as a response to excitation-potential E_t in a linear system.

EIS possesses the ability to study any intrinsic material property or specific processes that could influence the conductivity/resistivity or capacitivity of an electrochemical system. Therefore, EIS is a useful tool in the development and analysis of materials for biosensor transduction. For electrochemical sensing, impedance techniques are useful to monitor changes in electrical properties arising from biorecognition events at the surface of modified electrodes in the form of so-called Nyquist plots (see Fig. 4.20). For example, changes in the conductance of the electrode can be measured as a result of protein immobilization and antibody–antigen reactions on the electrode surface. To additionally measure changes in capacitance with EIS, sometimes referred to as Faradaic impedance spectroscopy, one scheme as proposed by Katz and Willner is to construct an array of interdigitated electrodes and to monitor antibody–antigen reactions in the gaps between the electrodes. Binding events of complementary antibody–antigen components alter the electrical properties in the gap between two electrodes, where changes in gap conductivity correspond to changes in the real impedance component $Z_r(w)$ and changes in the gap capacitance correspond to changes in the imaginary impedance component $Z_i(w)$.

$$Z(\omega) = R_s + \frac{R_{ct}}{1+\omega^2 R_{ct}^2 C_{dl}^2} - \frac{j\omega R_{ct}^2 C_{dl}}{1+\omega^2 R_{ct}^2 C_{dl}^2} \tag{4.13}$$

$$Z'(\omega) = R_s + \frac{R_{ct}}{1+\omega^2 R_{ct}^2 C_{dl}^2} \tag{4.14}$$

$$Z''(\omega) = -\frac{\omega R_{ct}^2 C_{dl}}{1+\omega^2 R_{ct}^2 C_{dl}^2} \tag{4.15}$$

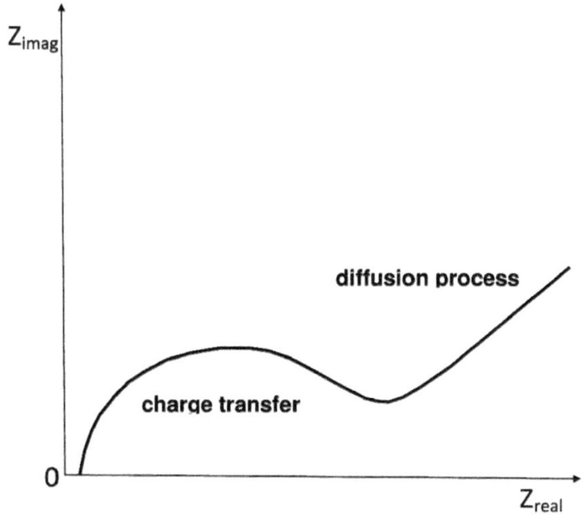

Figure 4.20 Nyquist plot for Randle's electrochemical cell model.

Since reactions are generally detected only in close proximity to the electrode surface, the electrodes themselves play a crucial role in the performance of electrochemical biosensors. Electrochemical sensing usually requires a reference electrode, a counter or auxiliary electrode and a working electrode, also known as the sensing or redox electrode. The reference electrode, commonly made from Ag/AgCl, is kept at a distance from the reaction site in order to maintain a known and stable potential. The working electrode serves as the transduction element in the biochemical reaction, while the counter electrode establishes a connection to the electrolytic solution so that a current can be applied to the working electrode. These electrodes should be both conductive and chemically stable.

The faradic biosensors detect biorecognition events which occur at the modified electrode by measuring the change in the Faradaic current (interfacial electron transfer resistance) owing to steric hindrance caused by the biomolecular interaction and/ or by the electrostatic repulsion between the free charges of the target molecules and the electroactive species in the supporting electrolyte. Redox probe selection depends on various parameters such as the charge, hydrophobicity/hydrophilicity, size of the redox couple, and the chemical and physical properties of the modified electrodes.

Amperometric devices are a type of electrochemical sensor, that continuously measures current resulting from the oxidation or reduction of an electroactive species in a biochemical reaction. Clark oxygen electrodes perhaps represent the basis for the simplest forms of amperometric biosensors, where a current is produced in proportion to the oxygen concentration. This is measured by the reduction of oxygen at a platinum working electrode in reference to a Ag/AgCl reference electrode at a given potential. Typically, the current is measured at a constant potential and this is referred to as *amperometry*. If a current is measured during controlled variations of the potential, this is referred to as *voltammetry*. Furthermore, the peak value of the current measured over a linear potential range is directly proportional to the bulk concentration of the analyte, i.e., the electroactive species.

Potentiometric devices measure the accumulation of a charge potential at the working electrode in comparison to the reference electrode in an electrochemical cell when zero or no significant current flows between them. In other words, potentiometry provides information about the ion activity in an electrochemical reaction. For potentiometric measurements, the relationship between the concentration and the potential is governed by the Nernst equation:

$$\text{EMF or } E_{\text{cell}} = E^{0}_{\text{cell}} - \frac{RT}{nF} lnQ, \tag{4.16}$$

where E_{cell} represents the observed cell potential at zero current. This is sometimes also referred to as the electromotive force or EMF. E^{0}_{cell} is a constant potential contribution to the cell, R the universal gas constant, T the absolute temperature in degrees Kelvin, n is the charge number of the electrode reaction, F is Faraday's constant and Q is the ratio of ion concentration at the anode to ion concentration at the cathode.

The direct determination of the analyte ion concentration with the Nernst equation is referred to as *direct potentiometry*. The lowest detection limits for potentiometric devices are currently often achieved with ion-selective electrodes (ISE). Therefore, by definition the detection limit is analyte specific and current devices have limits of detection in ranges between 10^{-8} to 10^{-11} M. Potentiometric sensors proved suitable for measuring low concentrations in tiny sample volumes, since they ideally offer the benefit of not chemically influencing a sample. The variety of ions, for which low detection limits are possible, is currently quite limited and is missing such important analytes as nickel, manganese, mercury and arsenate ions. A number of potentiometric devices are based on various forms of field-effect transistor (FET) devices to measure pH changes, selective ion concentrations and the kinetics of biocatalytic reactions involving enzymes. However, it can be said, that electrochemical biosensors are one of the most sensitive bioanalytical instruments with extremely low detection limits and possibly wide pool of different analytes for determination (see Table 4.5).

Table 4.5 Nanoparticle-based electrical methods for biosensing

Nanoparticle	Biomarker/analyte	LOD	Linear range
AuNPs	ApoE (Alzheimer biomarker)	80 pg mL^{-1}	100–12,500 pg mL^{-1}
AuNPs	Beta amyloid	19 pg mL^{-1}	20–12,500 pg mL^{-1}
AuNPs	Cortisol (biomarker of physiological conditions)	10 pg mL^{-1}	0.03–10,900 ng mL^{-1}
AuNPs	IgG	0.31 ng mL^{-1}	2–500 ng mL^{-1}
AuNPs	Breast cancer biomarkers	1.72 fM	0.05–1,000 pM
AuNPs	Muc1 (tumor biomarker)	0.1 nM	1–10 nM
AgNPs and AuNPs	Apo A-I (biomarker of risk of cardiovascular disease)	0.02 pg mL^{-1}	0.1 pg mL^{-1} – 50 ng mL^{-1}
AgNPs and CNTs	AFP (tumor biomarker)	0.061 pg mL^{-1}	0.1–5,000 pg mL^{-1}
AgNPs and CNTs	CEA (tumor biomarker)	0.093 pg mL^{-1}	0.1–5,000 pg mL^{-1}
Iridium oxide NPs	ApoE (Alzheimer biomarker)	7 nM	0.7–35 nM
CNTs	CEA (tumor biomarker)	0.018 ng mL^{-1}	0.05–5 ng mL^{-1}
SWCNTs	Glycoprotein from African Trypanosomes	4 fM	1–1,000 pM
Graphene	β-Lactoglobulin (milk allergen)	0.85 pg mL^{-1}	0.001–100 ng mL^{-1}
Graphene nanosheets	CEA (tumor biomarker)	40 fg mL^{-1}	0.0001–10 ng mL^{-1}
rGO	CEA and SCCA (tumor biomarkers)	CEA: 13 pg mL^{-1} SCCA: 10 pg mL^{-1}	CEA: 0.05–20 ng mL^{-1} SCCA: 0.03–20 ng mL^{-1}
rGO	CRP	0.08 ng mL^{-1}	2–1,000 ng mL^{-1}
Graphene and rGO	DNA from HIV virus	0.158 pM	$1–1\times10^5$ pM
ZrO_2–rGO	CYFRA-21-1 (tumor biomarker)	0.122 ng mL^{-1}	2–22 ng mL^{-1}
PtNPs	PSA (tumor biomarker)	1 ng mL^{-1}	1–30 ng mL^{-1}
PtNPs	Telomerase activity (tumor biomarker)	100 cells mL^{-1}	$5\times10^2–1\times10^7$ cells mL^{-1}

Note: AFP: a-fetoprotein; AgNPs: silver nanoparticles; ApoE: apolipoprotein E; Apo-A1: apolipoprotein A-1; AuNPs: gold nanoparticles; CNTs: carbon nanotubes; CEA: carcinoembryonic antigen; CRP: C-reactive protein; CYFRA-21-1: oral cancer biomarker: cytokeratin-19; IgG: human immunoglobulin; Muc 1: Mucin 1; PSA: prostate-specific antigen; PtNPs: platinum nanoparticles; GO: grapheme oxide; HIV: human immunodeficiency virus; rGO: reduced graphene oxide; SCCA: squamous cell carcinoma antigen; SWCNTs: single-walled CNTs; ZrO_2: zirconium dioxide.

Conductometric devices measure the ability of an analyte (e.g., electrolyte solutions) or a medium (e.g., nanowires) to conduct an electrical current between electrodes or reference nodes. Conductometric devices can be considered a subset of impedimetric devices. In most cases, conductometric devices have been strongly associated with enzymes, where the ionic strength, and thus the conductivity, of a solution between two electrodes changes as a result of an enzymatic reaction. Thus, conductometric devices can be used to study enzymatic reactions that produce changes in the concentration of charged species in a solution. The variable ionic background of clinical samples and the requirement to measure small conductivity changes in media of high ionic strength limit the applicability of such enzyme-based conductometric devices for biosensing. Another approach is to directly monitor the changes in conductance of an electrode as a result of the immobilization of, e.g., enzymes, complementary antibody–antigen pairs, etc. onto the electrode surface. The construction of multi-analyte conductance biosensors and conductive polymer-based devices has been made possible by the rapid development of semiconductor technology and by sensor integration with microelectronic devices, such as FET devices. Now there is an increased interest in conductometric immunosensors in combination with nanostructures, and especially nanowires, for biosensing. Although conductometric sensing has not been as extensively implemented as it could be, there are examples of successful development of these devices for practical application, such as drug detection in human urine and pollutant detection in environmental screening. Whole cells have also been used as a biorecognition element in conductometric biosensors for toxicity analysis by immobilizing the cells to a transducer of interdigitated electrodes.

4.4.2.5 Electrochemistry in combination with complementary biosensor techniques

Obviously, all sensing techniques demonstrate specific strengths in different, yet sometimes overlapping, areas of application. For example, both electrochemical and optical sensing techniques, can allow for real-time, in situ, non-destructive and label-free analysis of solutions, biolayers, surfaces, thin films, bulk materials and interfaces. On one hand, electrochemical techniques, such as CV, enable the in situ influence and monitoring of, e.g., redox reactions, system electrical response and reaction reversibility. On the other hand, optical techniques are known for their ability to measure mass adsorption kinetics (e.g., optical waveguide lightmode spectroscopy (OWLS), surface plasmon resonance (SPR) and ellipsometry). Gravimetric techniques like quartz crystal microbalance with dissipation monitoring (QCM-D) and imaging techniques like scanning probe microscopies, such as AFM, and fluorescent microscopies, such as confocal laser scanning microscopy (CLSM) can also be successfully combined with electrochemical techniques to enhance understanding of biointerfacial phenomena. These techniques provide high sensitivity close to the surface of the transducing element (electrode, waveguide, tip, etc.). The shared high interfacial sensitivity of electrochemical and other types of biosensor results in the simultaneous extraction of a richer set of initial data in addition to the benefit of increased control over the sensing environment.

4.4.3 Nanoparticles and Lectin-Based Biosensors

Our research group is focused on the design of cutting edge nanotechnology-based sensors and biosensors. These ultrasensitive (bio)sensors in combination with cost-effective and time-saving electrochemical methods should play an important role in diagnostic and prognostic applications. Based on this we place importance on the successful integration of self-assembled monolayers (SAM), biological molecules and nanomaterials in the area of glycomics. The application of metallic nanoparticles, such as AuNPs (Fig. 4.21), enables a significant improvement in the limit of impedimetric detection of prostate specific antigen (PSA—a biomarker for prostate cancer) as compared to a standard binary self-assembled monolayer aptasensor.

The lectin-based biosensors represent potential diagnostic and recognizing tools for cancer biomarker detection. Ideally these biosensors should exhibit three unique characteristics: (1) analysis of cancer biomarker with very low LOD; (2) ultrasensitive analysis of a glycan part of cancer biomarker and (3) both assays (i.e., cancer biomarker quantification and glycoprofiling) can be performed on the same interface due to label-free format of analysis.

Glycans are complex carbohydrates consisting of saccharide units that link together and are attached to proteins and lipids to form glycoproteins and glycolipids, respectively and play an important role in many cellular processes including immune response, tumor metastasis, infection by bacteria/viruses, inflammation, molecular recognition and cell signaling. Glycosylation is the most common co- and posttranslational modification of proteins, it is also increasingly recognized as a phenotype modulator of various pathological changes on cell surfaces, mostly during cancer development. Glycomics as a scientific discipline studying the structure and function of glycans is a younger sister of more developed genomics and proteomics.

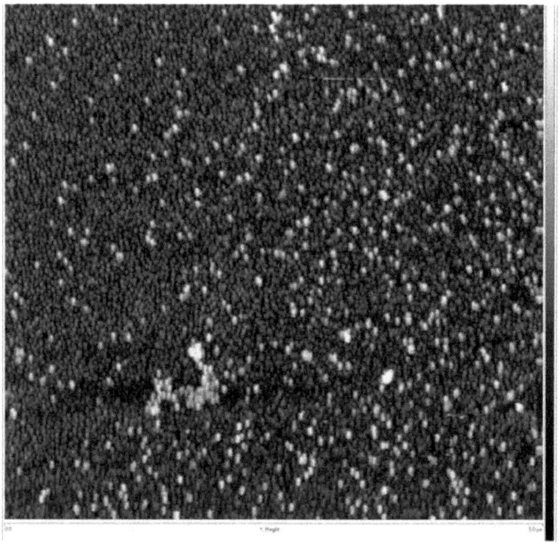

Figure 4.21 AFM image of gold nanoparticles with diameter of ~20 nm.

In most cases electrochemical impedance spectroscopy (EIS) is applied as a label-free, non-destructive and sensitive electrochemical method, which allows for the detection of analytes down to a single molecule level. After exposure of the biosensor to the analyte, R_{ct} changes thereby allowing for quantification of analyte concentration. The first EIS-based biosensor to evaluate glycan-lectin interactions was prepared by Joshi´s group and since then an increased interest can be seen to apply the approach especially in combination with immobilized lectins (Table 4.6) for analysis of intact glycoproteins and even various types of cancerous cells.

As follows from Table 4.7, a number of different analytes has been targeted by our group. We Have prepared MXene and oxidized MXene modified GCE for detection of O_2, NADH and H_2O_2. Specifically, the $Ti_3C_2T_x$ MXene was proven an excellent catalyst for reduction of H_2O_2, and the H_2O_2 sensor based on $Ti_3C_2T_x$ is our most sensitive device with a detection limit of 0.7 nM comparable to the best devices described so far (i.e., 0.3 nM) with a response time of ~10 s. It is conceivable that a further modification of $Ti_3C_2T_x$ by metallic nanoparticles could further enhance performance of modified $Ti_3C_2T_x$ to detect H_2O_2.

2D $Ti_3C_2T_x$ (MXene, where T: =O, –OH, –F) sheets modified with Pt nanoparticles (PtNPs) deposited on the surface of GCE showed much better and stable redox behavior in an anodic potential window as compared to the GCE patterned by pristine $Ti_3C_2T_x$. Detection of H_2O_2 on GCE modified by $Ti_3C_2T_x$/PtNP is possible at a potential value of 0 V vs. Ag/AgCl with LOD of 448 nM. Furthermore, the $Ti_3C_2T_x$/PtNP sensor could detect small redox molecules such as ascorbic acid (AA), acetaminophen (APAP), dopamine (DA) and uric acid (UA) and at a potential higher than +250 mV with high selectivity and LOD down to nM level. The redox behavior of small analytes can be effectively modulated by formation of outer membranes.

In another of our proof-of-principle studies, an anti-PSA DNA aptamer was co-immobilized with either 6-mercapto-1-hexanol (MCH) or 6-(ferrocenyl)hexanethiol (FcSH) for both impedimetric or amperometric detection, respectively. A PSA detection of as low as 10 pg mL^{-1} was achieved with a dynamic range from 10 pg mL^{-1} to 10 ng mL^{-1}, well within the clinically relevant values, whilst retaining high specificity of analysis.

We have demonstrated glycan biosensor using surface chemistry based on a mixed SAM composed of thiols bearing oligoethyleneglycol (OEG) moieties resisting nonspecific interactions. The biosensor has been applied for analysis of influenza viruses H3N2 with LOD of 13 viral particles in 1 μl, what is the lowest LOD for analysis of influenza viral particles by the glycan-based device achieved so far. Additional examples of what has been achieved within our group are given in Table 4.7.

Table 4.6 Lectin properties

Lectin	Common abbreviation	Source	MW (kDa)	Preferred sugar specificity
Aleuria aurantia	AAL	*Aleuria aurantia* mushrooms	72	Fucα6GlcNAc
Concanavalin A	Con A	*Canavalia ensiformis* (Jack Bean) seeds	104	αMan, αGlc
Dolichos biflorus	DBA	*Dolicos biflorus* (Horse Gram) seeds	111	αGalNAc
Griffonia (Bandeiraea) simplicifolia I	GSL I, BSL I	*Griffonia (Bandeiraea) simplicifolia* seeds	114	αGal, αGalNAc
Griffonia (Bandeiraea) simplicifolia I Isolectin B_4	GSL I-B_4	*Griffonia (Bandeiraea) simplicifolia* seeds	114	αGal
Griffonia (Bandeiraea) simplicifolia II	GSL II, BSL II	*Griffonia (Bandeiraea) simplicifolia* seeds	113	α or βGlcNAc
Lotus tetragonolobus	LTL	*Lotus tetragonolobus, Tetragonolobus purpurea* (Winged Pea, Asparagus Pea) seeds	107	αFuc

Lectin	Common abbreviation	Source	MW (kDa)	Preferred sugar specificity
Maackia amurensis I	MAL I, MAL	*Maackia amurensis* seeds	130	Galβ4GlcNAc
Maackia amurensis II	MAL II, MAH	*Maackia amurensis* seeds	130	Neu5Acα3Galβ3GalNAc
Phaseolus vulgaris Erythroagglutinin (PHA-E)	PHA-E	*Phaseolus vulgaris* (Red Kidney Bean) seeds	126	Galβ4GlcNAcβ2Manα6 (GlcNAcβ4) (GlcNAcβ4Manα3) Manβ4
Ricinus communis I	RCA I, RCA_{120}	*Ricinus communis* (Castor Bean) seeds	120	Gal
Ricinus communis II, ricin	RCA II, RCA_{60}, ricin	*Ricinus communis* (Castor Bean) seeds	0	Gal, GalNAc
Sambucus nigra	SNA, EBL	*Sambucus nigra* (Elderberry) bark	140	Neu5Acα6Gal/GalNAc
Ulex europaeus I	UEA I	*Ulex europaeus* (Furze Gorse) seeds	63	αFuc
Wheat Germ	WGA	*Triticum vulgaris* (wheat germ)	36	GlcNAc

Sugar abbreviations: Fuc: L-Fucose, Gal: D-Galactose, GalNAc: N-Acetylgalactosamine, Glc: D-Glucose, GlcNAc: N-Acetylglucosamine, Man: Mannose, Neu5Ac: N-Acetylneuraminic acid (sialic acid)

Table 4.7 SAMs, lectin and nanomaterials-based sensors

Nanoparticle/SAM	Lectin	Analyte	LOD	Detection technique
SAM consisting of 11-MUA and 6-MH	SNA I	FET and ASF	FET: 0.33 fM ASF: 0.54 fM	EIS/Au polycrystalline electrodes
AuNPs/SAM of 11-MUA and 6-MH	SNA I	FET and ASF	≈ 1 aM	EIS/Au polycrystalline electrodes
SAM of 11-MUA and DPS	Con A, RCA, SNA I	FET and ASF	(Con A): 1 fM (RCA): 10 fM (SNA I): 100 fM	EIS/Au polycrystalline electrodes
SAM consisting of 11-MUA and 6-MH (2D) 1st SAM/AuNPs/2nd SAM (MUA:MH) (3D)	SNA I	FET	FET (2D): 1 fM FET (3D): 10 aM	EIS/Au polycrystalline electrodes
SAM consisting of 11-MUA and 6-MH	MAA, DSL	H1N1, H5N1	H1N1: 50 aM H1N5: 50 zM	EIS/Au polycrystalline electrodes
CB SAM consisting of 11-MUA and SB	SNA I, RCA	Galactose, sialic acid	FET: 73 pM ASF: 2.7 nM	EIS/Au polycrystalline electrodes
SAM composed of MUA:MCH	SNA I, MAA	PSA	4 aM	EIS/Au polycrystalline electrodes

Nanoparticle/SAM	Lectin	Analyte	LOD	Detection technique
SAM composed of OEG-COOH and OEG	MAA I	H3N2	glycan binding lectin: 5 aM H3N2: 13 viral particles in 1 μl	EIS/Au polycrystalline electrodes
MCH/AuNPs/thiolated DNA aptamer with either FcSH or MCH	—	PSA	10 pg/ml, dynamic range up to 10 ng/ml	EIS/Au disc electrodes
SAM consisting of OEG-COOH:OEG-OH, HSA (3D) SAM consisting of 11-MUA and 6-MH (2D)	GOD-2C4 DBA	Con A	3D: 1.4 aM 2D: 270 aM	EIS/Au polycrystalline electrodes
$Ti_3C_2T_x$ MXene	—	H_2O_2	0.7 nM (−0.5 V)	CA/GCRDE
$Ti_3C_2T_x$ MXene {PtNP	—	H_2O_2	448 nM (0 V)	CA/GCRDE

Note: ASF: asialofetuin (N-acetylneuraminic acid); CB SAM: self-assembled monolayer (SAM) deposited from a solution containing a carboxybetaine-terminated thiol; DBA: lectin from *Dolichos biflorus*; DSL: *Datura stramonium* lectin; DPS: sulfobetaine derivates; FcSH: 6-(ferrocenyl)hexanethiol; FET: fetuin (N-acetylneuraminic acid), antibody GOD-2C4, a mouse IgG1κ antibody not cross-reacting with the GalNAc-β-O epitope or the blood group A antigen; GCERDE: glassy carbon rotating disc electrode H3N2: intact influenza viruses; 6-MH: 6-mercaptohexanol; HSA: human serum albumin; MAA: *Maackia Amurensis* lectin; MCH: 6-mercapto-1-hexanol; 11-MUA: 11-mercaptoundecanoic acid; OEG: oligoethylene glycol; PSA: prostate specific antigen; SB: thiolated sulphobetaine; SNA I: lectin from Sambucus nigra; RCA: *Ricinus communis* agglutinin; Con A: concanavalin A; 11-aminoundecanethiol.

References

Recommended literature

1. Hughes, G., Westmacott, K., Honeychurch, K. C., Crew, A., Pemberton, R. M., Hart, J. P. (2016). Recent advances in the fabrication and application of screen-printed electrochemical (bio)sensors based on carbon materials for biomedical, agri-food and environmental Analyses. *Biosensors*, 6, p. 50.
2. Taleat, Z., Khoshroo, A., Mazloum-Ardakani, M. (2014). Screen-printed electrodes for biosensing: A review (2008–2013). *Microchimica Acta*, 181, pp. 865–891.
3. Arduini, F., Micheli, L., Moscone, D., Palleschi, G., Piermarini, S., Ricci, F., Volpe, G. (2016). Electrochemical biosensors based on nanomodified screen-printed electrodes: Recent applications in clinical analysis. *Trends in Analytical Chemistry*, 79, pp. 114–126.
4. Domíýnguez Renedo, O., Alonso-Lomillo, M. A., Arcos Martínez, M. J. (2007). Recent developments in the field of screen-printed electrodes and their related applications. *Talanta*, 73, pp. 202–219.
5. Grieshaber, D., MacKenzie, R., Vörös, J., Reimhult, E. (2008). Electrochemical biosensors: Sensor principles and architectures. *Sensors*, 8, pp. 1400–1458.
6. Tang, Y., Ouyang, M. (2007). Tailoring properties and functionalities of metal nanoparticles through crystallinity engineering. *Nature Materials*, 6, pp. 754–759.
7. Sanvicens, N., Marco, M. P. (2008). Multifunctional nanoparticles: Properties and prospects for their use in human medicine. *Trends in Biotechnology*, 26, pp. 425–433.
8. Akbarzadeh, A., Samiei, M., Davaran, S. (2012). Magnetic nanoparticles: Preparation, physical properties, and applications in biomedicine. *Nanoscale Research Letters*, 7, p. 144.
9. Zamora-Gálvez, A., Morales-Narváez, E. C., Mayorga-Martinez, C., Merkoçi, A. (2017). Nanomaterials connected to antibodies and molecularly imprinted polymers asbio/receptors for bio/sensor applications. *Applied Materials Today*, 9, pp. 387–401.
10. Cheeveewattanagul, N., Morales-Narváez, E. H. A., Hassan, A.-R., Francisco Bergua, J., Surareungchai, W., Somasundrum, M., Merkoçi, A. (2017). Straightforward immunosensing platform based on graphene oxide-decorated nanopaper: A highly sensitive and fast

biosensing approach. *Advanced Functional Materials*, 27, 1702741. DOI: 10.1002/adfm.201702741.

11. Zamora Gálvez, A. C., Mayorga-Matinez, C., Parolo, C., Pons, J., Merkoçi, A. (2017). Magnetic nanoparticle-molecular imprinted polymer: A new impedimetric sensor for tributyltin detection. *Electrochemistry Communications*, 82, pp. 6–11.
12. Satheeshkumar, E., Makaryan, T., Melikyan, A., Minassian, H., Gogotsi, Y., Yoshimura, M. (2016). One-step solution processing of Ag, Au and Pd@MXene hybrids for SERS. *Scientific Reports*, 6. DOI: 10.1038/srep32049.
13. Sarycheva, A., Makaryan, T., Maleski, K., Satheesh Kumar, E., Melikyan, A., Minassian, H., Yoshimura, M., Gogotsi, Y. (2017). Two-dimensional titanium carbide (MXene) as surface-enhanced Raman scattering substrate. *The Journal of Physical Chemistry C*, 121, pp. 19983–19988.
14. Jiang, Y., Zhang, X., Pei, L., Yue, S., Ma, L., Zhou, L., Huang, Z., He, Y., Gao, J. (2018). Silver nanoparticles modified two-dimensional transition metal carbides as nanocarriers to fabricate acetycholinesterase-based electrochemical biosensor. *Chemical Engineering Journal*. DOI: 10.1016/j.cej.2018.01.111.
15. Kurbanoglu, S., Ozkan, S. A., Merkoçi, A. (2017). Nanomaterials-based enzyme electrochemical biosensors operating through inhibition for biosensing applications. *Biosensors and Bioelectronics*, 89, pp. 886–898.
16. Morales-Narváez, E., Baptista-Pires, L., Zamora-Gálvez, A., Merkoçi, A. (2017). Graphene-based biosensors: Going simple. *Advanced Materials*, 29, 1604905. DOI: 10.1002/adma.201604905.
17. Chamorro-Garcia, A., Merkoçi, A. (2016). Nanobiosensors in diagnostics. *Nanobiomedicine*, 3, pp. 1–26.
18. Rabti, A. C., Mayorga–Martinez, C., Baptista–Pires, L., Raouafi, N., Merkoçi, A. (2016). Ferrocene-functionalized graphene electrode for biosensing applications. *Analytica Chimica Acta*, 926, pp. 28–35.
19. Morales-Narváez, E., Golmohammadi, H., Naghdi, T., Yousefi, H., Kostiv, U., Horak, D., Pourreza, N., Merkoçi, A. (2015). Nanopaper as an optical sensing platform. *ACS Nano*, 9, pp. 7296–7305.
20. Bertok, T., Gemeiner, P., Mikula, M., Gemeiner, P., Tkac, J. (2013). Ultrasensitive impedimetric lectin based biosensor for glycoproteins containing sialic acid. *Microchimica Acta*, 180, pp. 151–159.
21. Bertok, T., Sediva, A., Katrlik, J., Gemeiner, P., Mikula, M., Nosko, M., Tkac, J. (2013). Label-free detection of glycoproteins by the lectin

biosensor down to attomolar level using gold nanoparticles. *Talanta*, 108, pp. 11–18.

22. Bertok, T., Klukova, L., Sediva, A., Kasak, P., Semak, V., Micusik, M., Omastova, M., Chovanová, L., Vlcek, M., Imrich, R., Vikartovska, A., Tkac, J. (2013). Ultrasensitive impedimetric lectin biosensors with efficient antifouling properties applied in glycoprofiling of human serum samples. *Analytical Chemistry*, 85, pp. 7324–7332.
23. Bertok, T., Sediva, A., Vikartovska, A., Tkac, J. (2014). Comparison of the 2D and 3D nanostructured lectin-based biosensors for in situ detection of sialic acid on glycoproteins. *International Journal of Electrochemical Science*, 9, pp. 890–900.
24. Hushegyi, A., Bertok, T., Damborsky, P., Katrlik, J., Tkac, J. (2015). An ultrasensitive impedimetric glycan biosensor with controlled glycan density for detection of lectins and influenza hemagglutinins. *Chemical Communications*, 51, pp. 7474–7477.
25. Bertok, T., Sediva, A., Filip, J., Ilcikova, M., Kasak, P., Velic, D., Jane, E., Mravcová, M., Rovenský, J., Kunzo, P., Lobotka, P., Smatko, V., Vikartovska, A., Tkac, J. (2015). Carboxybetaine modified interface for electrochemical glycoprofiling of antibodies isolated from human serum. *Langmuir*, 31, pp. 7148–7157.
26. Pihíková, D., Belický, Š., Kasák, P., Bertok, T., Tkac, J. (2016). Sensitive detection of a prostate specific antigen (PSA) using impedimetric assays with in-situ PSA´s glycan analysis. *Analyst*, 141, pp. 1044–1051.
27. Hushegyi, A., Pihíková, D., Bertok, T., Adam, V., Kizek, R., Tkac, J. (2016). Ultrasensitive detection of influenza viruses with a glycan-based impedimetric biosensor. *Biosensors and Bioelectronics*, 79, pp. 644–649.
28. Jolly, P., Zhurauski, P., Hammond, J. L., Miodek, A., Liébana, S., Bertok, T., Tkáč, J., Estrela, P. (2017). Self-assembled gold nanoparticles for impedimetric and amperometric detection of a prostate cancer biomarker. *Sensors and Actuators B: Chemical*, 251, pp. 637–643.
29. Kveton, F., Blsakova, A., Hushegyi, A., Damborsky, P., Blixt, O., Jansson, B., Tkac, J. (2017). Optimization of small glycan presentation for binding of a tumor-associated antibody. Application for construction of an ultrasensitive glycan biosensor. *Langmuir*, 33, pp. 2709–2716.
30. Lorencova, L., Bertok, T., Dosekova, E., Holazova, A., Paprckova, D., Vikartovska, A., Sasinkova, V., Filip, J., Kasak, P., Jerigova, M., Velic, D., Mahmoud, K. A., Tkac, J. (2017). Electrochemical performance

of $Ti_3C_2T_x$ MXene in aqueous media: Towards ultrasensitive H_2O_2 sensing. *Electrochimica Acta*, 235, pp. 471–479.

31. Lorencova, L., Bertok, T., Filip, J., Jerigova, M., Velic, D., Kasak, P., Mahmoud, K. A., Tkac, J. (2018). Highly stable $Ti_3C_2T_x$ (MXene)/Pt nanoparticles-modified glassy carbon electrode for H_2O_2 and small molecules sensing applications. *Sensors and Actuators B: Chemical*, 263, pp. 360–368.

Books

1. J. Bard, A., Stratmann, M. (2002). *Encyclopedia of Electrochemistry: vol. 9: Bioelectrochemistry*, eds., A. J. Bard, M. Stratmann, G. S. Wilson (Wiley-VCH Verlag GmbH), pp. 1–626.
2. Dasari, A., Njuguna, J. (2016). *Functional and Physical Properties of Polymer Nanocomposites*, Cadevall, M., Ros, J. and Merkoçi, A., Chapter 8 "Bismuth-Based Nanomaterials and Platforms for Sensing and Biosensing Applications" (John Wiley & Sons, Ltd.), pp. 159–181.
3. Escarpa, A., Cristina González, M., Ángel López, M. (2015). *Agricultural and Food Electroanalysis*, (John Wiley & Sons, Ltd.) pp. 1–479.

Chapter 5

Simulation of Different Biosensors Using MatLab

Eduard Jane

Department of Glycobiotechnology, Institute of Chemistry, Slovak Academy of Sciences, Bratislava 84538, Slovak Republic

eduard.jane58@gmail.com

Nowadays all fields of science are influenced with computers. This chapter focuses on two topics: data processing and biosensor modeling. It follows the "hands-on modeling" approach, i.e., how to get results from equations. Therefore, the problem is only briefly described and the focus is on the numerical solution and MatLab implementation. Readers are strongly encouraged to refer to the references included in the chapter for a better understanding of the physics involved and for the mathematical analysis. Note that the MatLab scripts can be written in several ways. The presented scripts are a compromise between the script readiness, physics clarity, and the MatLab most effective case. For easy read, some MatLab knowledge is assumed.

Glyconanotechnology: Nanoscale Approach for Novel Glycan Analysis and Their Medical Use
Edited by Tomáš Bertók

ISBN 978-981-4800-16-7 (Hardcover), 978-0-429-44849-2 (eBook)
www.jennystanford.com

5.1 Data Processing

This section focuses on data classification of measured data. There are two main types of data classification: unsupervised and supervised. In this example, PCA (unsupervised method) and neural networks (supervised method) will be presented.

5.1.1 Principal Component Analysis (PCA)

Principal component analysis is based on the eigenvalue analysis. The result of PCA is to transform data to a new set of variables and reduce the dimensionality of data. Consider the medical data that are used to differentiate benign and malignant breast cancer. MatLab dataset for breast cancer will be used.

The input/markers data are

1. Clump thickness
2. Uniformity of cell size
3. Uniformity of cell shape
4. Marginal adhesion
5. Single epithelial cell size
6. Bare nuclei
7. Bland chromatin
8. Normal nucleoli
9. Mitoses

The results are binary outputs: benign and malignant. The total count of the patient is 699. The results are already scaled, so there is no need for preprocessing.

To start covariance, a matrix should be formed—in this case, with nine markers [1]:

$$\mathbf{S}=\begin{pmatrix} \mathbf{(variance(marker1,marker1)} & \cdots & \mathbf{covariance(marker1,marker9)} \\ \vdots & \ddots & \vdots \\ \mathbf{covariance(marker9,marker1)} & \cdots & \mathbf{variance(marker9,marker9))} \end{pmatrix}$$

where

$$\mathbf{covariance}\ (m_1,m_2)=\frac{1}{N-1}\sum_{i=1}^{N}(m_{1i}-\overline{m_1})(m_{2i}-\overline{m_2}) \qquad (5.1)$$

$$\textbf{variance } (m_1) = \frac{1}{N-1}\sum_{i=1}^{N}(m_{1i} - \overline{m_1})^2 \tag{5.2}$$

The eigenvalues can be found solving the equation

$$|S - \lambda I| = 0, \tag{5.3}$$

where I is the identity matrix. Afterward when λ is the solved the eigenvector **a** is obtained by solving

$$(S - \lambda_j I)\ a_j = 0 \tag{5.4}$$

The new transformed variables can be found using

$$Z = m \cdot A \tag{5.5}$$

where A is the eigenvector from Eq. 5.4

```
clear all; clc; close all;
load cancer_dataset.MAT
data=cancerInputs';
datar=cancerTargets';
for ii=1:2
poz1{ii}=find(datar(:,ii)==1); % find position of benign and malignant
end
meand=mean(data); % computation of means
for ii=1:size(data,2)
td(:,ii)=data(:,ii)-meand(ii);
end
covmat=cov(td); % computation of covariance matrix
[V,D]=eigs(covmat); % first 6 eigval and eigvec
val=diag(D);
findata=data*V(:,1:2);
figure(1) % one principal axis
hold on
plot(findata(poz1{1},1),zeros(size(poz1{1},1),1),' k.')
plot(findata(poz1{2},1),zeros(size(poz1{2},1),1),' ko')
xlabel('pc1')
figure(2) % two principal axis
hold on
plot(findata(poz1{1},1),findata(poz1{1},2),' k.')
```

```
plot(findata(poz1{2},1),findata(poz1{2},2),' ko')
xlabel('pc1')
ylabel('pc2')
figure(3) % eigenvalues not shown
stem(val,'k')
xlim([0.5,6.5])
```

The MatLab output should be two plots (Fig. 5.1). On plot (a) only one principal component is assumed. A closer look reveals that dotted markers are located in the interval (0.3;0.8)—benign and open circles are the malignant. The plot (b) shows two principal axis plots. From the picture it is clear that second component does not differentiate the data.

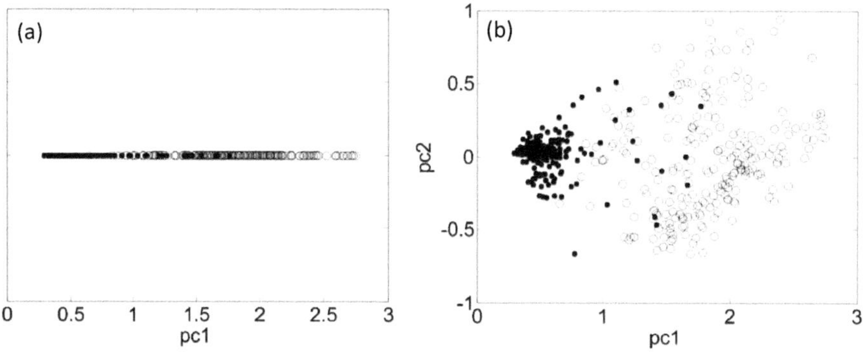

Figure 5.1 One principal component (a), two principal components (b).

5.1.2 Neural Networks

Neural network are computing systems inspired by the biological neural networks that constitute animal brains. Neural networks and deep learning currently provide the best solutions to many problems in image recognition, speech recognition, and natural language processing. The neural network is built from perceptrons. A perceptron takes several inputs and produce a single output. Obviously, one perceptron cannot model an animal brain. The perceptrons are formed in neural network. In this example, breast cancer dataset will be used. The input is divided in training, testing, and validation data set. The result from neural network is confusion matrix and ROC curve. The MatLab breast cancer dataset is used as the input data [2].

```
clear all; clc ;close all;
load cancer_dataset.MAT
posb=find(cancerTargets(1,:)==1); % benign datapart
posm=find(cancerTargets(2,:)==1); % malignant datapart
net = patternnet(10);
net.trainParam.showWindow = false;
net.divideFcn = 'dividerand';
net.divideParam.trainRatio = 70/100; % 70% of sample as training data
net.divideParam.valRatio = 15/100; % 15% of sample as validating data
net.divideParam.testRatio = 15/100; % 15% of sample as testing data
[net1,tr] = train(net,cancerInputs,cancerTargets);
y = sim(net1,cancerInputs); % simulate output
nzt=tr.trainInd;
nzv=tr.valInd;
nztest=tr.testInd;
valtrain=sim(net1,cancerInputs(:,tr.trainInd));
valvalid=sim(net1,cancerInputs(:,tr.valInd));
valtest=sim(net1,cancerInputs(:,tr.testInd));
[tprtrain,fprtrain,thresholds]=roc(cancerTargets(:,tr.trainInd),valtrain);
[tprval,fprval,thresholds]=roc(cancerTargets(:,tr.valInd),valvalid);
[tprtest,fprtest,thresholds]=roc(cancerTargets(:,tr.testInd),valtest);
[tprall,fprall,thresholds]=roc(cancerTargets,y);
%%
auc=@(x,y) trapz(x,y);
close all
subplot(2,2,1)
plot(fprtrain{1},tprtrain{1},'linewidth',1.5)
AUC1=auc(fprtrain{1},tprtrain{1});
xlabel('1-specificity')
ylabel('sensitivity')
title(strcat('Train AUC=',num2str(AUC1)))
set(gca,'box','off', 'fontsize',8)
subplot(2,2,2)
hold on
plot(fprval{1},tprval{1},'linewidth',1.5)
AUC2=auc(fprval{1},tprval{1});
% plot(opt2(1),opt2(2),'o','MarkerSize',4,'linewidth',1.5)
xlabel('1-specificity')
ylabel('sensitivity')
 title(strcat('Val AUC=',num2str(AUC2)))
```

```
set(gca,'box','off', 'fontsize',8)
subplot(2,2,3)
hold on
plot(fprtest{1},tprtest{1},'linewidth',1.5)
AUC3=auc(fprtest{1},tprtest{1});
% plot(opt3(1),opt3(2),'o','MarkerSize',4,'linewidth',1.5)
xlabel('1-specificity')
ylabel('sensitivity')
title(strcat('Test AUC=',num2str(AUC3)))
set(gca,'box','off', 'fontsize',8)
subplot(2,2,4)
hold on
plot(fprall{1},tprall{1},'linewidth',1.5)
AUC4=auc(fprall{1},tprall{1});
% plot(opt4(1),opt4(2),'o','MarkerSize',4,'linewidth',1.5)
xlabel('1-specificity')
ylabel('sensitivity')
 title(strcat('All AUC=',num2str(AUC4)))
set(gca,'box','off', 'fontsize',8)
naz={'','train','val','test'};
figure(2)
plotconfusion(cancerTargets,y,'all')
```

The result from MatLab is shown in Fig. 5.2. On the left, ROC curves are plotted for train dataset, validating dataset, test dataset and all data. The AUC (area under curve) values are in titles of Fig. 5.2a. On right side Confusion matrix is plotted. On the diagonal number of correctly classified data (true positives and true negatives green boxes). False positive and false negative data are shown on the red boxes. The overall machine learning prediction success is summarized in the blue square.

5.2 Biosensor Simulation

5.2.1 Methods of Lines, Diffusion Equation

This subsection deals with partial differential equations (PDEs) (especially parabolic). Typical parabolic PDEs are, for example, diffusion equation and heat equation. In some special cases, the equation can be solved analytically [3], but in practical cases, the numerical/semianalytical approach is needed [4, 5].

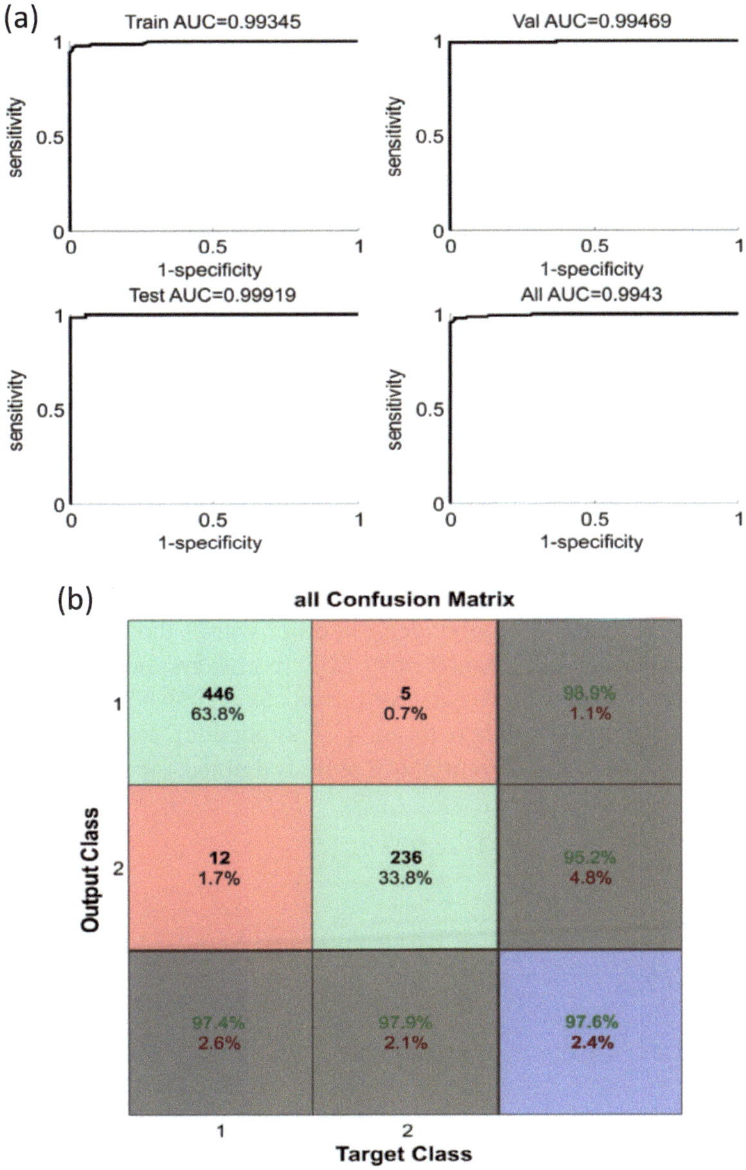

Figure 5.2 ROC curve for training, testing, validation and all data (a). Confusion matrix for all data (b).

For the sake of simplicity, only one-dimensional problem will be presented. One of the most popular semianalytical methods for this problem is the *method of lines* (MOL) [4, 6]. The principle will be much clearer after the next example.

Consider diffusion PDE in one dimension:

$$\frac{\partial c(x,t)}{\partial t} = D\frac{\partial^2 c(x,t)}{\partial c^2} \tag{5.6}$$

With diffusion coefficient D = 0,01 a.u. With boundary conditions

$$c(-1,t) = 0, \quad c(1,t) = 0 \tag{5.7}$$

and initial conditions

$$c(x,0) = \exp\left(\frac{-x^2}{0.01}\right) \tag{5.8}$$

The basic idea of this method is to discretize spatial differentiation with finite differences. Once this is done, the variable is spatially independent. In other words, the discretization of spatial variable "transforms" PDE to the system of ordinary differential equation. The discretization of x is shown in Fig. 5.3.

The finite difference formula is applied on the each node. The second order differential i-th point is defined.

$$\frac{\partial^2 c_i(x,t)}{\partial x^2} \cong \infty \frac{c_{i-1}(t) - 2c_i(t) + c_{i+1}(t)}{xL^2}, \tag{5.9}$$

where xL is the distance between nodes. Now it can be plugged in Eq. 1.6.

$$\frac{\partial c_i(t)}{\partial t} = D\frac{c_{i-1}(t) - 2c_i(t) + c_{i+1}(t)}{xL^2} \tag{5.10}$$

Using this expression, one can generate n (c_1 to c_n) equation according to Fig. 5.3. The undefined/fictitious nodes c_0 used first equation (for c_1) and c_{n+1} used in n-th equation (cn) are replaced with boundary conditions.

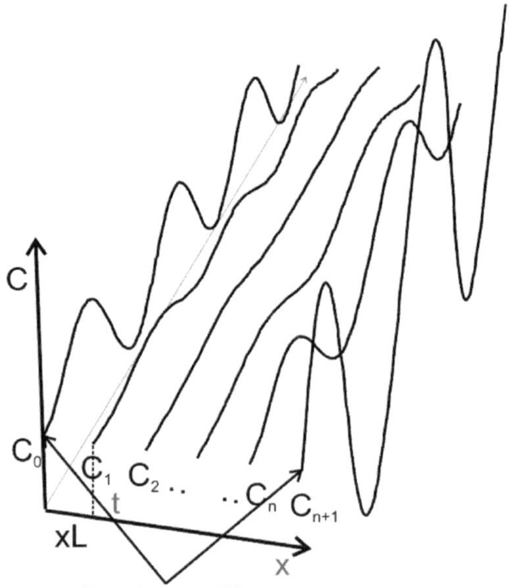

Figure 5.3 Method of line scheme. Ordinary differential equations are evaluated in time domain.

Now the complete list of first-order differential equations is formed. To solve it numerically initial condition are needed for the every equation, which are computed using Eq. 5.8. This can be done by plugging $i \cdot xL$ (i is the index of spatial discretization) in the initial condition.

```
function[]=dif1()
clear all;clc;
tspan=linspace(0,5,30); % time coordinate
nd=19; %number of nodes without boundary
xL=2/(nd+1); % length between nodes
xspat=-1:xL:1; % spatial coordinate
xr=-1:xL:1;
y0=exp(-xspat.^2/0.01); % IC
c0=0; % BC
cend=0; % BC
dif=0.01; % dif coeff
[T,Y]=ode15s(@q,tspan,y0(2:end-1)); % call ode solver
[TT,XX]=meshgrid(xr,tspan); % plotgrid and figure
```

```
mesh(TT,XX,[zeros(30,1),Y,zeros(30,1)])
colormap([0 0 0])
ylabel('t')
xlabel('x')
zlabel('c')
function[dzdt]= q(t,z)
  dzdt=zeros(nd,1);
    dzdt(1)=dif*(+(c0-2*z(1)+z(2))/(xL.^2)); % left boundary
  for ii=2:nd-1
   dzdt(ii)=dif*(+(z(ii-1)-2*z(ii)+z(ii+1))/(xL.^2)); % all middle nodes
  end
  dzdt(nd)=dif*(+(z(nd-1)-2*z(nd)+cend)/(xL.^2)); %right boundary
end
end
```

The MatLab output is a three-dimensional graph, with spatial axis, time axis, and "result/concentration" axis.

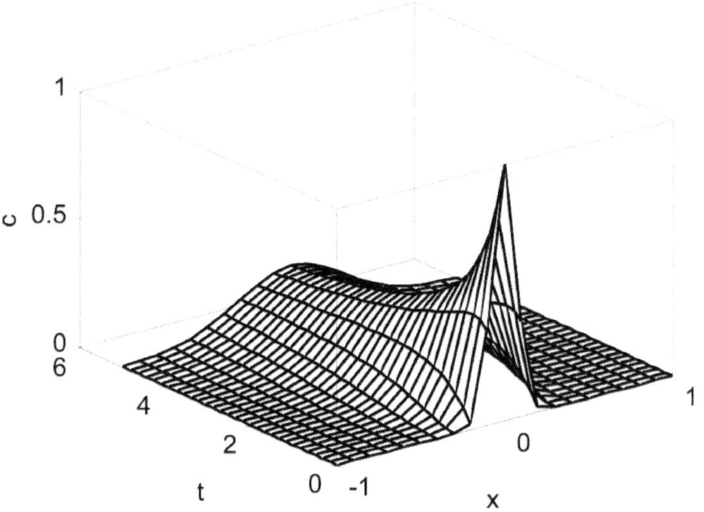

Figure 5.4 Diffusion equation solved with method of lines. Note the spiky peak is due to the low number of nodes.

5.2.2 Affinity Type of Biosensor

In this subsection, we present a theoretical model for the binding of an antigen in solution to single binding antibody on the surface. A simplified analytical solution can be found [7]. The focus will on numerical solution. On the surface, we expect the following chemical reaction [8, 9]:

$$Ab + Ag \xleftrightarrow{k_f, k_b} Ab.Ag \tag{5.11}$$

with reaction rate k_f and k_b for forward and backward reaction, respectively. The surface concentration of free antibodies is Γ_0. The kinetic of the bounded antigen is

$$\frac{d\Gamma_1(t)}{dt} = k_f c(0, t)(\Gamma_0 - \Gamma_1(t)) - k_b \Gamma_1(t), \tag{5.12}$$

where $c(0, t)$ is the concentration of antigen at the surface. The concentration is diffusion controlled.

$$\frac{\partial c(x,t)}{\partial t} = D \frac{\partial^2 c(x,t)}{\partial x^2} \tag{5.13}$$

The boundary conditions are

$$D \left. \frac{\partial c(x,t)}{\partial x} \right|_{x=0} = k_f c(0,t)(\Gamma_0 - \Gamma_1(t)) - k_b \Gamma_1(t)$$
$$c(L, t) = c_{\text{bulk}} \tag{5.14}$$

Initial conditions are set

$$c(x,0) = 0;\ c(L,0) = c_{\text{bulk}};\ \Gamma_1(0) = 0 \tag{5.15}$$

The result of simulation is coverage $\Gamma(t)$ on the surface. It can be compared with the result from SPR (surface plasmon resonance) [9], or the antigen can be marked fluorescence detection antigen with fluorescence [10].

Before we start writing the MatLab script, we should look closer at the first boundary condition, which can approximated:

$$D \frac{c(xL,t) - c(0,t)}{xL} = k_f c(0,t)(\Gamma_0 - \Gamma_1(t)) - k_b \Gamma_1(t) \tag{5.16}$$

From this equation, we can express $c(0, t)$:

$$c(0,t) = \frac{c(xL,t)D + \Gamma_1(t))k_b xL}{D + \Gamma_0 k_f xL - \Gamma_1(t))k_f xL} \tag{5.17}$$

The MatLab script begins with the definition of some constants: dif is the diffusion constant in cm^2s^{-1}, L is the is the thickness in centimeters, gama0 surface coverage in $mol\cdot cm^{-2}$, k_f and k_b are reaction rate constants in $s^{-1}\cdot mol^{-1}\cdot cm^3$ and s^{-1}, respectively, nd is the number of nodes, and cend is the bulk concentration in $mmol\cdot cm^{-3}$.

```
clear all; close all; clc;
dif=3e-5;L=0.06;gama0=6e-13;
kf=1e6;kb=1e1;
nd=101;cend=1e-6;
tspan=linspace(0,250,150);
[T,Y,Z,gamaq,xL,xspat]=rov(dif,nd,d,tspan,gama0,k1,k2,cend);%call
function
```

In next step function rov.m is called with the child function q. Length between nodes (xL), initial conditions ($y0$) are set in the function beginning. After the equations are solved with ode15s solver, the antigen concentration profile (stored in variable Z) and surface coverage are extracted. Note that the indexes with $nd + 1$ i.e., $Y(..., nd + 1)$ belongs to $\Gamma_1(t)$, so the gamaq is $\Gamma_1(t)$. xx is the spatial coordinate vector.

```
function [T,Y,Z,gamaq,xL,xx]=rov(dif,nd,d,tspan,gamma0,k1,k2,cend)
xL=d/(nd+1);
y0=[zeros(1,nd-1),cend,0];
[T,Y]=ode15s(@q,tspan,y0);
q0=(Y(:,1)*dif+Y(:,nd+1)*k2*xL)./(dif+gamma0*k1*xL-Y(:,nd+1)*k1*xL);
Z=[q0,Y(:,1:end-1),cend*ones(length(tspan),1)];
gamaq=Y(:,nd+1);
xx=xL.*(0:nd+1);
function[dzdt]= q(t,z)
dzdt=zeros(nd+1,1);
c0=(z(1)*dif+z(nd+1)*k2*xL)/(dif+gamma0*k1*xL-z(nd+1)*k1*xL);% eq 1.17
dzdt(1)=dif*(+(c0-2*z(1)+z(2))/(xL.^2)); % left boundary
for ii=2:nd-1
dzdt(ii)=dif*(+(z(ii-1)-2*z(ii)+z(ii+1))/(xL.^2));
```

```
end
dzdt(nd)=dif*(+(z(nd-1)-2*z(nd)+cend)/(xL.^2)); %right boundary
dzdt(nd+1)=c0*k1*(A0-z(nd+1))-k2*z(nd+1);% gama(t)
end
end
```

Child function *q*, called in the ode solver, is very similar to the diffusion problem solved earlier; c_0 is redefined with boundary condition and one more $z(:,nd+1)$ is added for the Γ_1, associated with Eq. 5.12.

Back in the main script, we get time *T*, concentration of antigen *Z*, and bounded antigen surface concentration gamaq. Some points are chopped off for the picture clarity:

```
% the main script continue here
[TT,XX]=meshgrid(xspat(1:5:end),tspan(1:50)); % grid
mesh(TT,XX,Z(1:50,1:5:end))
% plots
colormap([0 0 0])
ylabel('t[s]')
xlabel('x[cm]')
zlabel('c[\mol/dm^-^3]')
figure(2)
hold on
plot(tspan,gamaq,'k-','linewidth',1.5)
xlabel('t[s]')
ylabel('\Gamma [mol/cm^2]')
set(gca,'box','on')
```

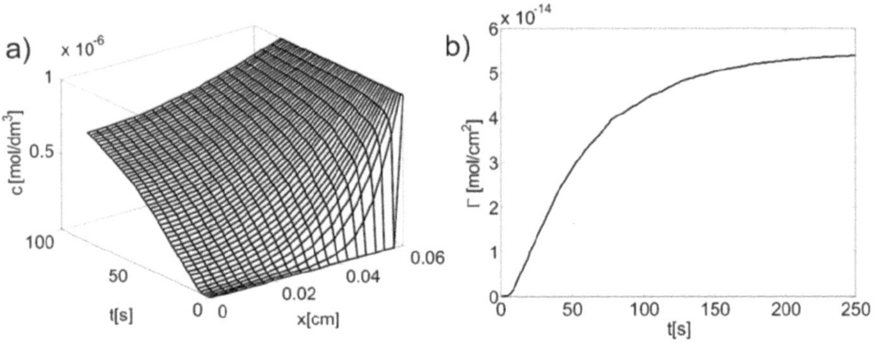

Figure 5.5 Concentration profile of antigen (a). Time evolution of bounded antigen on the surface (b).

5.2.3 Enzyme Biosensor

5.2.3.1 Monolayer enzyme biosensor

This chapter is inspired with Baronas et al.'s book [11]. Two physical processes are analyzed in simulation of enzyme biosensor: enzyme kinetics and diffusion. Enzyme kinetics is usually described with the Michaelis–Menten kinetics:

$$E + S \xleftrightarrow{k_f, k_f} ES \xrightarrow{k_2} P + E \tag{5.18}$$

The product reaction rate is defined as follows:

$$\frac{\partial P(t)}{\partial t} = k_2 ES = \frac{k_2 e_o S(t)}{\frac{k_b + k_2}{k_f} + S(t)} = \frac{V_M S(t)}{K_M S(t)}, \tag{5.19}$$

where V_M is the saturation reaction rate and K_M is the Michaelis–Menten constant.

In this model, we will consider the biosensor as the enzyme immobilized on the surface of electrode. The governing equations for this type of biosensor will be

$$\frac{\partial S(x,t)}{\partial t} = D_S \frac{\partial^2 S(x,t)}{\partial x^2} - \frac{V_M S(x,t)}{K_M + S(x,t)}, \tag{5.20}$$

$$\frac{\partial P(x,t)}{\partial t} = D_P \frac{\partial^2 P(x,t)}{\partial x^2} + \frac{V_M S(x,t)}{K_M + S(x,t)}, \tag{5.21}$$

where x is the spatial coordinate and is in the interval 0 to L, which is the thickness of the enzyme layer. D_s and D_P are the diffusion coefficients for the substrate and the product, respectively.

In case of the amperometric biosensor, due to the electrode polarization the concentration of the reaction product at the electrode surface is permanently reduced to zero. Therefore, the boundary condition should be set:

$$\left.\frac{\partial S(x,t)}{\partial x}\right|_{x=0} = 0 \tag{5.22}$$

$$P(0, t) = 0, S\,(L, t) = S_{BULK}, P(L, t) = 0$$

Concentration the substrate the membrane is constant S_{BULK}. The product is diluted in the bulk, so it is negligibly small and set to 0. In the case of amperometric biosensors, due to the electrode polarization, the concentration of the reaction product at the electrode surface ($x = 0$) is permanently reduced to zero.

The initial conditions are:

$$P(x, 0) = 0; S(x, 0) = 0 \cap S(x, L) = S_{BULK} \tag{5.23}$$

The amperometric biosensor response is current density defined as

$$j(t) = \frac{i(t)}{A} = nFD_P \left. \frac{\partial P(x,t)}{\partial t} \right|_{x=0}, \tag{5.24}$$

where n is the number of exchange electrons, F Faraday constant, D_P diffusion coefficient of the product.

Now we can switch to the MatLab script. In the first part, the constants are defined: F is the Faraday constant in $A \cdot s \cdot \mu mol^{-1}$, all concentrations are in $\mu mol \cdot cm^{-3}$, L, the thickness of the membrane layer, is in centimeters, and diffusion coefficients are in $cm^2 \cdot s^{-1}$.

```
clear all; close all; clc;
F=0.096485; difS=3e-6;difP=3e-6; Vmax=0.1;Km=0.1;
nd=101;d=0.01;s01=0.001;p01=0;se=0.001;pe=0.;
tspan=linspace(0,30,200);
[T,Y,subz,prodz,xspat,xL]=enzyme(difS,difP,Vmax,Km,...
nd,d,s01,p01,se,pe,tspan);
```

The main script calls the function enzyme.m with child function q. This function is similar to the previous examples. One change should be made in the boundary. The substrate at the electrode has a reflective boundary (i.e., first derivative equals to 0). Suppose that the first node is on the boundary. In this case, we use central first-order difference formula:

$$\left. \frac{\partial c(x,t)}{\partial x} \right|_{x=0} \cong \frac{c_2(t) - c_o(t)}{2xL} = 0 \tag{5.25}$$

From this equation, c_0 should be equal to c_2. Therefore

$$\left.\frac{\partial^2 c(x,t)}{\partial x^2}\right|_{x=0} \cong \frac{-2c_1(t)+2c_2(t)}{xL^2} = 0. \tag{5.26}$$

The second change in child function q consists of two variables: the substrate and the product. In this case *dzdt* has dimensions $z(:,2)$, with the index 1 for substrate and the index 2 for product. At the end of function $dzdt(:,2)$ is reshaped to $dydt(:,1)$, which is the correct form for ode solver ode15s:

```
function [T,Y,sub,prod,xspat,xL]=enzyme(difS,difP,Vmax,...
 Km,nd,d,s01,p01,se,pe,tspan)
s1=[zeros(1,nd-1),s01];
p1=[zeros(1,nd-1),p01];
y0=[s1',p1']; % initial conditions
xL=d/(nd+1);
xspat=(0:nd+1).*xL; % spatial coordinate
[T,Y]=ode15s(@q,tspan,y0);
send=linspace(s01,s01,length(tspan));
pend=linspace(p01,p01,length(tspan));
pb=linspace(p01,p01,length(tspan));
sub=[Y(:,1),Y(:,1:nd),send'];% one more line of Y(:,1) is added
prod=[pb',Y(:,nd+1:end),pend' ];

 function[dydt]= q(t,y)
  dzdt=zeros(nd,2);
  z=reshape(y,[],2);
  dzdt(1,1)=difS*(-2*z(1,1)+2*z(2,1))/(xL.^2)...
   -Vmax*z(1,1)/(Km+z(1,1));
  dzdt(1,2)=difP*(0-2*z(1,2)+z(2,2))/(xL.^2)...
   +Vmax*z(1,1)/(Km+z(1,1));
  for ii=2:nd-1
   dzdt(ii,1)=difS*(z(ii-1,1)-2*z(ii,1)+z(ii+1,1))/(xL.^2)...
    -Vmax*z(ii,1)/(Km+z(ii,1));
   dzdt(ii,2)=difP*(z(ii-1,2)-2*z(ii,2)+z(ii+1,2))/(xL.^2)...
    +Vmax*z(ii,1)/(Km+z(ii,1));
  end
  dzdt(nd,1)=difS*(z(nd-1,1)-2*z(nd,1)+se)/(xL.^2)...
   -Vmax*z(nd,1)/(Km+z(nd,1));
  dzdt(nd,2)=difP*(z(nd-1,2)-2*z(nd,2)+pe)/(xL.^2)...
   +Vmax*z(nd,1)/(Km+z(nd,1));
```

```
  dydt=reshape(dzdt,[],1);
 end
 %end of function enzyme

%Matlab main script continue here
 sub=subz*1000; % concentration in umol/dm3
 prod=prodz*1000; % concentration in umol/dm3
 didt=(prodz(:,2)-prodz(:,1))/xL;
jc=1e7*F*difP*didt; %current in nA/mm2
 g=1;g1=6;tend=50;% data reduction and cut
% Plotting results
figure(1) % Substrate
 [TT,XX]=meshgrid(xspat(1:g1:end),tspan(1:g:tend));
mesh(TT,XX,sub(1:g:tend,1:g1:end),'linewidth',1.5)
ylabel('t[s]')
xlabel('x[cm]')
zlabel('c[\mumol/dm^-^3]')
colormap([0 0 0])
 figure(2) % Product
mesh(TT,XX,prod(1:g:tend,1:g1:end),'linewidth',1.5)
ylabel('t[s]')
xlabel('x [cm]')
zlabel('c [\mumol/dm^-^3]')
colormap([0 0 0])
set(gca,'box','on')
tim=[2,3,5,10];
figure(3) % Substrate and product cuts
subplot(3,1,1)
hold on
for ii=1:length(tim)
plot(xspat,sub(tim(ii),:),'k', 'linewidth',1.5)
end
axis('tight')
xlabel('x [cm]')
ylabel('c [\mumol/dm^-^3]')
set(gca,'box','on')
subplot(3,1,2)
hold on
for ii=1:length(tim)
plot(xspat,prod(tim(ii),:),'k', 'linewidth',1.5)
```

```
end
axis('tight')
xlabel('x [cm]')
ylabel('c [\mumol/dm^-^3]')
set(gca,'box','on')
tspan(tim)
figure(4) % current density
plot(tspan,jc,'k', 'linewidth',1.5)
xlabel('t[s]')
ylabel('j[nA. mm^-^2]')
set(gca,'box','on')
```

After copying the main script in file and function in the second file, after the successful run, the graphs plotted in Fig. 5.6 are obtained.

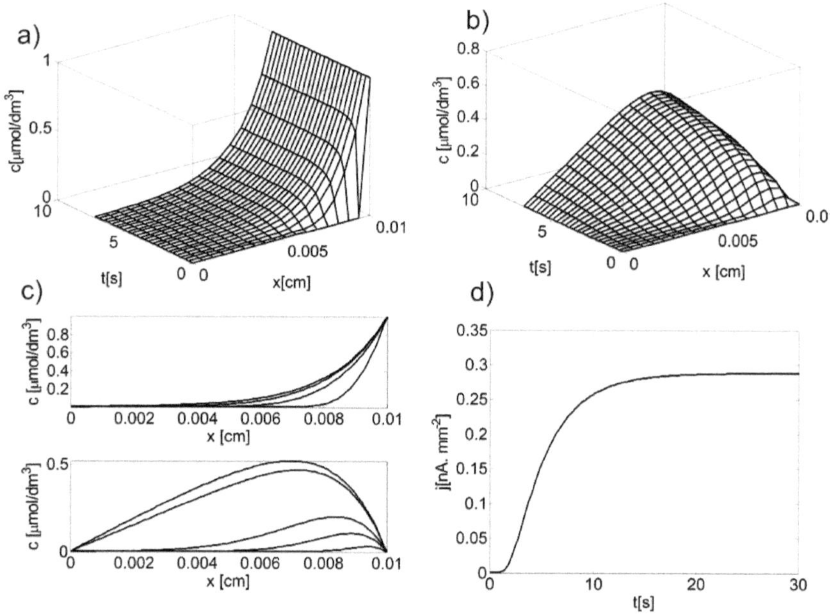

Figure 5.6 Substrate (a) and product (b) concentration evolution. Concentration profile cuts (c) are in 0.15s; 0.6s; 1.3s; 7.4s and 15s and current density evolution (d).

The results are the concentration profiles of the substrate, the product, and the current density. Note that concentrations were recomputed to μmol/dm^3. Current density has nA/mm^2 dimension. Simulation time was 30 s.

5.2.3.2 Steady-state solution of enzyme biosensor, biosensor sensitivity

The time evolution electric current density, substrate and product concentration were computed in the previous example. There are some cases where only maximal current (steady state) is needed. For example, an interesting parameter is biosensor sensitivity [11].

$$B(S_{\mathrm{Bulk}}) = \frac{S_O}{I(S_{\mathrm{BULK}})} \frac{dI(S_{\mathrm{BULK}})}{dS_{\mathrm{BULK}}} \tag{5.27}$$

Other parameter is dependence of current density on layer thickness.

The variables become steady state at long time (at the infinity). In this case, we can assume that the substrate and the product concentration changes in time are approximately 0. In this case, the governing equations in the previous model simplify to the system of ordinary differential equations:

$$\frac{d^2S(x)}{dx^2} = \frac{V_{\max}}{K_{\mathrm{M}} + S(x)} \tag{5.28}$$

$$\frac{d^2P(x)}{dx^2} = \frac{V_{\max}}{K_{\mathrm{M}} + S(x)} \tag{5.29}$$

For each equation, we need two boundary conditions, same as in the previous example:

$$\left.\frac{dS(x)}{dx}\right|_{x=0} = 0;\ P(0) = 0, S(L) = S_{\mathrm{BULK}},\ P(L) = 0; \tag{5.30}$$

To solve this problem, we can use bvp4c solver implemented in MatLab.

The steady state computation is less time consuming, than time dependent computation.

The MatLab script consist again from defined constants and boundary conditions:

```
clear all; close all; clc;
F=0.096485;
difS=3e-6;
difP=3e-6;
Vmax=[0.05,0.1,0.15];
Km=0.1;
d=0.01; % length of enzyme layer
se=0.005; % concentration of substrate at bulk boundary umol/
cm3==mmol/dm3
pe=0.00; % concentration of product at bulk boundary
```

Current densities for different V_{max} and concentration are computed. Therefore two for loops are added to the main script.

```
for jj=1:3 % Loop for different Vmax
for ii=1:35% Loop for different concentration
[Y,sub,prod,xspat,xL]=enzyme_steady(difS,difP,d,Vmax(jj),Km,...
se*10^(ii/10),pe);
koncsub(ii,jj)=se*10^(ii/10);
 didt=(prod(:,2)-prod(:,1))/xL;
jc(ii,jj)=1e7*F*difP*didt;
prodf{ii,jj}=prod';
subf{ii,jj}=sub';
end
end
```

The main script calls function enzyme_steady in this case. Bvp4c solver can be used only for first-order ode. The equation should be rewritten to the form:

$$\frac{dS(x)}{d(x)} = sx(x) \tag{5.31}$$

$$\frac{dp(x)}{dx} = px(x) \tag{5.32}$$

$$\frac{dsx(x)}{dx} = \frac{1}{D_s}\frac{V_{max}}{K_M + S(x)} \tag{5.33}$$

$$\frac{dpx(x)}{dx} = \frac{1}{D_P}\frac{V_{max}}{K_M + S(x)} \tag{5.34}$$

Two new support variables *sx* and *px* are added to the equations. In the next script, $x(1)$ is for S, $x(3)$ is for P, $x(2)$ is for sx, and $x(4)$ is for px. Child function equ is for equation definition, function eqbc defines a set of boundary conditions, and the last function is the initial condition. The result in this case is not sensitive to the initial condition. In other words, for all initial conditions, the result should be same. The reason for setting initial conditions is the use of "collocation method" in bvp4c solver. For details, see ref. [12].

```
function [Y,sub,prod,xcor,xL]=enzyme_steady(difS,difP,d,Vmax,Km,se,pe)
xspat=linspace(0,d,50);
xL=mean(diff(xspat));
solinit = bvpinit(xspat,@initc);
sol = bvp4c(@equ,@eqbc,solinit);
xcor=unique(xspat);
Y=deval(sol,xcor);
sub=Y(1,:);
prod=Y(3,:);
% ------------------------------------------------------------
function dx = equ(t,x) %equations 1.31-1.34
 dx=zeros(4,1);
 dx(1)=x(2);
 dx(3)=x(4);

 dx(2)=1/difS*Vmax*x(1)./(Km+x(1));
 dx(4)=-1/difP*Vmax*x(1)./(Km+x(1));

end
% boundary conditions
function res = eqbc(XL,XR)
res = [ XL(2)
 XL(3)-pe
 XR(1)-se
 XR(3)
 ];
end
% -------------
function yinit = initc(x)
 yinit = [0 0 0 0 ];
```

```
end
end
% end of function enzyme_steady
```

Back in the main script, only plots are done. Note that three different V_M are used in simulation,

```
%The main script continue here
figure(1)
shape='osv';
hold on
for jj=1:3
 fc=(koncsub(1:end-1,jj)./jc(1:end-1,jj)).*diff(jc(:,jj))./diff(koncsub(:,jj));
 plot(koncsub(1:end-1,jj),fc,strcat('-k',shape(jj)),'linewidth',1.5)
end
xlabel('c(S) [mmol/dm^3]')
ylabel('B [a.u]')
set(gca,'box','on','xscale','log')
axis('tight')
```

The second example shows the current dependence on the enzyme layer thickness. The function enzyme_steady.m is identical to the previous example. Only minor changes are done in MatLab script.

```
clear all; close all; clc;
F=0.096485;
difS=3e-6;
difP=3e-6;
Vmax=[0.05,0.1,0.2];
Km=0.1;
d=0.0001;
se=0.001;
pe=0.00;
for jj=1:3 % Loop for different Vmax
for ii=1:40 % Loop for different concentration
[Y,sub,prod,xspat,xL]=rov(difS,difP,d*10^(ii/15),Vmax(jj),Km,se,pe);
 didt=(prod(:,2)-prod(:,1))/xL;
jc(ii,jj)=1e7*F*difP*didt;
prodf{ii,jj}=prod';
subf{ii,jj}=sub';
```

```
dcor(ii,jj)=d*10^(ii/15);
end
end
shape='osv';
hold on
for jj=1:3
plot(dcor(:,jj),jc(:,jj),strcat('-k',shape(jj)),'linewidth',1.5)
end
xlabel('x[cm]')
ylabel('j[nA.mm^-^2]')
set(gca,'box','on','xscale','log')
axis('tight')
```

On the Fig. 5.7a is the plot of current density dependence on the enzyme layer thickness. Different markers are for the different V_M. The current density increases with the V_M. Figure 5.7b shows the sensitivity biosensor on the concentration of substrate. The higher value V_M increases the interval of concentration with sensitivity close to 1.

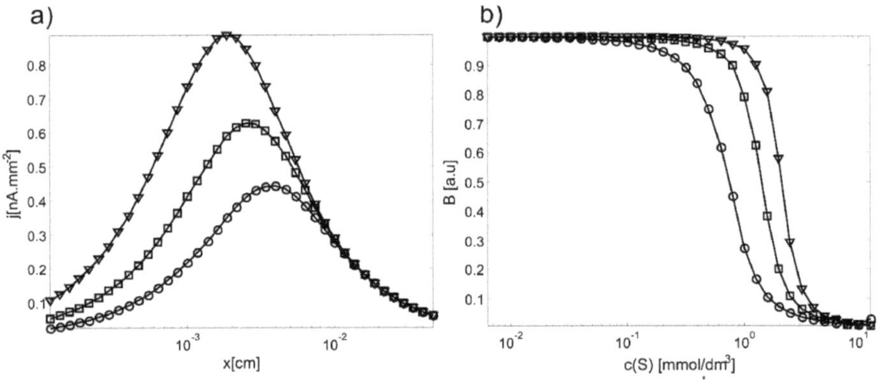

Figure 5.7 Current density dependence on the layer thickness (a). Sensor sensitivity (b) V_{max} = 0.05 (circles); 0.1 (squares); 0.15 (triangles). Note that the *x* axis is in the logarithmic scale.

5.2.3.3 Flow injection

The Previous mathematical model described biosensors in the batch mode. The substrate bulk concentration was constant during the whole time. Suppose we add a finite amount of substrate to the biosensor. This change will affect the right boundary (in the L distance) in some time interval.

To model flow injection rectangle/window function can be used. The rectangle function is noncontinuous, and for numerical purposes, it is better to replace it with the function

$$S_{\text{BULK}}(t) = \frac{S_{\text{BULK}}}{2}\left(\tanh\left(\frac{t_{\text{end}} - t}{w}\right) - \tanh\left(\frac{t_{\text{begin}} - t}{\text{eps}}\right)\right), \tag{5.35}$$

where, tanh is the hyperbolic tangent w is coefficient of "edge sharpness," eps is the floating-point relative accuracy. MatLab eps value is 2^-52.

If w increases, the edge becomes "softer," as shown in Fig. 5.8. The left edge is sharp, the substrate was added to the sensor in a short time and in the next step buffer was added. The flow injection model mimics the final amount of substrate and it's decreasing.

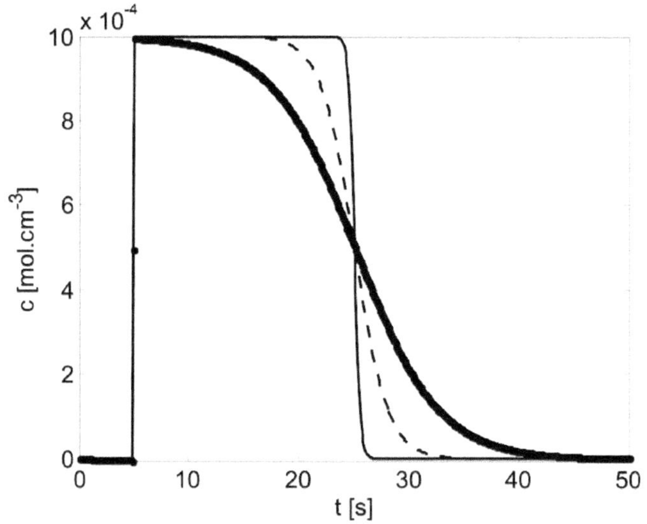

Figure 5.8 Different enzyme profiles on outer boundary w = 0.5 (solid line); 3 (dashed line); 7 (dotted-thick line).

Only one line for the S bulk (send variable in the MatLab script) should be replaced from constant to Eq. 5.35.

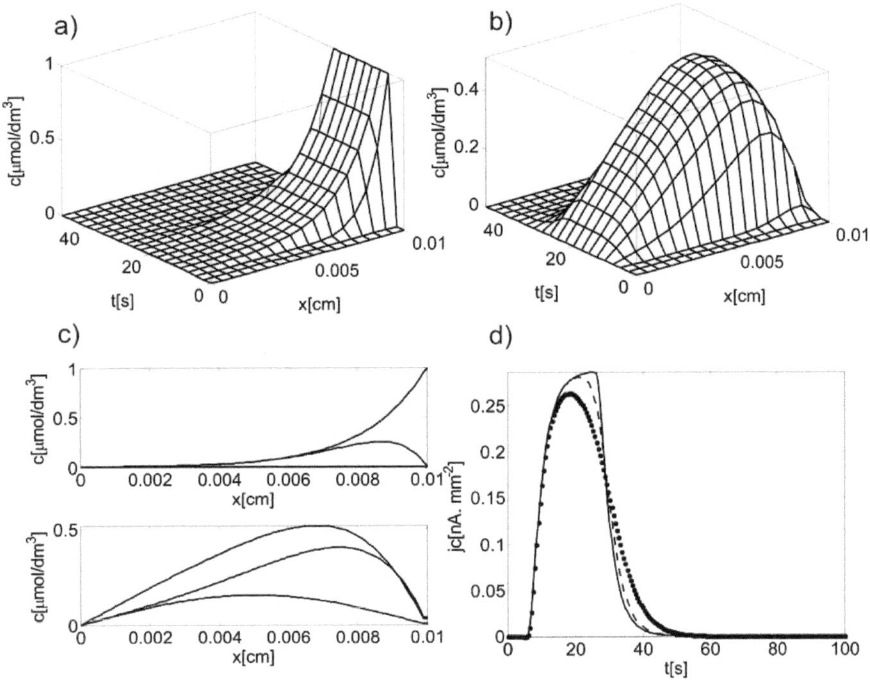

Figure 5.9 Concentration profiles for substrate (a) and product (b) and cuts (c) at w = 7. The current density (d) w = 0.5 (solid line); 3 (dashed line); 7 (dotted line).

5.2.3.4 Cyclovoltammetric simulation of enzyme electrode

The response of the sensor was computed in the previous examples. In the voltammetric experiment, electrical perturbation is applied. In this case, voltammetric simulation on electrochemically mediated enzyme is presented. The cyclic voltammograms offer information about the diffusive or adsorptive nature of the electrode process, its kinetic and thermodynamic parameters, as well as the existence and characteristics of coupled homogeneous chemical reactions [13]. In a typical cyclic voltammetry experiment, the potential is swept linearly with time from some starting potential, E_i, where species A is stable (i.e., not electroreduced), to some other, more negative potential, E_v, at which electron transfer between species A and the electrode is rapid, and species B is formed.

Quasi steady-state approach is assumed (*d intermediate/dt~0)*. *A* and *B* are the concentrations of reduced and oxidized forms of an enzyme, respectively. Enzyme is uniformly distributed in the layer and the diffusion of reduced and oxidedized enzyme is neglected [14].

$$A \leftrightarrow B + e^-$$
$$B + S \xleftrightarrow{k_1, k_{-1}} \text{intermediate} \xrightarrow{k_2} P + A \tag{5.36}$$

Governing equations for this example are

$$\frac{\partial A(x,t)}{\partial t} = D_A \frac{\partial^2 A(x,t)}{\partial x^2} + k_1 k_2 \frac{B(x,t)S(x,t)}{k_2 + k_{-1}} \tag{5.37}$$

$$\frac{\partial B(x,t)}{\partial t} = D_B \frac{\partial^2 B(x,t)}{\partial x^2} + k_1 k_{-1} \frac{B(x,t)S(x,t)}{k_2 + k_{-1}} - k_1 B(x,t)S(x,t) \tag{5.38}$$

$$\frac{\partial S(x,t)}{\partial t} = D_S \frac{\partial^2 S(x,t)}{\partial x^2} + k_1 k_{-1} \frac{B(x,t)S(x,t)}{k_2 + k_{-1}} - k_1 B(x,t)S(x,t) \tag{5.39}$$

$$\frac{\partial P(x,t)}{\partial t} = D_P \frac{\partial^2 P(x,t)}{\partial x^2} + k_1 k_2 \frac{B(x,t)S(x,t)}{k_2 + k_{-1}} \tag{5.40}$$

With boundary condition on the electrode surface ($x = 0$)

$$D_B \left.\frac{\partial B(x,t)}{\partial x}\right|_{x=0} = (k_{\text{red}} B(0,t) - k_{\text{ox}} A(0,t)) \tag{5.41}$$

$$D_A \left.\frac{\partial A(x,t)}{\partial x}\right|_{x=0} = -D_B \left.\frac{\partial B(x,t)}{\partial x}\right|_{x=0} \tag{5.42}$$

$$\left.\frac{\partial S(x,t)}{\partial x}\right|_{x=0} = 0; \quad \left.\frac{\partial P(x,t)}{\partial x}\right|_{x-0} = 0 \tag{5.43}$$

Boundary conditions on the enzyme layer end (L)

$$A(L,t) = A_{\text{BULK}};\ B(L,t) = 0;\ S(L,t) = S_{\text{BULK}};\ P(L,t) = 0 \tag{5.44}$$

And finally the initial conditions

$$A(x, 0) = A_{\text{BULK}};\; B(x, 0) = 0; S(x, 0) = 0 \cap S(L, 0) = S_{\text{BULK}};$$

$$P(x, 0) = 0 \tag{5.45}$$

For reduction and oxidation rates, the Butler–Volmer equation is used:

$$k_{\text{red}} = k_0 \exp\left(-\frac{\alpha F}{RT}\theta\right); k_{\text{ox}} = k_0 \exp\left((1-\alpha)\frac{F}{RT}\theta\right); \tag{5.46}$$

Finally, the electric current density can be computed:

$$j(t) = \frac{I(t)}{A} = -F(k_{\text{red}}B(0,t) - k_{\text{ox}}A(0,t)) \tag{5.47}$$

The implementation of the mediator boundary condition on the electrode is the hardest part. The following coupled equations should be solved:

$$D_B \frac{B(xL,t) - B(0,t)}{xL} = (k_{\text{red}}B(0,t)) - k_{\text{ox}}A(0,t)) \tag{5.48}$$

$$D_A \frac{A(xL,t) - B(0,t)}{xL} = -D_B \frac{B(xL,t) - B(0,t)}{xL} \tag{5.49}$$

$A(0, t)$ and $B(0, t)$ are the searched variables. For solving, the symbolic toolbox can be used, or it can be done classically with a pencil and paper.

All variables are defined with "syms.". Variable sol is the structure with $A0$ and $B0$.

```
clear all
syms A A0 xL kred A0 kox B B0 difA difB
eq=[difB*(B-B0)/xL==kred*B0-kox*A0,...
 difA*(A-A0)/xL==-difB*(B-B0)/xL];
sol=solve(eq,A0,B0);
sol.A0
sol.B0
```

After this preparation, the MatLab script is straightforward. Again, a script begins with the definitions of constants. The variable thet contains the vector of the "V" signal.

```
clear all; close all; clc;
difA=3e-7;difS=3e-7;difP=3e-7;difB=3e-7;
nd=101;d=0.01;can=1e-3;cbn=0*1e-3; ];% umol/cm^3
k0=1e-3; % cm/s
k1=1e6;k2=1*1e2;km1=1*1e2;p01=0;
s01=[0,3e-4,8e-4,2e-3,5e-3,7e-3,1.2e-2,15e-4];% umol/cm^3
thet=1*([0.3:-0.01:-0.3, -0.301:0.01:0.301 ]); % potential
tspan=linspace(0,50,length(thet)); % time coordinate
for ii=1:8
[T,Y,p(ii,:),sub,A0,B0,prod,xspat,xL,eso]=rovcyclo(difA,difB,difS,difP,nd,d,ca
n,cbn,thet,k0...
k1,km1,k2,tspan,p01,s01(ii));
end
```

The MatLab calls function rovcyclo.m. Eight different substrate concentrations are simulated in the for loop.

```
function[TT,Y,p,sub,A0,B0,prod,xspat,xL,es0]=rovcyclo(difA,difB,difS,difP,n
d,d,can,...
cbn,thet,k0,k1,km1,k2,tspan,p01,s01)
A=can*[ones(1,nd)];
B=cbn*[ones(1,nd)];
s1=s01*[ones(1,nd)];
p1=[zeros(1,nd-1),p01];
y0=[A',B',s1',p1'];
xL=d/(nd+1);
xspat=(0:nd+1).*xL;
fac=96500/8.314/300;
F=96.4685; mA.s/umol
alfa=0.5;
[TT,Y]=ode15s(@q,tspan,y0);
kred0=k0.*exp(-alfa.*fac.*thet)';
kox0=k0.*exp((1-alfa)*fac.*thet)';
ca00=-((-difA.*difB.*Y(:,1)-difA.*kred0.*xL.*Y(:,1)-difB.*kred0.*xL.*Y(:,nd
+1))...
./(difA*difB+difB.*kox0.*xL+difA*kred0*xL));
cb00=-((-difA.*difB.*Y(:,nd+1)-difA.*kox0.*xL.*Y(:,+1)-difB.*kox0.*xL.*Y(:,
nd+1))...
./(difA*difB+difB.*kox0.*xL+difA.*kred0*xL));

p=-F*(kred0.*cb00-kox0.*ca00););%current density
```

```
send=linspace(s01,s01,length(tspan)); %boundary conditions
pend=linspace(p01,p01,length(tspan)); %boundary conditions
pb=linspace(p01,p01,length(tspan)); %boundary conditions
aa=linspace(can,can,length(tspan)); %boundary conditions
bb=linspace(cbn,cbn,length(tspan)); %boundary conditions
A0=[ca00,Y(:,1:nd),aa']; % variable extracting conditions
B0=[cb00,Y(:,nd+1:2*nd),bb'];
sub=[Y(:,2*nd+1),Y(:,2*nd+1:3*nd),send'];
prod=[Y(:,3*nd+1),Y(:,3*nd+1:4*nd),pend' ];
es0=k1.*B0.*sub./(km1+k2);

function[dydt]= q(t,y) % Function with equations

theta= interp1(tspan,thet,t);
dzdt=zeros(nd,4);
z=reshape(y,[],4);
kred=k0.*exp(-alfa.*fac*theta);
kox=k0.*exp((1-alfa)*fac*theta);
ca0=-((-difA*difB*z(1,1)-difA*kred*xL*z(1,1)-difB*kred*xL*z(1,2))...
./(difA*difB+difB*kox*xL+difA*kred*xL));
cb0=-((-difA*difB*z(1,2)-difA*kox*xL*z(1,1)-difB*kox*xL*z(1,2))...
./(difA*difB+difB*kox*xL+difA*kred*xL));

dzdt(1,1)=difA*(ca0-2*z(1,1)+z(2,1))/(xL.^2)...
+(k1*k2*z(1,2)*z(1,3))/(k2+km1);
dzdt(1,2)=difB*(cb0-2*z(1,2)+z(2,2))/(xL.^2)...
+(k1*km1*z(1,2)*z(1,3))/(k2+km1)-k1*z(1,2)*z(1,3);
dzdt(1,3)=difS*(-2*z(1,3)+2*z(2,3))/(xL.^2)...
+(k1*km1*z(1,2)*z(1,3))/(k2+km1)-k1*z(1,2)*z(1,3);
dzdt(1,4)=difP*(-2*z(1,4)+2*z(2,4))/(xL.^2)...
+(k1*k2*z(1,2)*z(1,3))/(k2+km1);

for ii=2:nd-1
dzdt(ii,1)=difA*(z(ii-1,1)-2*z(ii,1)+z(ii+1,1))/(xL.^2)...
+(k1*k2*z(ii,2)*z(ii,3))/(k2+km1);
dzdt(ii,2)=difB*(z(ii-1,2)-2*z(ii,2)+z(ii+1,2))/(xL.^2)...
+(k1*km1*z(ii,2)*z(ii,3))/(k2+km1)-k1*z(ii,2)*z(ii,3);
dzdt(ii,3)=difS*(z(ii-1,3)-2*z(ii,3)+z(ii+1,3))/(xL.^2)...
+(k1*km1*z(ii,2)*z(ii,3))/(k2+km1)-k1*z(ii,2)*z(ii,3);
dzdt(ii,4)=difP*(z(ii-1,4)-2*z(ii,4)+z(ii+1,4))/(xL.^2)...
```

```
+(k1*k2*z(ii,2)*z(ii,3))/(k2+km1);
end
dzdt(nd,1)=difA*(z(nd-1,1)-2*z(nd,1)+can)/(xL.^2)...
+(k1*k2*z(nd,2)*z(nd,3))/(k2+km1);
dzdt(nd,2)=difB*(z(nd-1,2)-2*z(nd,2)+0)/(xL.^2)...
+(k1*km1*z(nd,2)*z(nd,3))/(k2+km1)-k1*z(nd,2)*z(nd,3);
dzdt(nd,3)=difS*(z(nd-1,3)-2*z(nd,3)+s01)/(xL.^2)...
+(k1*km1*z(nd,2)*z(nd,3))/(k2+km1)-k1*z(nd,2)*z(nd,3);
dzdt(nd,4)=difP*(z(nd-1,4)-2*z(nd,4)+0*p01)/(xL.^2)...
+(k1*k2*z(nd,2)*z(nd,3))/(k2+km1);

dydt=reshape(dzdt,[],1);
end
end
%end of function rovcyclo.m
```

The main script continues with plotting. Note that cyclovoltagramms are plotted in two figures for plot clarity.

```
g=5;
g1=6;
tend=50;
close all
figure(1)
[TT,XX]=meshgrid(xspat(1:g1:end),tspan(1:g:end));
mesh(TT,XX,1000*A0(1:g:end,1:g1:end),'linewidth',1.5) %A
ylabel('t[s]','FontSize' ,16)
xlabel('x[cm]','FontSize' ,16)
zlabel('c[\mumol/dm^3]','FontSize' ,16)
colormap([0 0 0])
set(gca, 'FontSize' ,16)
figure(2)
mesh(TT,XX,1000*B0(1:g:end,1:g1:end),'linewidth',1.5) %B
ylabel('t[s]','FontSize' ,16)
xlabel('x[cm]','FontSize' ,16)
zlabel('c[\mumol/dm^3]','FontSize' ,16)
colormap([0 0 0])
set(gca, 'FontSize' ,16)
figure(3)
mesh(TT,XX,1000*sub(1:g:end,1:g1:end),'linewidth',1.5) % Substrate
ylabel('t[s]','FontSize' ,16)
```

```
xlabel('x[cm]','FontSize' ,16)
zlabel('c[\mumol/dm^3]','FontSize' ,16)
colormap([0 0 0])
set(gca, 'FontSize' ,16)
 figure(4)
 [TT,XX]=meshgrid(xspat(1:g1:end),tspan(1:g:end));
mesh(TT,XX,1000*prod(1:g:end,1:g1:end),'linewidth',1.5) % Product
ylabel('t[s]','FontSize' ,16)
xlabel('x[cm]','FontSize' ,16)
zlabel('c[\mumol/dm^3]','FontSize' ,16)
colormap([0 0 0])
set(gca, 'FontSize' ,16)
figure(5) % Cyclovoltammogram
hold on
plot(thet,1e3*p(1,:),'k-','linewidth' ,1.5)
plot(thet,1e3*p(2,:),'k--','linewidth' ,1.5)
plot(thet,1e3*p(3,:),'k.-','linewidth' ,1.5)
plot(thet,1e3*p(8,:),'k-','linewidth' ,1.3)
xlabel('\theta [V]','FontSize',16)
ylabel('I [mA/cm^2]','FontSize',16)
set(gca,'box','on')
set(gca, 'FontSize' ,16)
axis('tight')
figure(6) % Cyclovoltammogram
hold on
plot(thet,1e3*p(4,:),'k-','linewidth' ,1.5)
plot(thet,1e3*p(5,:),'k--','linewidth' ,1.5)
plot(thet,1e3*p(6,:),'k.-','linewidth' ,1.5)
plot(thet,1e3*p(7,:),'k-','linewidth' ,1.3)
 xlabel('\theta [V]','FontSize',16)
ylabel('j [mA/cm^2]','FontSize',16)
set(gca,'box','on')
set(gca, 'FontSize' ,16)
set(gca,'box','on' )
axis('tight')
```

The MatLab output is shown in Figs. 5.10 and 5.11. The upper row (a,b) of Fig. 5.10 shows the concentration profiles of the reduced and oxidized forms of the enzyme. From the figure, it is clear that the mediator reacts with the substrate near the electrode. At the higher distance, it remains constant. The concentration profiles of the substrate and the product have similar profiles.

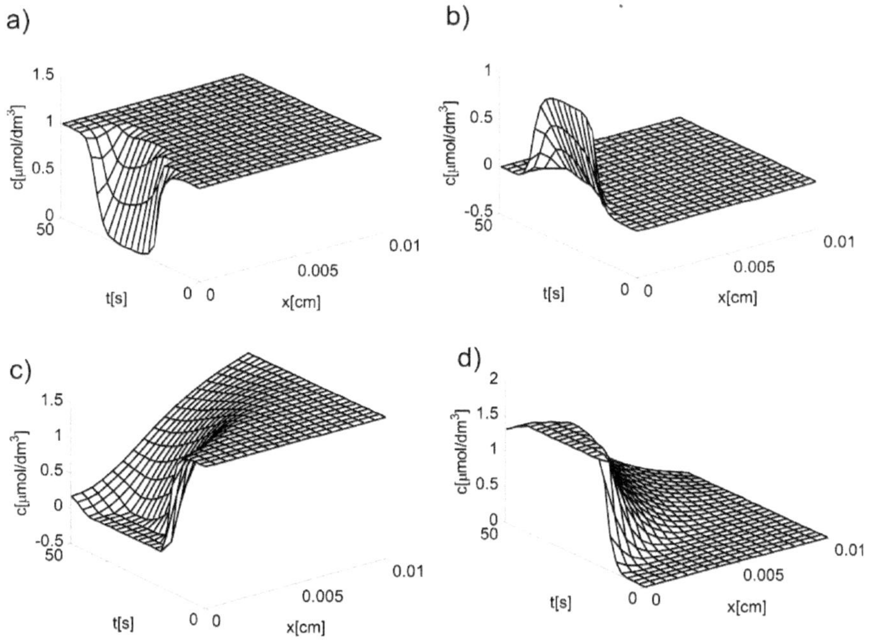

Figure 5.10 Concentration profile of reduced (a) and oxidized form of an enzyme (b). Concentration profile of substrate (c) and product (d) at the substrate concentration 1.5e-3 mol/dm^3.

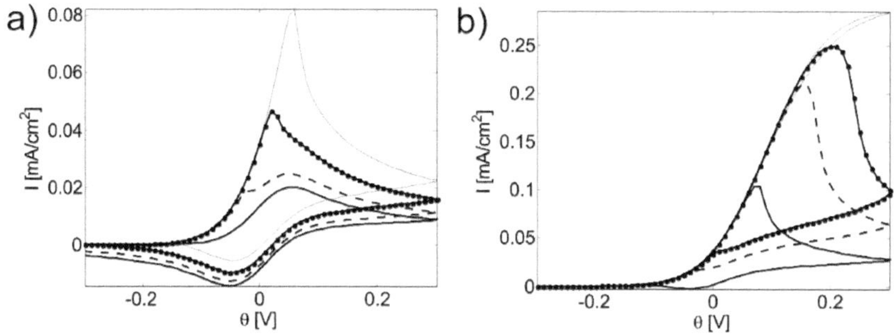

Figure 5.11 Cyclovoltammetric signal. Thick solid line without substrate, dashed, dotted, and thin solid is with 3e-4, 8e-4, and 1.5e-3 substrate concentration, respectively, on the left figure. On the right, the concentrations of the substrate are 2e-3, 5e-3, 7e-3, 1.2e-2 in the order solid dashed, dotted, and thin solid.

The cyclovoltagrams are shown on Fig. 5.11. The thick solid line shows the process without the substrate; only the mediator is present. Increasing the concentration of substrate has an effect on the cathodic current.

5.2.3.5 Fourier transform linear sweep ac voltammetry

In this section, we will continue with voltage perturbation and its effect on current. Only a linear function of voltage in time will be used [15]. Moreover, the small ac modulation will be added. The purpose of this experiment is to avoid the capacitance and to measure only the Faradaic current [16].

The output signal is more complicated in analysis than in the cyclovoltammetric case. The fast Fourier transform and the concept of narrow band gap filter will shed some light [17]. The script computes the results for the experiment with and without the substrate.

The main MatLab script:

```
clear all; close all; clc;
difA=3e-7;difS=3e-7;difP=3e-7;difE=3e-7;
difB=3e-7;
nd=101;d=0.01;can=1e-3;cbn=0*1e-3;
k0=1e-3;k1=1e6;k2=1*1e2;km1=1*1e2;p01=0;
s01=2e-3;
 thet0=1*([-0.3:0.00005:0.3 ]);
 % step is lowered
tspan=linspace(0,100,length(thet0));
omega=5; % frequency
ac=0.025*sin(omega.*tspan);
thet=thet0+ac;
%%
[T,Y,p,sub,A0,B0,prod,xspat,xL,eso]=rovcyclo(difA,difB,difS,...
difP,nd,d,can,cbn,thet,k0,k1,km1,k2,tspan,p01,0);
[T,Y,p1,sub1,A01,B01,prod1,xspat,xL,eso]=rovcyclo(difA,difB,...
difS,difP,nd,d,can,cbn,thet,k0,k1,km1,k2,tspan,p01,s01);
%To this point the MatLab script is similar to cyclovoltammetric
close all
figure(1)
hold on
```

```
plot(thet0,1e3*p,'k-','linewidth',1.5)
set(gca,'Ydir','reverse','Xdir','reverse','box','on','linewidth',1.5)
xlabel('\Theta [V]','FontSize',16)
ylabel('I [mA/cm^2]','FontSize',16)
set(gca,'Ydir','reverse','Xdir','reverse','box','on')
set(gca, 'FontSize' ,16)
axis('tight')
```

The MatLab script ffte.m is used to the Fourier transform of the computed signal. The found frequencies correspond to the dc signal ($w = 0$) and first harmonics ($w = 5$) and the second harmonics ($w = 10$):

```
% The MatLab script continues
figure(2)
hold on
plot(thet0,1e3*p1,'k-','linewidth',1.5)
set(gca,'box','on','linewidth',1.5)
xlabel(' \theta [V]','FontSize',16)
ylabel('I [mA/cm^2]','FontSize',16)
set(gca,'box','on')
set(gca, 'FontSize' ,16)
axis('tight')

 [w,pp0,pp]=ffte(tspan',1e3*p);
[w,pp01,pp1]=ffte(tspan',1e3*p1);
figure(3)
hold on
plot(w,pp,'k--','linewidth',1.5)
plot(w,pp1,'k-','linewidth',1.5)
xlim([-2.5*omega,2.5*omega])
xlabel('\omega [Hz]','FontSize',16)
ylabel('int [a.u]','FontSize',16)
set(gca, 'FontSize' ,16)
set(gca,'box','on' )
```

The narrow-band filter is used to extract the signal around the specified frequency.

```
figure(4)
[t1,f2a,w,F]=iffte(w,pp0,0,2,3);
```

```
[t1,f2a1,w,F]=iffte(w,pp01,0,2,3);
 poz1=find(t1>100,1);
 if isempty (poz1)
   poz1=length(t1);
 end
hold on
plot(t1(floor(length(t1)/2):10:poz1),f2a(floor(length(t1)/2):10:poz1),'k--
','linewidth',1.5)
plot(t1(floor(length(t1)/2):10:poz1),f2a1(floor(length(t1)/2):10:poz1),'k-
','linewidth',1.5)
xlabel('t [s]','FontSize',16)
ylabel('I [a.u]','FontSize',16)
set(gca, 'FontSize' ,16)
set(gca,'box','on' )
set(gca,'box','on')
axis('tight')
figure(5)
[t1,f2a,w,F]=iffte(w,pp0,omega,1.5,3);
[t1,f2a1,w,F]=iffte(w,pp01,omega,1.5,3);
subplot(2,1,1)
plot(t1(floor(length(t1)/2):10:poz1),(f2a(floor(length(t1)/2):10:poz1)),'k-')
xlabel('t [s]','FontSize',16)
ylabel('I [a.u]','FontSize',16)
set(gca, 'FontSize' ,16)
set(gca,'box','on' )
axis('tight')
subplot(2,1,2)
plot(t1(floor(length(t1)/2):10:poz1),(f2a1(floor(length(t1)/2):10:poz1)),'k-')
xlabel('t [s]','FontSize',16)
ylabel('I [a.u]','FontSize',16)
set(gca, 'FontSize' ,16)
set(gca,'box','on' )
axis('tight')
figure(6)
[t1,f2a,w,F]=iffte(w,pp0,2*omega,1.5,3);
[t1,f2a1,w,F]=iffte(w,pp01,2*omega,1.5,3);
subplot(2,1,1)
plot(t1(floor(length(t1)/2):1:poz1),(f2a(floor(length(t1)/2):1:poz1)),'k-')
xlabel('t [s]','FontSize',16)
ylabel('I [a.u]','FontSize',16)
```

```
set(gca, 'FontSize' ,16)
set(gca,'box','on' )
axis('tight')
subplot(2,1,2)
plot(t1(floor(length(t1)/2):10:poz1),(f2a1(floor(length(t1)/2):10:poz1)),'k-')
xlabel('t [s]','FontSize',16)
ylabel('I [a.u]','FontSize',16)
set(gca, 'FontSize' ,16)
set(gca,'box','on' )
axis('tight')
```

In this script, two new functions are used. Ffte.m is used for the fast Fourier transform of the computed signal. The results are the variables for the power spectrum.

```
function[w,F1,F2]=ffte(t,f)
ts=mean(diff(t));
t=[t;zeros(100000,1)]; % Zero padding
f=[f;zeros(100000,1)]; % Zero padding
nn=length(t);
ws=1/ts/length(t);
ws=2*pi/ts/length(t);
w=ws*(-floor(nn/2):1:floor(nn/2));
F=fftshift(fft(f));
F1=(fftshift(fft(f)));
end
```

The function iffte.m is the inverse fast Fourier transform and "Gaussian type" narrow band filter. The narrow band filter is defined with the function

$$g(\omega)=\exp\left(-\left(\frac{\omega-\omega_0}{f}\right)^{2*\mathrm{round}(n)}\right), \tag{5.50}$$

where ω_c is the center frequency of the band filter and f is the width of the filter. The n parameter determines the filter edge sharpness.

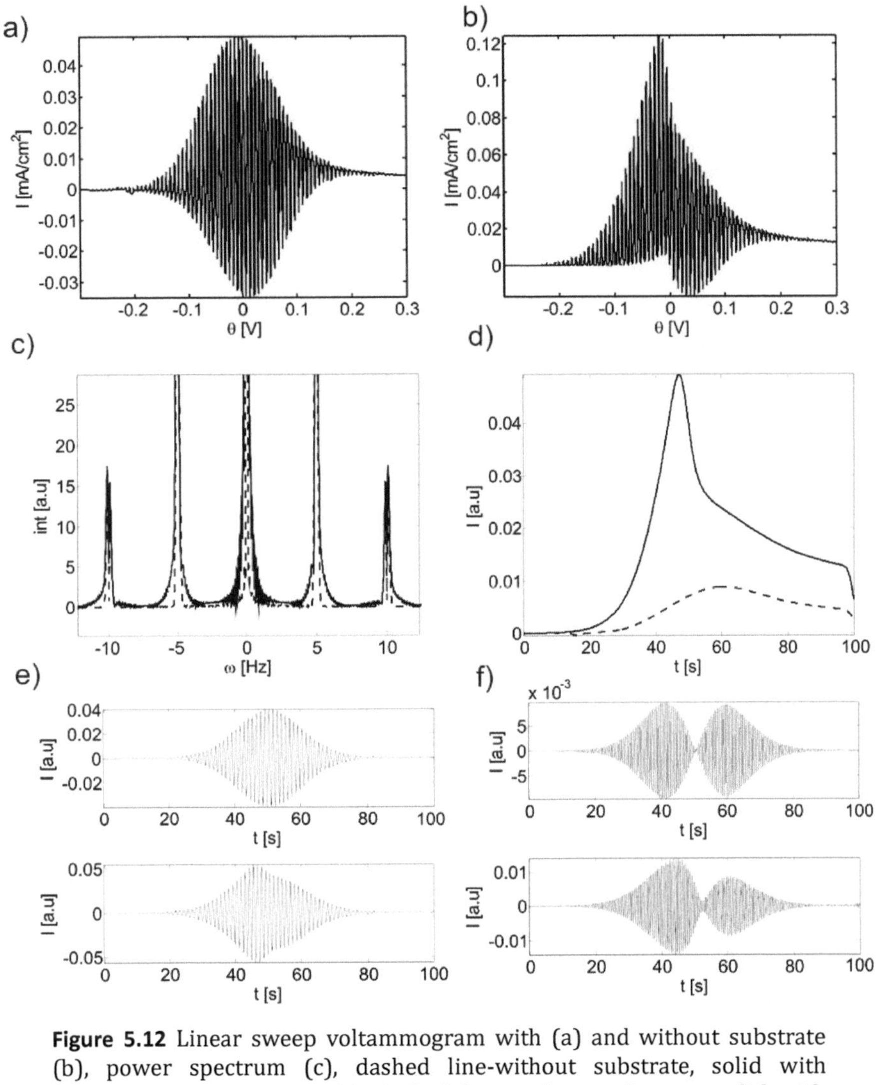

Figure 5.12 Linear sweep voltammogram with (a) and without substrate (b), power spectrum (c), dashed line-without substrate, solid with substrate, dc component (d) dashed line-without substrate, solid with substrate, first (e) and second harmonics (f) without substrate and with substrate are on top and down respectively.

```
function[t1,f2a,w,F]=iffte(w,F1,c,f,nx)
poz=find(w>=0);
q=lp(w(poz));
```

```
F=q'.*F1;
ws=mean(diff(w));
nn=length(w);
ts=2*pi/ws/length(w);
t=ts*(-floor(nn/2):1:floor(nn/2));
f1=fftshift(ifft(ifftshift(F)));
f2=real(f1);
t1=t(floor(length(t)/2):end);
f2a=f2(floor(length(t)/2):end);
t1=t(1:end);
f2a=f2(1:end);
 function[y]= lp(x)
  v=exp(-((x-c)/f).^2*round(nx));
  y=[wrev(v), v(2:end)]; % negative frequencies
 end
end
```

The MatLab script generates six figures. On the first graph (a) is the current vs. potential in the experiment without the substrate. On the second graph (b) is the current vs. potential in the experiment with the substrate. The third (c) figure is the power spectrum. On the fourth (d) figure is the dc signal and on the last two graphs are the first and second harmonics, respectively. On the top are the signals without the substrate.

5.2.3.6 Two-layer model of enzyme biosensor (advanced)

In this last example, we considered the simulation model sketched in Fig. 5.13. A perfect membrane layer was assumed in the previous subsections, and we deal only with the enzyme layer.

According to Baronas, there are various reasons for applying a multilayer approach to the modeling of biosensors. Multilayer models are usually used in the following cases:

(a) The bulk solution is assumed to be slightly stirred or non-stirred. This assumption leads to two-compartment models.
(b) The enzyme layer is covered with an inert outer membrane.
(c) The membrane stabilizes the enzyme layer and creates a diffusion limitation to the substrate, i.e., lowers the substrate concentration in the enzymatic layer and thereby prolongs the calibration curve of the biosensor.
(d) The electrode is covered with a selective membrane.

Selective membranes are usually impermeable to certain molecules and permeable to a desired substance. This arrangement can notably increase the biosensor selectivity. The selective layer can also protect the metal interface of the electrode.

(e) In multienzyme systems, enzymes are often immobilized separately in different active layers packed in a sandwich-like multilayer arrangement. This approach seems to be a rather fast and cheap method to design biosensors for different purposes [11, 18].

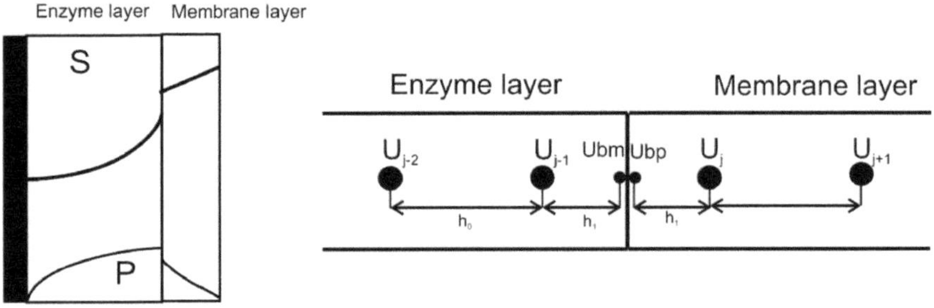

Figure 5.13 Two-layered model. On the interface, two points are added to implement jump boundary condition.

In this case, the physical analysis for the enzyme layer is same. It contains Michaelis–Menten kinetics and diffusion process. In the membrane layer, only the diffusion process is present. Moreover, the membrane is not perfectly permeable for the substrate or the product, so on the interface of enzyme layer and membrane layer is the concentration "jump."

The governing equations for enzyme layer are

$$\frac{dS_1(x,t)}{\partial t} = D_{S1}\frac{\partial^2 S_1(x,t)}{\partial x^2} - \frac{V_M S_1(x,t)}{K_M + S_1(x,t)} \tag{5.51}$$

$$\frac{\partial P_1(x,t)}{\partial t} = D_{P1}\frac{\partial^2 P_1(x,t)}{\partial x^2} + \frac{V_M S_1(x,t)}{K_M + S_1(x,t)} x \in [0, L_1) \tag{5.52}$$

For the membrane layer

$$\frac{\partial S_2(x,t)}{\partial t} = D_{S2}\frac{\partial^2 S_2(x,t)}{\partial x^2} \tag{5.53}$$

$$\frac{dP_2(x,t)}{\partial t} = D_{P2}\frac{\partial^2 P_2(x,t)}{\partial x^2} x \in [L, L_1) \tag{5.54}$$

Boundary condition should be set on electrode surface $x = 0$ and on the membrane $x = L$ are same:

$$\left.\frac{dS_1(x,t)}{\partial x}\right|_{x=0} = 0 \tag{5.55}$$

$P(0, t) = 0, S\,(L, t) = S_{\text{BULK}}, P(L, t) = 0$

Moreover, the interfacial boundary condition should be added for concentration:

$$S_1 = k_1 S_2, \quad P_1 = k_2 P_2 \tag{5.56}$$

And for flow

$$D_{S1}\left.\frac{\partial S_1(x,t)}{\partial x}\right|_{x=L_1} = D_{S2}\left.\frac{\partial S_2(x,t)}{\partial x}\right|_{x=L_1} \tag{5.57}$$

$$D_{P1}\left.\frac{\partial P_1(x,t)}{\partial x}\right|_{x=L_1} = D_{P2}\left.\frac{\partial P_2(x,t)}{\partial x}\right|_{x=L_1} \tag{5.58}$$

The solution is similar to one layer problem. Some changes should be done one interface. Consider the layer interface with nodes sketched on the figure. Nodes with index $j-2$ and $j-1$ are in first layer. Nodes with index j, $j+1$ belong to the second layer.

Two more points/nodes(Ubm and Ubp) are added on interface to boundary condition.

If the boundary is in the middle of $j-1$ and j node one can express node with Taylor series [19]:

$$U_{j-2} = Ubm - \frac{dUbm}{dx}\frac{3}{2}L + \frac{d^2Ubm}{dx^2}\left(\frac{3}{2}L\right)^2 \tag{5.59}$$

$$U_{j-1} = Ubm - \frac{dUbm}{dx}\frac{1}{2}L + \frac{d^2Ubm}{dx^2}\left(\frac{1}{2}L\right)^2 \tag{5.60}$$

$$U_j = Ubp - \frac{dUbp}{dx}\frac{1}{2}L + \frac{d^2Ubp}{dx^2}\left(\frac{1}{2}L\right)^2 \tag{5.61}$$

$$U_{j+1} = Ubp - \frac{dUbp}{dx}\frac{3}{2}L + \frac{d^2Ubp}{dx^2}\left(\frac{3}{2}L\right)^2 \tag{5.62}$$

and the discontinuity in the concentration and flux at interfaces between the layers

$$Ubm = k_1 Ubp \tag{5.63}$$

$$D_1\frac{dUbm}{dx} = D_2\frac{dUbp}{dx} \tag{5.64}$$

Now there are six equations and six variables (*Ubm*, *Ubp*, first differentials and second differentials). The equations have an analytical solution and again pencil and paper or the MatLab symbolic toolbox can be used. All variables should be declared with the "syms" command. Variable eq contains a list of equations. MatLab script is written in more general case. Therefore, *dx* variables are replaced within the next step.

```
clear all; clc;close all;
syms U1m U2m U U1p Dx Dx2 dudxm du2dxm dudxp...
  du2dxp Ubm Ubp k1 difS difS2 xL ...
  h1 h0 k2 difP difP2

 eq=[-U1m+Ubm-Dx*dudxm+Dx^2/2*du2dxm;...
-U2m+Ubm-Dx2*dudxm+Dx2^2/2*du2dxm;...
-U+Ubp+Dx*dudxp+Dx^2/2*du2dxp;...
-U1p+Ubp+Dx2*dudxp+Dx2^2/2*du2dxp;...
-Ubp+k1*Ubm;-difS*dudxm+difS2*dudxp];
eq1=subs(subs(eq,Dx,xL/2),Dx2,3*xL/2);
vys=solve(eq1,dudxp, dudxm, Ubm, Ubp, du2dxm, du2dxp);
```

The solution of the equation is plugged in the finite difference scheme for nonuniform grid:

$$\frac{d^2U_{j-1}}{dx^2} = \frac{h_1 Ubm - (h_1 + h_0)U_{j-1} + h_0 U_{j-2}}{h_0 h_1 (h_0 + h_1)} \tag{5.65}$$

$$\frac{d^2U_j}{dx^2} = \frac{h_0U_{j+1} - (h_1 + h_0)U_j + h_1Ubp}{h_0h_1(h_0 + h_1)} \tag{5.66}$$

```
% the script continues
rig=2*(h1*U1p-(h1+h0)*U+h0*vys.Ubp)/(h0*h1*(h0+h1));
lef=2*(h0*vys.Ubm-(h1+h0)*U1m+h1*U2m)/(h0*h1*(h0+h1));
r=simplify(subs(subs(rig,h0,xL),h1,xL/2))
l=simplify(subs(subs(lef,h0,xL),h1,xL/2))
```

The solution of this equation will be shown and used in the next example. In the beginning, the constants are defined. Note that number of nodes increased:

```
clear all; close all; clc;
F=0.096548;
difS=3e-6;difP=3e-6; difS2=3e-6; difP2=3e-6;
k1=1.5; k2=2;
Vmax=0.1;Km=0.1;
nd=202;
d=0.01; % length of enzyme layer
s01=0.; p01=0. ; % Initial conditions
se=0.001; % Boundary condition
pe=0.00; % Boundary condition
tspan=linspace(0,50,200);
```

The main script calls the function rov2enz.m. There is a new variable ndx. It defines a node where the interface is located.

```
[T,Y,subz,prodz,xspat,xL]=rov2enz(difS,difP,difS2,difP2,k1,k2,Vmax,Km,nd,
d,s01,p01,se,pe,tspan);
```

```
function [T,Y,sub,prod,xspat,xL]=rov2enz(difS,difP,difS2,difP2,k1,k2,Vmax,K
m,nd,d,s01,p01,se,pe,tspan)
s1=[zeros(1,nd-1),s01];
p1=[zeros(1,nd-1),p01];
y0=[s1',p1'];
xL=d/(nd+1);
xspat=(0:nd+1).*xL;
 ndx=floor(8.5*nd/10);
[T,Y]=ode15s(@q,tspan,y0);
send=linspace(s01,s01,length(tspan));
```

```
pend=linspace(p01,p01,length(tspan));
sub=[Y(:,1),Y(:,1:nd),send'];
prod=[zeros(length(tspan),1),Y(:,nd+1:end),Y(:,end)];
 function[dydt]= q(t,y)
  dzdt=zeros(nd,2);
  z=reshape(y,[],2);
  p0=0;
  dzdt(1,1)=difS*(-2*z(1,1)+2*z(2,1))/(xL.^2)-Vmax*z(1,1)/(Km+z(1,1));
  dzdt(1,2)=difP*(p0-2*z(1,2)+z(2,2))/(xL.^2)+Vmax*z(1,1)/(Km+z(1,1));

  for ii=2:ndx-2
   dzdt(ii,1)=difS*(z(ii-1,1)-2*z(ii,1)+z(ii+1,1))/(xL.^2)-Vmax*z(ii,1)/
   (Km+z(ii,1));
   dzdt(ii,2)=difP*(z(ii-1,2)-2*z(ii,2)+z(ii+1,2))/(xL.^2)+Vmax*z(ii,1)/
   (Km+z(ii,1));
  end
% -----------------------------------------------------------------------------------------
```

Node *ndx*–1 belongs to the enzyme layer and the node ndx belongs to the membrane layer. The formula defined in the previous example is used for the nodes. Note that the formula for the substrate and product differs only in jump constants and diffusion coefficients.

```
% ----------------------------------------------------------------------------------------
---------------
  dzdt(ndx-1,1)=(9*z(ndx,1)*difS2 - 3*z(ndx-1,1)*difS + 3*z(ndx-2,1)*difS ...
-z(ndx+1,1)*difS2 - 12*z(ndx-1,1)*difS2*k1 + 4*z(ndx-2,1)*difS2*k1)/...
  (3*xL^2*(difS + difS2*k1))-Vmax*z(ndx-1,1)/(Km+z(ndx-1,1));

  dzdt(ndx,1)=-(12*z(ndx,1)*difS - 4*z(ndx+1,1)*difS +
  3*z(ndx,1)*difS2*k1...
  - 9*z(ndx-1,1)*difS*k + z(ndx-2,1)*difS*k1 - 3*z(ndx+1,1)*difS2*k1)/...
  (3*xL^2*(difS + difS2*k1));

  dzdt(ndx-1,2)=(9*z(ndx,2)*difP2 - 3*z(ndx-1,2)*difP + 3*z(ndx-2,2)*difP...
  - z(ndx+1,2)*difP2 - 12*z(ndx-1,2)*difP2*k2+ 4*z(ndx-2,2)*difP2*k2)/...
  (3*xL^2*(difP + difP2*k2))+(+Vmax*z(ndx-1,1)/(Km+z(ndx-1,1)));

  dzdt(ndx,2)=-(12*z(ndx,2)*difP - 4*z(ndx+1,2)*difP +
  3*z(ndx,2)*difP2*k2...
  - 9*z(ndx-1,2)*difP*k2+ z(ndx-2,2)*difP*k2 - 3*z(ndx+1,2)*difP2*k2)/...
```

```
(3*xL^2*(difP + difP2*k2));
% --------end of interface
```

The MatLab function then continues as in monolayer case.

```
% ----------membrane layer
  for ii=ndx+1:nd-1
    dzdt(ii,1)=difS2*(z(ii-1,1)-2*z(ii,1)+z(ii+1,1))/(xL.^2);
    dzdt(ii,2)=difP2*(z(ii-1,2)-2*z(ii,2)+z(ii+1,2))/(xL.^2);
  end
  dzdt(nd,1)=difS2*(z(nd-1,1)-2*z(nd,1)+se)/(xL.^2);
  dzdt(nd,2)=difP2*(z(nd-1,2)-2*z(nd,2)+pe)/(xL.^2);
  dydt=reshape(dzdt,[],1);
 end
end
%end of subfunction
```

In the main script, the data are chopped and plotted, which is similar to the monolayer example:

```
sub=1000*subz; % change the concentration to umol/dm3
prod=1000*prodz; % change the concentration to umol/dm3
didt=(prodz(:,2)-prodz(:,1))/xL;
jc=1e7*F*difP*didt; % compute current density
g=1;
g1=6;
tend=50;
figure(1) % substrate profile
[TT,XX]=meshgrid(xspat(1:g1:end),tspan(1:g:tend));
mesh(TT,XX,sub(1:g:tend,1:g1:end),'linewidth',1.5)
ylabel('t[s]')
xlabel('x[cm]')
zlabel('c[\muM]')
colormap([0 0 0])
figure(2) % product profile
mesh(TT,XX,prod(1:g:tend,1:g1:end),'linewidth',1.5)
ylabel('t[s]')
xlabel('x[cm]')
zlabel('c[\muM]')
colormap([0 0 0])
set(gca,'box','on')
tim=[2,3,5,10];
```

```
figure(3) % substrate and product  profile cuts
subplot(2,1,1)
hold on
for ii=1:length(tim)
plot(xspat,sub(tim(ii),:),'k', 'linewidth',1.5)
end
axis('tight')
xlabel('x[cm]')
ylabel('c[\muM]')
set(gca,'box','on')
subplot(2,1,2)
hold on
for ii=1:length(tim)
plot(xspat,prod(tim(ii),:),'k', 'linewidth',1.5)
end
axis('tight')
xlabel('x[cm]')
ylabel('c[\muM]')
set(gca,'box','on')
tspan(tim)
figure(4) % current density
plot(tspan,jc,'k', 'linewidth',1.5)
xlabel('t[s]')
ylabel('jc[nA. mm^-^2]')
set(gca,'box','on')
```

The output of the script should result in the Fig. 5.14. The concentration profiles clearly show the concentrations jumps at the interface. The membrane interface shows decrement on the substrate site and in increment on the product side.

5.3 Outlook

This chapter gives the readers some ability to solve the one-dimensional diffusion equation and modify it to solve "biosensor simulations." Some examples can be solved with the pdepe solver. See the MatLab documentation. In the examples discussed in this chapter, only planar electrode was assumed. The other geometry examples are cylindrical geometry and hemisphere geometry. The diffusion equation for hemispherical electrodes:

$$\frac{\partial c(r,t)}{\partial t} = D\left(\frac{\partial^2 c(r,t)}{\partial r^2} + \frac{2}{r}\frac{\partial c(r,t)}{\partial r}\right) \tag{5.67}$$

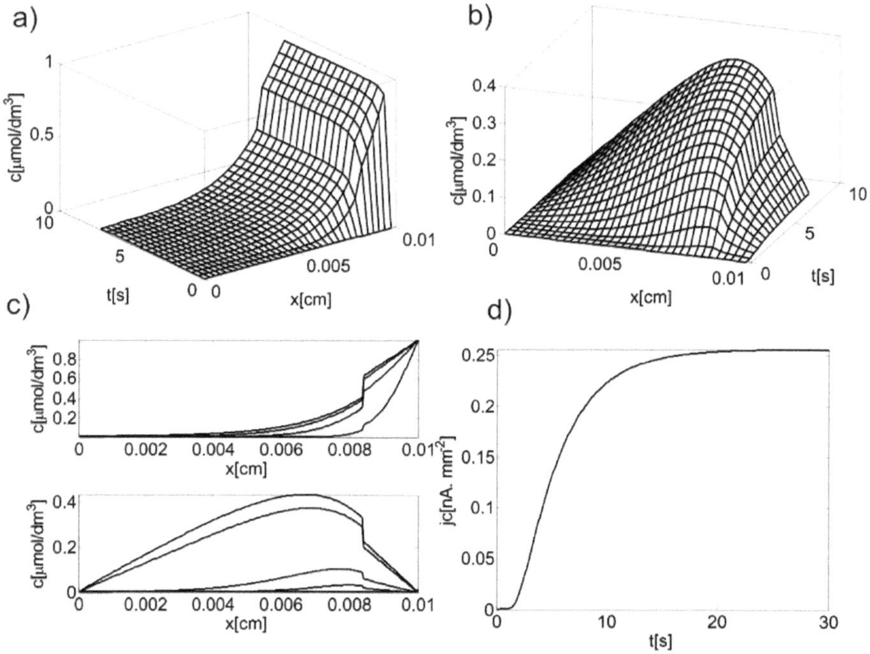

Figure 5.14 Substrate (a) and product (b) concentration evolution in two-layer model. Concentration profiles cuts (c) are in 0, 15; 0.6; 1.3; 7.4; 15 second and current density evolution (d).

And for cylindrical electrodes:

$$\frac{\partial c(r,t)}{\partial t} = D\left(\frac{\partial^2 c(r,t)}{\partial r^2} + \frac{1}{r}\frac{\partial c(r,t)}{\partial r}\right) \tag{5.68}$$

Note that the equations were transformed to spherical and cylindrical and r is the new spatial variable.

Cyclic voltammetry and ac voltammetry are a type of electrochemical method. Other types, e.g., chronoamperometry and differential pulse voltammetry, etc., can also be implemented. Readers are advised to use their experimental data and try to fit it.

References

1. Martines W., Moonjung C. (2014): *Statistics in Matlab. A Primer* (CRC press).
2. Marvin L. (2016): *Neural Networks with MatLab* (CreateSpace Independent Publishing Platform).
3. Miura T., Seki K. (2015): Diffusion influenced adsorbtion kinetics, *J. Phys. Chem.. B*, 2015, 119(34), 10954 10961
4. Schiesser W. E., Griffiths G. W. (2009): *A Compendium of Partial Differential Equation Models, Method of Lines Analysis with Matlab* (Cambridge University press).
5. Britz D., Strutwolf J. (2014): *Digital Simulation Electrochemistry* (Springer International Publishing Switzerland).
6. White R. E., Subramanian V. R. (2010): *Computational Methods in Chemical Engineering with Maple* (Springer).
7. Sadana A. (2002): *Engineering Biosensors: Kinetics and Designs Applications* (Academic press).
8. Davis M. E. (1984): *Numerical Methods and Modeling For Chemical Engineers* (John Wiley and Sons).
9. Gao Y., Zhou Q., Matharu Z., Liu Y., Kwa T., Rezvin A. (2014): A mathematical method for extracting cell secretion rate from affinity biosensors continuously monitoring cell activity, *Biomicrofluidics*, 8, 021501.
10. Bartlett P. (2008): *Bioelectrochemistry, Fundamentals, Experimental Techniques and Applications* (John Wiley).
11. Baronas R., Ivanauskas F., Kulys J. (2009): *Mathematical Modeling of Biosensors: An Introduction for Chemist and Mathematicians* (Springer).
12. Shampine L. F., Gladwell I., Thompson S. (2003): *Solving ODES with Matlab* (Cambridge University press).
13. Compton, R. G. (2014): *Understanding Voltammetry: Simulation of Electrode Processes* (Imperial College Press London, GB).
14. Yokoyama K., Kayanuma Y. (1998): Cyclic voltammetric simulation for electrochemically mediated enzyme reactions and determination of enzyme kinetic constant, *Anal. Chem.*, 70, 3368–3376.
15. Cavaghan D. J., Bond A. M. (2000): A complete numerical simulation of the techniques of alternating current linear sweep and cyclic

voltammetry: Analysis of a reversible process by conventional and fast Fourier transform method, *J. Electroanal. Chem.*, 480, 133–149.

16. Adamson H., Bond A. M., and Parkin A., (2017): Probing biological redox chemistry with large amplitude Fourier transformed ac voltammetry, *Chem. Commun.*, 53, 9519.
17. Bracewell R. (1999): *Fourier Transform and Its Applications* (McGraw-Hill Science).
18. Aseris V., Baronas R., Petrauskas K. (2016): Computational modelling of three-layered biosensor based on chemically modified electrode, *Comp. Appl. Math.*, 35, 405–421.
19. Hickson R. I., Barry S. I., Mercer G. N., Sidhu H. S. (2011): Finite difference schemes for multilayer diffusion, mathematical and computer modeling, 54, 210–220.

Chapter 6

Self-Assembled Monolayers for Surface Modification

Peter Kasak

Centre for advanced materials, Qatar University, P.O. Box 2713, Doha, Qatar

peter.kasak@qu.edu.qa

Self-assembled monolayers (SAM) have recently been recognised as core elements for improving technology in biosensing, separation molecules, development of drugs, nanoelectronics and nanotechnology. A number of strategies have been developed for immobilisation of biological elements like glycan, protein, DNA onto SAM, leading to enhanced robustness, specificity and sensitivity in tailored applications. Approaches have been devised for the construction of SAM, preventing non-specific interactions such as an essential interface element with the environment. This chapter summarises the construction of self-assembled monolayers, approaches for immobilising biological elements,

Glyconanotechnology: Nanoscale Approach for Novel Glycan Analysis and Their Medical Use
Edited by Tomáš Bertók

ISBN 978-981-4800-16-7 (Hardcover), 978-0-429-44849-2 (eBook)
www.jennystanford.com

especially glycan, and the applications of SAM in nanotechnology, nanoelectronics, and biosensing.

6.1 Introduction

Organic molecules that organise on a substrate surface on a nanoscale into a highly structured film with a high density of entire molecules move toward molecular self-assembly. Whitesides and Grzybowski defined self-assembly as "the autonomous organisation of elements into patterns and structures, without human intervention" [1]. A primary characteristic of nature is self-assembly; any structure showing distinctive scales in length higher than those of their individual element can be deemed to be engaged in self-assembly. In nature, self-assembly is a ubiquitous process in which larger systems are spontaneously constructed from suitable building blocks in a dynamic manner. This, entails a change from a disordered state to a well-organised state. Self-assembly involves the co-existence of covalent and non-covalent interactions which lead to the development of hierarchically complex structures. Self-assembly largely relies on multiple non-covalent interactions, which consist of electrostatic interactions, hydrogen bonding, dipole–dipole interaction, and steric repulsion, van der Waals forces and hydrophobic interactions (Fig. 6.1).

In nature, the building blocks for the creation of complex structures are predominantly based on glycans—sugars, amino acids, nucleotides and lipids. These building blocks self-assemble into complex structures with the involvement of elements such as phospholipid bilayers in cell membranes, micelles from amphiphilic molecules, double helical structures of nucleic acids, the folding and complexes of proteins and scaffolds. The potential for exploiting the recognition properties of chemically designed building units has led scientists to pursue the bottom-up construction of functional assemblies and complex machinery.

Self-assembled monolayers (SAM) represent one of the most rigorously and extensively studied examples of inanimate self-assembling systems [2]. The first assays investigating the self-assembly of an organic compound from a solution on a surface can be traced back to 1946, in a study focusing on the adsorption and oleophobicity of a long aliphatic compound comprising acid and amine on Pt surface [3]. In 1980, J. Savig defined a self-assembled monolayer as a proposed application of a mixed monolayer for the investigation of oleophobic behaviour.

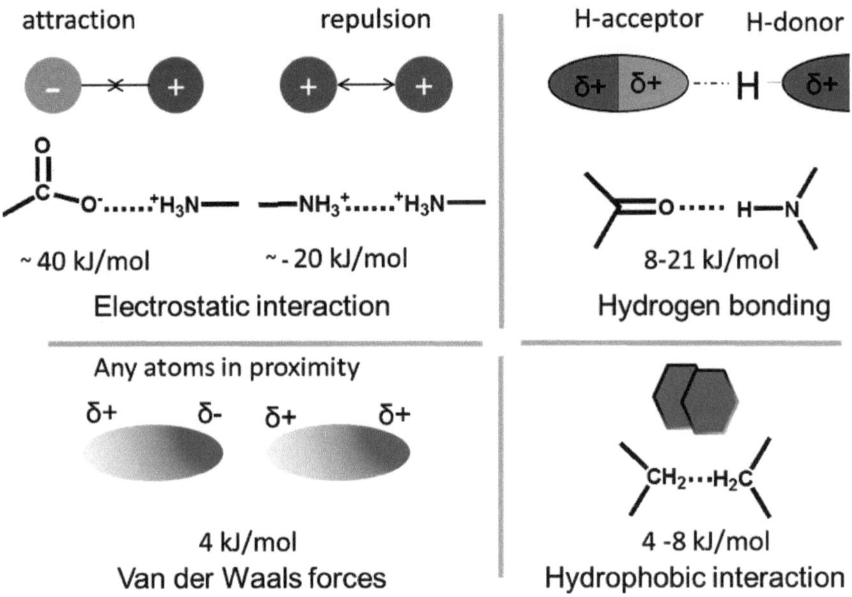

Figure 6.1 Non-covalent interaction related to self-assembly: schematic presentation; the example approximates binding energy and type of bond.

6.2 SAM Elements

SAM is formed from organic molecules with an anchoring group, linker and functional terminal group; the anchoring group binds onto the substrate surface, as shown in Fig. 6.2. SAM can be prepared either from a solution containing these compounds or from a gas phase. The main factors that affect the final order and quality of the monolayer formed can be summarised as follows:

(1) Cleanliness, smoothness, and purity of the original substrate and its pre-treatment
(2) Crystallinity of the substrate for the compaction
(3) Type and character of the substrate and anchored group
(4) Type, purity and concentration of the assembled molecules and their solution

(5) Length and chemical character of the linker
(6) Type of the terminal group (size, bulk and chemical properties)
(7) Time and temperature for monolayer assembly formation

During SAM formation, these molecules are usually chemisorbed on the substrate surface and subsequently assemble into large, well-organised molecular domains. This process leads to two-dimensional materials and affords major advantages such as simple preparation, relative stability under ambient conditions, and availability for use in modulation of interfacial characteristics. It should be noted that the mechanical and chemical robustness of monolayer systems is limited in comparison with polymeric modification.

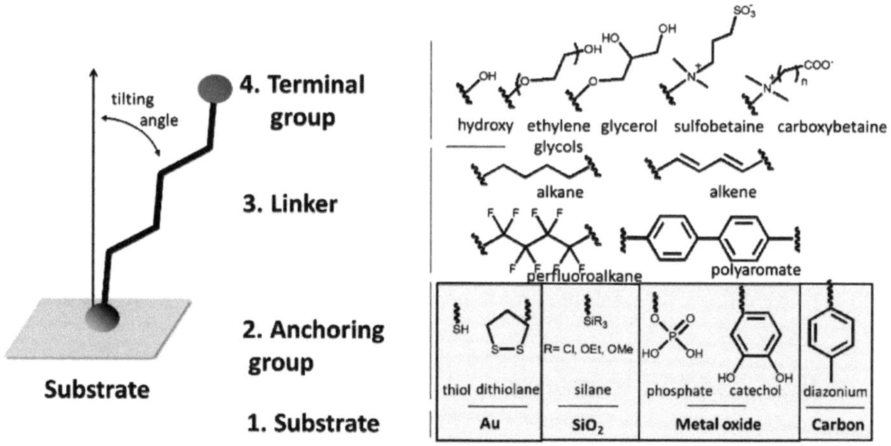

Figure 6.2 (Left) Schematic presentation of typical aliphatic self-assembling molecule on substrate: the substrate and the three main parts of the molecule (anchoring group, linker and terminal group) are indicated. (Right, from top to bottom) Some of the most common terminal groups for SAM serve to prevent non-specific interaction, linker type used in SAM, and anchored group with related substrates.

6.2.1 Substrate in SAM

The substrate surface usually requires preparation: in the case of noble metals such as gold on the surfaces of commercially available electrodes by physical vapour deposition (PVD). Pre-treatment of the gold surface, quartz or silica oxide is usually carried out with hot piranha solution (3:1 concentrated H_2SO_4: 30% H_2O_2). For most surfaces, such as ITO (Indium tin oxide), glassy carbon or metal surfaces, the application of chemomechanical and electrochemical polishing pre-treatment can also lead to substrates with clean and flat surfaces. Another method for cleaning gold substrate is the combination of immersion into KOH + H_2O_2 and potential sweep in the potassium hydroxide solution. In the formation of a gold substrate on a silicon wafer adhesion, first a layer of titanium, chromium or nickel of several nanometres (1–5 nm) needs to be applied and then a 100–300 nm layer of gold is deposited. It should be noted that Ag and Cu are susceptible to oxidation and have more defects than Au. Moreover, gold can be easily processed and is available as a thin film and as a colloid.

The substrate metal interface stabilises the surface atoms and tunes the electronic state. With Au or Ag, the most frequently used substrate and anchoring group combinations are thiol or dithiolane functionality.

6.2.2 Anchoring Group to Surface

Monolayers from thiols or disulphides can be prepared simply by the immersion of a sample into an ethanolic or aqueous solution and these molecules are spontaneously organised in trans conformation at an angle between 20° and 30°. Thiols are less stable than dithiolane due to possible oxidation. A dithiolane moiety bound to the Au substrate is more stable due to the presence of two binding S groups in the molecules. The density of these molecules is, however, lower than that of the thiols due to steric hindrance.

Organosilane or trichlorosilane derivatives in contact with the substrate lead to irreversible condensation with the surface –OH group from silicone oxide (SiO_2), quartz or ITO to form siloxane O–Si–O. This binding is much stronger and localised, and the tilting of the molecule is less than the thiol-based anchoring. Moreover, siloxane chemistry can be combined with silicone-based technology to exploit the resulting chemical and physical stability of this binding and modification. The metal oxide as a substrate exhibits a strong interaction with the phosphate or catecholate group to create a stable monolayer. The organophosphate forms a highly organised monolayer on metal oxide and alloy and their stability and durability in an aqueous environment is superior to the siloxane-based anchoring.

In the case of a carbon-based substrate, diazonium salt can be attached to the surface by decomposition of a diazonium ion and insertion of the aryl part to the substrate. In another approach, π-π interaction or hydrophobic interaction of the aromatic and aliphatic structure with anchoring capability based on π-π or hydrophobic interaction is used. Carbon as a glassy carbon electrode or hydrophobic polymer-based surface can be modified through non-covalent immobilisation via long aliphatic or perfluorated alkyl tails or through large aromatic systems such as pyrene or similar derivatives.

Recently, the application of 2D material from the graphene-based family has attracted attention and gained in relevance as a monolayer. These structures consist of a large aromatic surface enabling the hydrophobic interaction with the carbon substrate and an edge functionality for further modification and immobilisation of the biological element. One example is a monolayer on a glassy carbon electrode from graphene oxide with immobilised lectin for the detection of glycoprotein down to 1 aM without any polymers (Fig. 6.3) [4].

The binding ability and selectivity in binding of the anchoring group is a major limitation in the SAM approach. Differences in the binding of the element to the surface make it possible even to prepare a different SAM from a solution with different elements.

The systems with the greatest number of studies and applications are based on alkanethiolates on Au surfaces or alkyl-siloxanes on hydroxyl-terminated surfaces. In general, the adsorption process of SAM formation starts immediately upon immersion in the solution; however, the subsequent slow reorganisation to achieve the densest and most defect-free surface requires a number of hours. A typical procedure for SAM preparation entails the 1–10 mM aqueous or ethanolic solution of a derivative and takes 10–16 h.

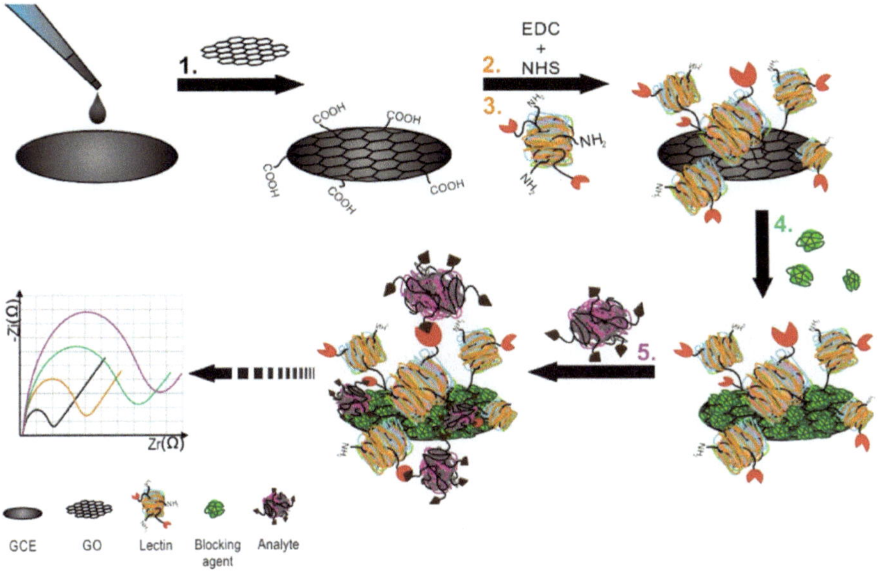

Figure 6.3 Graphene oxide as 2D material for interaction with electrode and immobilisation of lectin molecules for detection of glycoprotein (analyte). Reproduced from Ref. 4 with permission from The Royal Society of Chemistry.

6.2.3 Linker between Anchoring and Terminal Groups

The linker in SAM is responsible for a well-defined thickness, which is 1–3 nm for organic molecules. In addition, it acts as a physical barrier and tunes the electric conductivity and local optical properties [5]. Below, applications are proposed on how to choose an efficient and effective linker. The most widely used linkers are linear aliphatic chains containing 10 or more methylene groups which form a compact and dense monolayer. A compact SAM but with a less dense linker can be achieved by applying lipoic acid derivatives with steric hindrance in the dithiolane anchoring unit. The linker is predominantly alkane in its chemical character. Alkene, polyfluoroalkane, polyaromatic linkers are also employed, but are mainly used in nanoelectronic devices to modulate the electronic properties of (semi-)conductor and dielectric surfaces.

6.2.4 Terminal Group Preventing Non-specific Interaction on SAM

The terminal functional group at the interface governs the surface properties and character of the chemical functionality of a surface.

The terminal group regulates the surface contact with complex biological samples in terms of intactness to the biological species that would impede transduction of the signal. In most biosensing and biotechnological applications, a system is required with an interface between the SAM and the biological environment so as to select only the target object. Such a system has to prevent the adsorption, adhesion or fouling of any other species apart from the target object. Much effort is expended in pursuing a surface with no or extremely low fouling properties. Whitesides' group pioneered the investigation of SAM to prevent fouling and, after testing and screening of large functionally diverse SAM surfaces, they postulated four main features that prevent fouling at SAM [6]: (i) the SAM should contain a hydrophilic group, (ii) it should contain hydrogen-bond acceptor groups, but (iii) not contain hydrogen-bond donors, and (iv) its overall electrical charge should be neutral without a net charge. SAMs which meet these criteria provide an additional hydration layer for preventing non-specific protein adsorption and also subsequent cell adhesion.

One of the classes of compounds used for the prevention of fouling is the hydroxyl group or oligo-(ethylene glycol) (OEG). The SAM from them is formed from derivatives bearing a thiol-anchoring group. However, PEG under physiological conditions is subjected to oxidation or degradation mainly at an elevated temperature and such a SAM also requires a long linkage to an element accessible for sensing. Polyglycerols are more thermally stable and less prone to oxidation and show protein resistance similar to PEG. Other promising candidates for protein resistance at SAM are zwitterions. Zwitterions consist of positive and negative groups in the same molecules, hence their overall charge is balanced. Hydration is more pronounced than in former candidates since the ion-ion and dipole-ion interactions with water molecules are stronger than the weak hydrogen binding-based hydration interactions in PEG and PG. Zwitterion materials also display excellent blood compatibility, stabilise the protein after bioconjugation and can be employed in implanted biosensors, in the encapsulation of living cells and in smart systems [7, 8]. Accordingly, zwitterion materials have attracted attention and are becoming a benchmark for protein, bacteria and cell-repellent materials.

Apart from the formation of SAM, other techniques are used for the prevention of biofouling, such as adjacent layers to existing SAM by biological or synthetic elements such as sugars, proteins or polymers.

6.3 Immobilisation of Glycan and Biological Elements on SAM

In addition to fouling prevention, some applications require the regulated adsorption of proteins or cell adhesion on the surface; this involves ligation to specific biological species such as glycan or polysaccharide through a linker [9].

Besides regulating the adsorption of proteins or cell adhesion, the surface of SAM in biosensing requires the presence of an element targeting biological species. These elements can be introduced in four different ways, as depicted in Fig. 6.4. The first two approaches are based on physical adsorption of the element. The first approach proceeds either through an electrostatic ion-ion interaction between the charged terminated group and the element (Fig. 6.4A) and the second through interaction of the hydrophobic surface with low energy and the hydrophobic part of the element (Fig. 6.4B).

The third approach proceeds via direct adsorption of the element with the anchored group on the substrate (Fig. 6.4C). Such an approach is suitable only for small molecules, due to possible complications in preserving the element functionality during the synthetic strategy. The fourth approach is a chemical reaction of the terminal group on the monolayer with the element by immobilisation of the element on the surface (Fig. 6.4D). This approach is used most frequently in the biological sciences and entails a SAM surface which is usually formed from a mixed monolayer consisting of molecules with a functionality preventing biofouling and molecules with a functionality capable of ligation with the element.

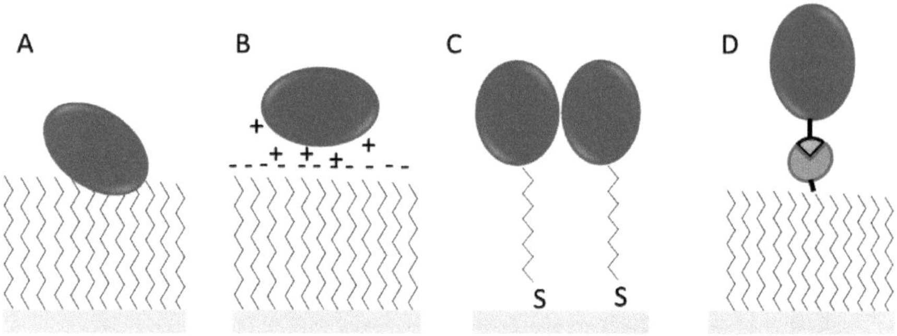

Figure 6.4 Different strategies for immobilisation of biological elements as red globule (glycan, polysaccharide, protein, DNA): (a) physisorption through hydrophobic interaction on SAM, (b) electrostatic complexation (c) direct covalent link, (d) covalent binding to the SAM terminal group.

The first two approaches are non-covalent and can be divided into non-specific and specific, as depicted in Figs. 6.5a,b.

Chemical reaction and final covalent binding can also be divided into covalent non-specific (Fig. 6.5c) and specific (Fig. 6.5d) by the site applied.

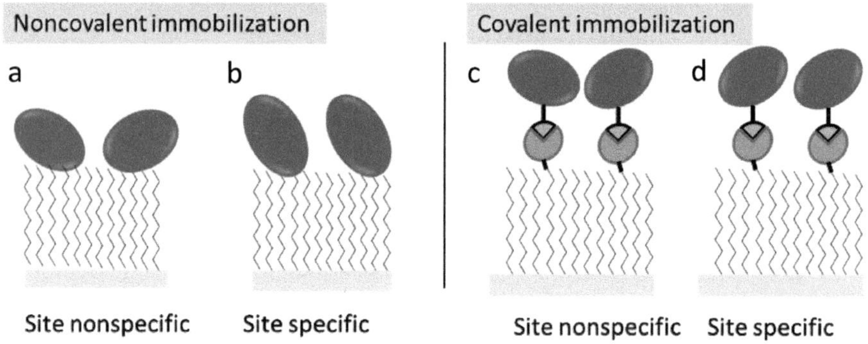

Figure 6.5 Different strategies for immobilisation of biological element. (a) Non-covalent, site-non-specific immobilisation; (b) non-covalent, site-specific immobilisation; (c) covalent, site non-specific immobilisation, and (d) covalent, site-specific immobilisation of biological elements on the SAM.

6.3.1 Non-Specific Non-Covalent Immobilisation

Typical non-specific non-covalent immobilisation is by adsorption of the biological element onto different SAM or even modified polymeric surfaces. This immobilisation often proceeds via a hydrophobic patch in the biological element, as shown in Fig. 6.6a. Due to a weak adsorption interaction, such an approach is usually applied to sufficiently large molecules. For strongly charged molecules, another non-specific non-covalent immobilisation strategy is the formation of SAM bearing chargeable moieties such as amino or carboxylic acid groups. Electrostatic interaction with polycations (polyethylene imine, polylysine...) depicted in Fig. 6.6b or polyanions (heparin, DNA, cell membrane...) illustrated in Fig. 6.6c, affords the effective immobilisation of such molecules.

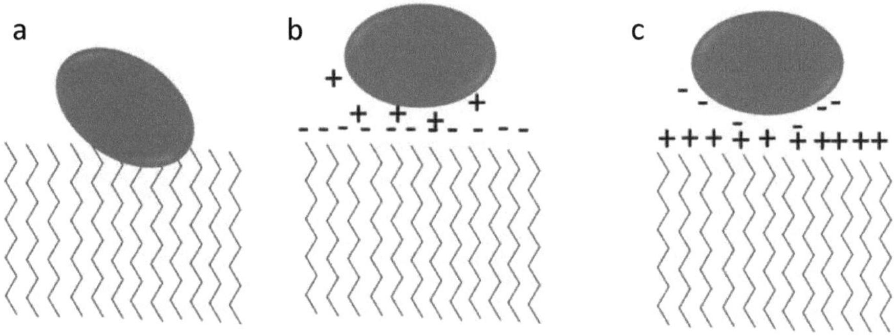

Figure 6.6 Non-covalent specific immobilisation based on (a) hydrophobic interaction and (b) and (c) electrostatic interactions.

6.3.2 Non-Covalent Site-Specific Immobilisation

Strategies for the non-covalent site-specific immobilisation of glycans, oligosaccharides or biological elements are mostly developed for microarrays with relatively simple molecules (Fig. 6.7). As an example, glycans conjugated with lipid, so-called neoglycolipids, are immobilised through hydrophobic interactions with aliphatic thiol on gold or a different type of membrane based on nitrocellulose, polycarbonate or polyvinylidenedifluoride (Fig. 6.7a).

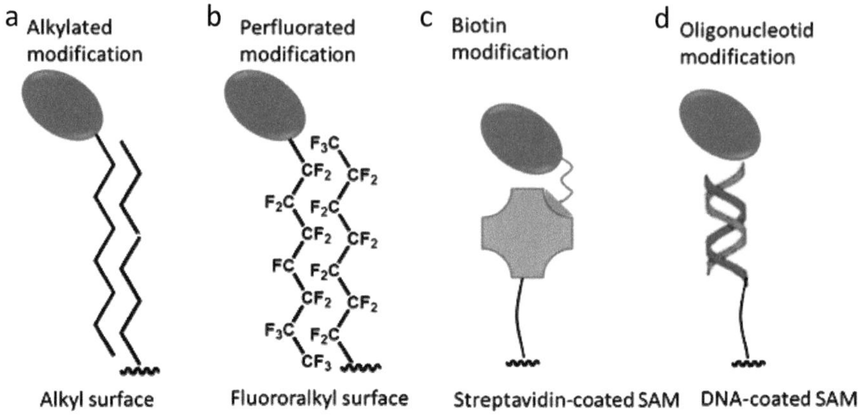

Figure 6.7 Strategies for non-covalent site-specific immobilisation (a) alkyl–alkyl hydrophobic interaction; (b) fluorous-fluorous interaction; (c) streptavidin-biotin interaction and (d) DNA hybridisation.

Similarly, an immobilisation strategy via fluorous-fluorous interaction of perfluoated alkylthiol or phosphate on SAM with fluoroalkyl modification of glycan or biological element was developed (Fig. 6.7b) This carries the advantage of the relatively simple isolation of perfluoro-modified glycans during preparation by fluorous chromatography.

A widely used approach, not only for immobilisation but for conjugation in general, is employment of the strong interaction between biotin–streptavidin, where K_d for the biotin–streptavidin complex is approximately 10^{-15} M with four binding sites for biotin on streptavidin molecules. This immobilisation was applied to SAM containing streptavidin for glycan microarrays (Fig. 6.7c). Another specific non-conjugated immobilisation utilises DNA hybridisation (Fig. 6.7d). In the application of such an approach, glycan units are conjugated with oligonucleotide and a complementary fragment of oligonucleotide is immobilised on the surface of SAM. During the immobilisation, a double helix is formed from the complementary oligonucleotide that leads to sequenced DNA.

6.3.3 Covalent Non-Specific Immobilisation

The most straightforward, simplest and easiest covalent immobilisation strategy for glycan structures and a biological element is based on light activation. This approach can be used spatially or patterning can be applied. For the immobilisation of glycans, immobilisation can be applied via photolabile groups such as aryltrifluoromethyldiazirine (Fig. 6.8a) or 4-azido-2,3,5,6-tetrafluorophenyl (Fig. 6.8b). After irradiation, these groups are transformed into highly active carbine or nitrene, respectively, then this group is instantly inserted in C–H bonding.

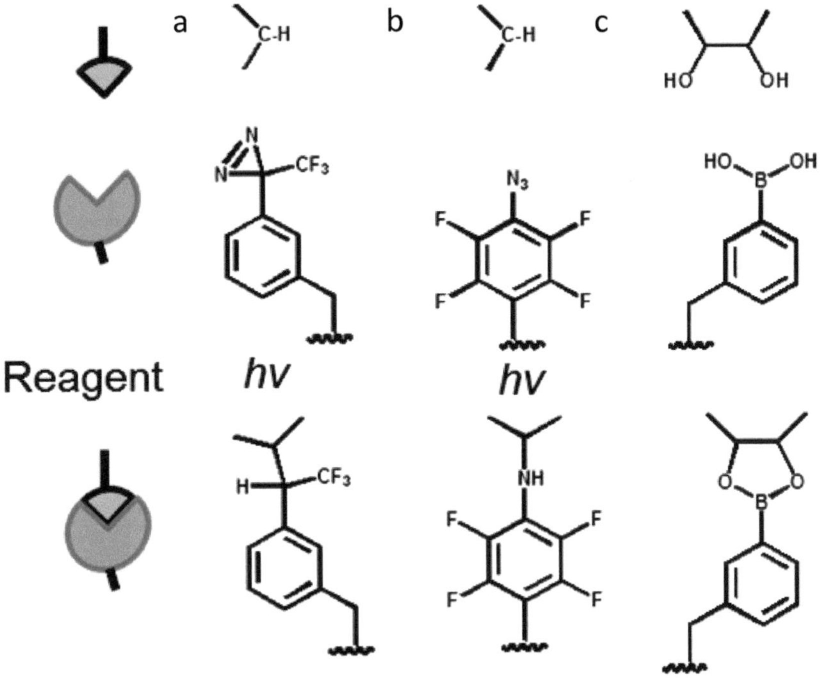

Figure 6.8 Covalent and site-non-specific immobilisation. Immobilisation of biological element through (a) aryltrifluoromethyldiazirine, (b) 4-azido-2,3,5,6-tetrafluorophenyl group, and (c) phenylboronic acid-coated surfaces.

Recently, a non-specific immobilisation strategy was developed for the recognition of 1,2- or 1,3-diols fragments in glycan units by aromatic boronic acids (Fig. 6.8c). During this reaction, a SAM-immobilised boronic ester or boronate group is formed, depending on the pH, with 1,2- or 1,3-diols fragments of glycan or biological species.

6.3.4 Specific Covalent Immobilisation

Prototypical synthetic strategies for specific covalent immobilisation with terminated groups and a counterpart from elements are summarised in Figs. 6.9 and 6.10.

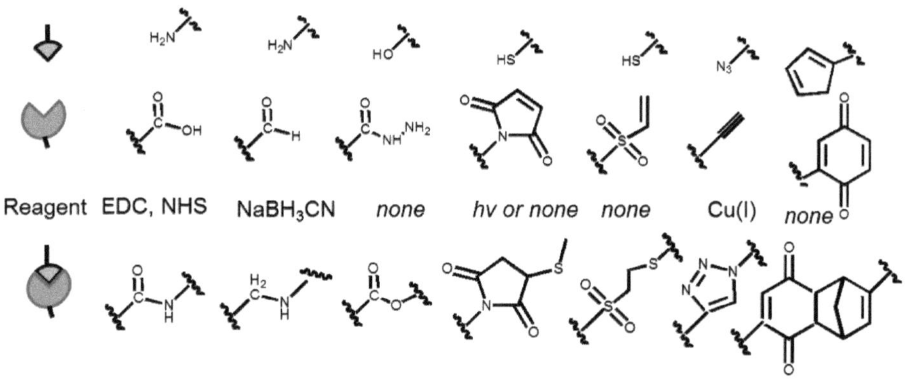

Figure 6.9 Strategies for specific covalent immobilisation.

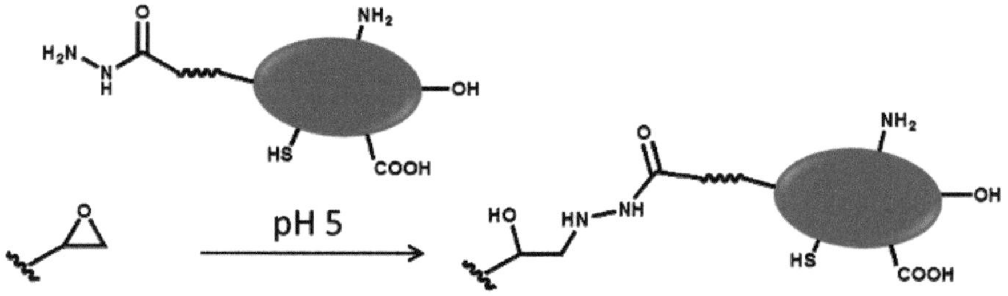

Figure 6.10 Specific covalent immobilisation between epoxide and hydrazido group at pH 5.

Immobilisation through carboxylic, carbonyl, amino or thiol groups usually does not require any further synthetic modification of the element. These are present as pendant groups from amino acid in (glyco)proteins and glycans [10, 11].

A frequently used synthetic strategy is the employment of N-ethyl-N′-(3-dimethylaminopropyl) carbodiimide (EDC)/N-hydroxysuccinimide (NHS) coupling chemistry. This involves activation of the carboxylic terminal groups on SAM or in the biological element followed by the reaction of the exposed primary amine groups from the element or SAM, respectively (Fig. 6.9a). It is worth noting that, during the activation and coupling stages, the pH should be adjusted so that the activated reaction centres are predominantly in the form of carboxylic acid and, in the coupling stage, the pH is adjusted so that the coupled group is in the form of a primary amine, not an ammonium salt [12].

Another simple approach is through the reaction of carbaldehyde and the exposed amino group from the biological element or vice versa. Using this approach, Schiff's base is formed, which can subsequently be reduced with a mild reducing agent such as $NaBH_3CN$ to a stable secondary amine (Fig. 6.9b). One disadvantage of the reductive amination of sugars is the destruction of the reducing end of the sugar segment, which might disrupt the recognition or interaction of glycans; however, this can be overcome by using milder aminooxy or hydrazido groups instead of the amino group in the reaction with an alcohol group (Fig. 6.9c).

Click reactions intensify applications in bioconjugation, as these are defined as reactions with a wide scope, simple to perform, specific to a reagent and insensitive to an aqueous environment. One type of click reaction is thiol-ene coupling (Figs. 6.9d,e). One example is the application of this reaction to a maleineimide-modified SAM and the SH group from the exposed cysteine of proteins structure (Figs. 6.9d) [13]. This reaction proceeds spontaneously or by the application of UV light. With the aid of UV radiation, a radical is formed in the thiol part that enhances addition of the thiol group to the activated double bond in SAM.

Immobilisation approaches that involve engineering the element with another functionality require chemical modification prior to the immobilisation process. However, the process affords a more specific and spatially localised linkage element to SAM. The best-known engineered click reaction is Huisgen's cycloaddition between organic azides and alkynes in the presence of Cu(I) (Fig. 6.9f). In this reaction, Cu(I) is readily oxidised to Cu(II); to avoid this poisoning the catalyst, a simple reduction reagent, such as ascorbic acid, electrical current or light source, can be used. As regards the synthesis, it is easier to introduce an azido group to the biological element. This click reaction was demonstrated in the immobilisation on simple saccharides, oligosaccharides or sugar-shelled dendrimers.

The Diels-Alder reaction also offers an alternative to controlled and engineered immobilisation (Fig. 6.9g). The advantage of this type of reaction is that it can be tailored by a redox transformation of an activated ketone to an inactive dihydroxyphenyl derivative upon the application of an electrical potential [14].

Epoxide-modified SAM or epoxide-coated surfaces are also widely used for immobilisation with a variety of biological elements. A specific ligation reaction was discovered for glycan microarrays. Here, the epoxide functionality on the surface can preferably react with hydrazide under mild acidic conditions in a sodium phosphate buffer at pH 5 (Fig. 6.10). In the absence of a hydrazide functionality, the epoxide coating can also be applied to the immobilisation of biological species which also bear strong nucleophiles, such as thiols or amines.

The immobilisation process of biological species can even be monitored, as was demonstrated in atomic force microscopy (AFM) studies on carboxybetaine-based SAM on a gold surface. Initially, SAM has a roughness of 0.39 nm then, during immobilisation of lectin *Sambucus nigra agglutin,* the roughness first increases up to 1.4 nm and then, after full confluency of the lectin on the surface, decreases to 1.2 nm due to reduction voids on the surface, as depicted in Fig. 6.11 [15].

Figure 6.11 AFM images of progression of *Sambucus nigra agglutin* (SNA) lectin immobilisation over time; surface roughness factor is depicted beneath. Reprinted with permission from [15]. Copyright 2015 American Chemical Society.

6.4 Mixed SAM

Mixed SAM is a simple approach which can minimise non-specific interactions and at the same time provide a different linkage density of elements on the SAM surface. It is recognised that bulky molecules such as proteins or complex carbohydrates need sufficient space for free movement, separation and proper orientation after immobilisation to achieve optimal conditions for the required performance. Additionally, multivalent binding and orientation elements are often substantial for strong recognition between elements on the SAM and in the environment. As an example, the separation and orientation of glycans have to conform with the binding site to ensure the high degree of complexation required, as depicted in Fig. 6.12.

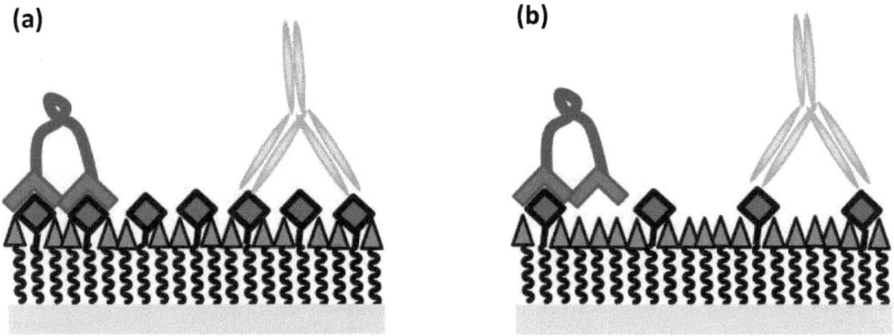

Figure 6.12 Schematic presentation of immobilised (bio)analyte showing that multivalent binding, orientation and density is crucial for recognition to occur. (a) An immobilised recognition element strongly binds with analyte with high density between binding sites and (b) other analyte cannot bind strongly to the immobilised element with low density on the surface.

Formation of a mixed monolayer can be achieved by different approaches, as shown in Fig. 6.13.

Usually, a mixed monolayer requires the preparation of SAM with two distinct linker molecules (Fig. 6.13a). One linker molecule consists of the element or permits linking with the biological element (see the linkage terminal group in Fig. 6.3). The other is a molecule which lacks a linkage with the terminal group to ensure separation and distribution of the element and, at the same time, prevent non-specific interaction of the protein. Another approach to mixed SAM formation is by introducing to the already formed SAM surface new thiolated molecules via the thiol-thiol surface exchange mechanism (Fig. 6.13b). The strategy for mixed

SAM formation with an equal number of different linkers includes the use of asymmetric disulphide (Fig. 6.13c) or the hydrolysis of anhydride from linear disulphide with amine or alcohols and leading to an acid and an acid derivative (amide or ester) terminal group mixed SAM (Fig. 6.13d).

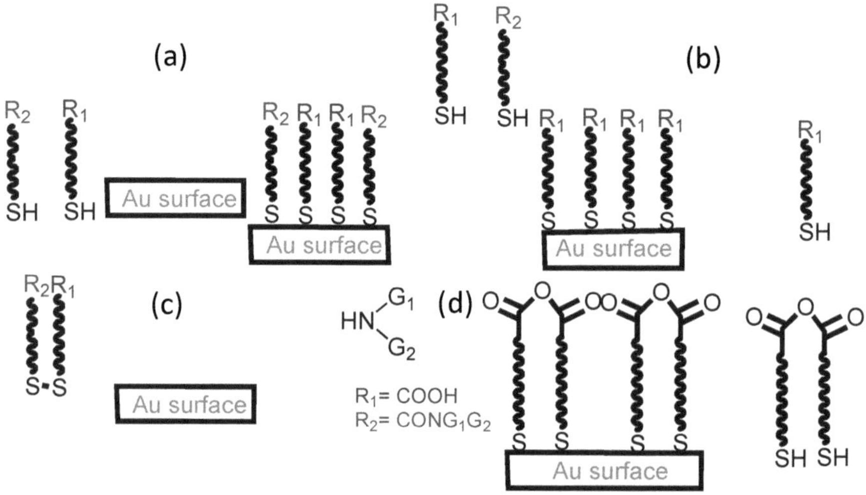

Figure 6.13 Different approaches to mixed SAM formation on gold surface (a) direct mixture of two elements; (b) surface exchange; (c) hetero disulphide application and (d) anhydride aminolysis.

Frequently, an oligoethyleneoxide derivative with thiol together with another thiol molecule with linkable functionality was used in mixed monolayers; however, in this case, a long linkage for the immobilisation of glycans and dendrimers is required, as shown in Fig. 6.14 [16].

Another example for mixed SAM is use of a zwitterionic sulphobetaine derivative (SB) in combination with a carboxybetaine derivative (CB). The terminal carboxylic group of CB derivative possesses a higher pKa value than the sulphate group in SB. This difference makes possible selective conjugation of the carboxylic acid with EDC and NHS chemistry and subsequent immobilisation while the SB remains intact in this treatment (Fig. 6.15a). An advantage of this approach is that it does not require derivatives with a long spacer for ligation and there is a controlled distribution of the biological element for immobilisation [17]. This approach was applied to glycomic analyses of human IgG samples in an array format.

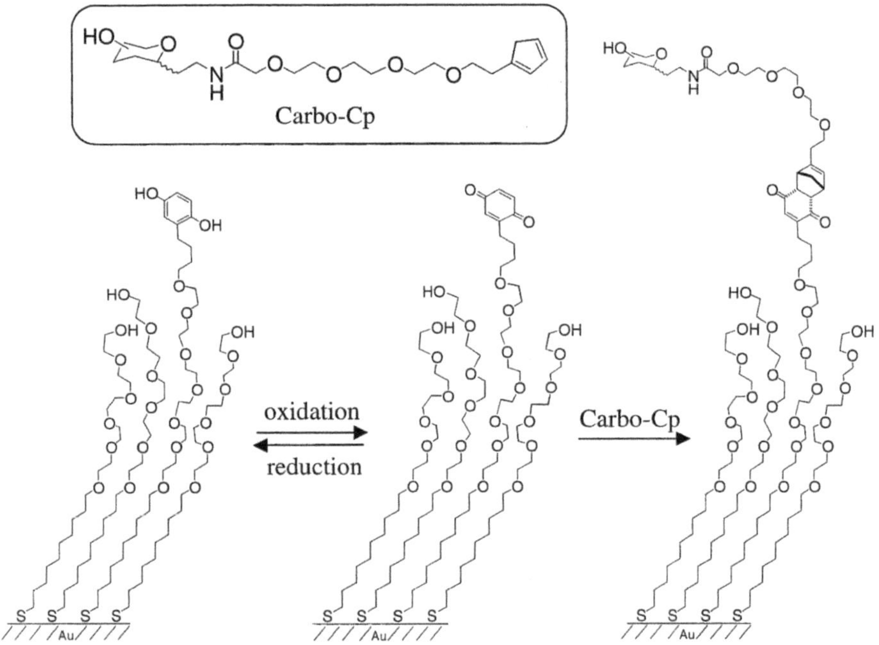

Figure 6.14 Chemical Approach to the Preparation of Carbohydrate Chips with mixed polyethylene glycol (PEG) derivatives. The strategy is based on the Diels-Alder-mediated immobilisation of carbohydrate-cyclopentadiene conjugates to a monolayer that presents benzoquinone groups. Because the Diels-Alder reaction is rapid, selective, and quantitative, all carbohydrates within the array are presented at a uniform density. Reprinted from [16], Copyright 2002 with permission from Elsevier.

It should be noted that the resultant ratio of the two different compounds for mixed SAM need not necessarily reflect the initial concentration of these elements in the solution. As shown in Fig. 6.15b for due compound distribution an analysis of the surface is required. In this case, reductive desorption by applying cyclic voltammetry for calculating the density of thiols present on an Au electrode surface showed that the thiol density of a mixed SAM was higher than in SAM composed of a single element. Density and the presence of different functionalities also have an influence on adsorption properties, as shown for adsorbed human serum albumin (HSA) in Fig 6.15c.

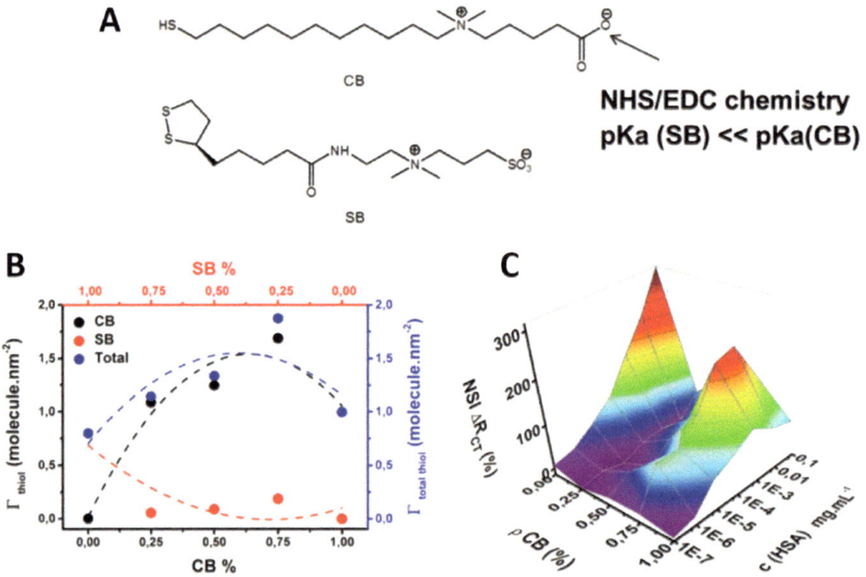

Figure 6.15 (a) Structure of carboxy-(CB) and sulphobetaine (SB) derivatives; (b) A reductive desorption procedure employing cyclic voltammetry for calculation of density of thiols present on an Au electrode surface. The experiment was performed by running CV in de-aerated 100 mM NaOH solution in a potential window from 500 to 1,500 mV at a scan rate of 100 mV. Blue line = total thiols, black line = CB, red line = SB; (c) Investigation of an interaction of human serum albumin (HSA) with an interface modified by a mixed SAM composed of CB and SB using EIS. NSI = interactions expressed as a relative R_{CT} change. Reprinted with permission from [17]. Copyright 2016 American Chemical Society.

6.5 Characterisation of SAM

As nanotechnology has been boosted over the last three decades, there have been significant advances in characterisation techniques for investigation of the mechanism of SAM formation as well as the qualitative and quantitative determination of the interaction between SAM and other biological and chemical elements. The conventional characterisation techniques, their advantages and disadvantages, are summarised in Table 6.1. The principle of measurement and detailed description of particular methods is beyond the scope of this chapter and some particular techniques are described in other chapters in this book and several reviews [18, 19].

Table 6.1 Methods for SAM characterisation

Methods	Measured characteristic	Advantages	Disadvantages
Contact angle	Wettability, surface energy	Simple and fast performance	Uncertainty in value. Extra analysis required
XPS	Chemical composition	Quantitative and qualitative chemical composition of the surface	Expensive equipment, sample contamination, complex data
Surface plasmon resonance (SPR)	Binding properties	Very good sensitivity; real-time and *in situ* measurements	Expensive chip, limited substrate and thickness
Quartz crystal microbalance	Frequency of quartz crystal, mass calculation	Different substrate available, good sensitivity; real-time and in situ analysis	Limited to hydrated mass; influence from energy dissipation for mass calculation
Fluorescence microscopy	Fluorescence	Good sensitivity, ability to measure competitive adsorption	Expensive and labelled reagent, quenching of fluorescence
Infrared reflection-absorption spectroscopy (IRAS)	Functional groups	Prove for well-ordered SAM	Difficult, expensive set-up, flat surface, complex data
GI Fourier-transform infrared spectroscopy (GI FTIR)	Functional groups	Fast performance	Difficult set-up, complex data analysis
Ellipsometry	Thickness	Fast and relatively cheap	Only on flat surface and homogenous layer

(*Continued*)

Table 6.1 (*Continued*)

Methods	Measured characteristic	Advantages	Disadvantages
Neutron reflectivity	Surface structure	Thickness, density, and roughness information	Complex and difficult data analysis
Cyclic voltammetry	Redox processes	Rapid measurement	Conductive substrate and redox probe required, peak, current reach plateau
Electrochemical impedance spectroscopy	Resistance of system	Label-free method for biological element, sensitive	Insufficient surface passivation for conductive substrate
ToF-SIMS Time-of Flight Secondary Ion Mass Spectrometry (ToF-SIMS)	Fragment of molecules of SAM elements	Orientation and conformation; very high surface sensitivity	Expensive equipment; limited to low molecular mass fragments
Atomic Force Microscopy (AFM)	Roughness, thickness	Resolution to single element	Not fully quantitative analysis

Characterisation and mechanism of SAM formation as well as the coverage can be achieved by different techniques. Usually, a combination of techniques is required to fully characterise SAM qualitatively and quantitatively.

6.6 Nanotechnology and SAM

6.6.1 Nanopatterning Processes

A large number of inventive and advanced nano- and micropatterning fabrication processes have been investigated for SAM. They have gained in attraction for academic and technological applications as well as for progressive industrial innovations. Moreover, introducing a biological element for interaction makes these approaches even more beneficial for biological and biomedical applications such as (nano)biochips, (nano)biosensors, drug development and fundamental studies of molecular biology. Nanopatterning techniques and their combinations are schematically depicted in Fig. 6.16.

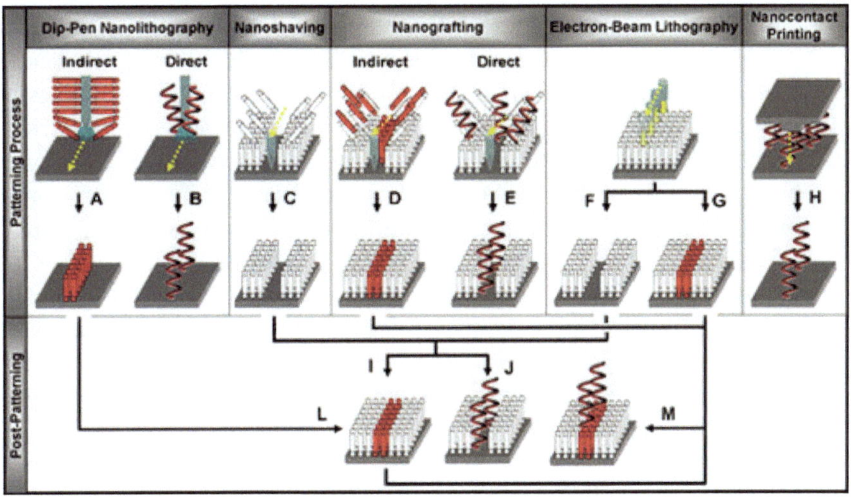

Figure 6.16 Schematic representation of the different lithographic techniques employed to indirectly or directly immobilise biomolecules on surfaces at nanometre scale resolution. In the indirect approach, the nanopatterns created by the different lithographic techniques are used in a second stage (i.e. post-patterning process) as templates to immobilise the biomolecules on surfaces. Reproduced with permission [20] from Springer.

Dip-pen nanolithography forms SAM from molecules which are released by diffusion from the AFM tip to the substrate via the meniscus of the solvent formed between the tip and substrate [21]. This approach provides for the fabrication of patterned biomolecules arrays. This strategy was used in a biological immunoassay with anti-p24 antibody electrostatically immobilised carboxylate as the terminated group which was evenly distributed by dip-pen nanolithography on a 100 nm scale. Passivation by an oligoethylene derivative was used on bare gold and the array was used for detecting HIV-1 p24 antigen and surpassed the limit of detection of a conventional ELISA by 100-fold [22].

Nanoshaving facilitates the formation of a small patterned area by mechanically removing SAM by AFM tip [23], which was even used for the controlled deposition of individual single-walled carbon nanotubes. Another technique is nanografting with a combination of nanoshaving in the presence of a solution with other molecules capable of immediately adsorbing onto the surface and forming a patterned surface which was applied to the adsorption of well-ordered DNA nanostructures [24].

Electron beam (e-beam) lithography deals with mask-free e-beam-focused technology at sub-10 nm resolution. It makes possible the release of an electron beam-sensitive part and creates a patterned surface on SAM. This process can tune the density between the patterns and complex structured arrays for biomolecules that have been created [25]. Another use of e-beam technology is the initiation of chemical reactions on the surface. P. Mendes et al. showed that e-beam was capable of reducing the nitro terminated group in 3-(4-nitrophenoxy)-propyltrimethoxysilane to amino group on the 100 nm scale and the amine group thus formed afforded the subsequent patterning with negatively charged gold nanoparticles [26].

Usually, in nano-contact printing, a polydimethylsiloxane (PDMS) patterned stamp is immersed in a solution with molecules for SAM formation and is subsequently contact-printed onto a substrate, where SAM is formed only in the positively patterned area from the stamp on the surface.

Some of these techniques can be used successively in post-modification by filling or backfilling the bare region with other (bio)molecules introduced into the bulk solution, thereby increasing the attraction of these approaches.

6.6.2 Switchable SAM

Molecular systems that have the capacity to reversibly or irreversibly change their properties by applying external stimuli or under different environmental conditions are denoted as "smart" materials [27–29]. The external triggers for this change are, typically, electric fields, optical, thermal, magnetic fields, mechanical force, chemical or biological molecules. Applying molecular self-assembly to smart materials is an excellent approach due to the cooperation effect of all the elements present in the monolayer, which leads to amplification of the switching process. Switchable SAM can be divided into reversible and irreversible switches.

One example of an irreversible switch on SAM is hydrolysis of carboxybetaine ester as a lipoic acid derivative (Fig. 6.17). This carboxybetaine ester-based SAM exhibits a cationic character and makes possible complexation for negatively charged species such as polyphosphate backbone from DNA or negatively charged citrate-capped Au nanoparticle. By contrast, after hydrolysis at a higher pH of 9, the ester group from carboxybetaine ester is transformed to carboxylate and forms a carboxybetaine derivative with an overall balanced charge that leads to the releasing of DNA molecules or else no adsorption of the Au nanoparticles [30].

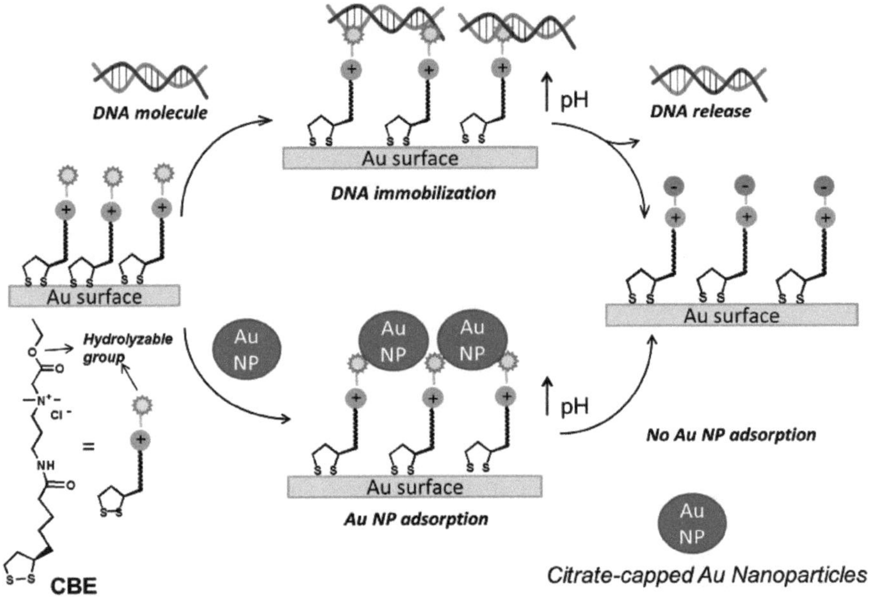

Figure 6.17 Chemical structure of irreversible carboxybetaine ester derivative and schematic illustration of interaction of SAM from this ester with DNA and Au nanoparticles (AuNPs) and change in interaction after pH-induced hydrolysis to carboxybetaine character. Reproduced with permission [30] from American Chemical Society.

An interesting example of a reversible switchable surface is a combination of so-called low density SAM (LD SAM) and an electrical field as an external trigger. LD SAM on a gold surface was prepared from SAM formed with a linear ester with a bulk ester group derivative. Subsequent to hydrolysis, a terminated negative carboxylate group with low density of LD SAM was formed. This assembly can change from straight conformation with the negative carboxylate head group of a hydrophilic character in the case of a negative potential to a bent conformation in the case of a positive potential applied. In this event, the substrate attracts the carboxylate-terminated ion resulting in its hydrophobic character. This transition is reversible, exhibiting tunability in wettability [31] (Fig. 6.18) and impedance [32] and can potentially be applied to an electrically modulated linkage through a carboxylate group.

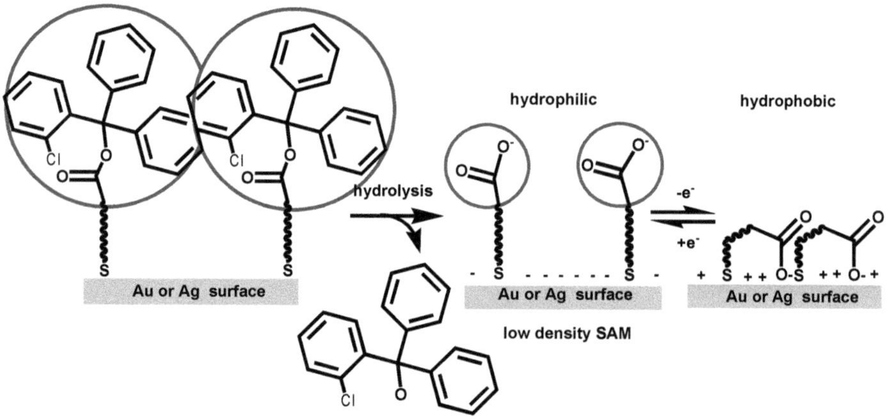

Figure 6.18 Schematic presentation of low density SAM (LD SAM) preparation and reversible switch based on the electrical field applied.

6.6.3 Nanostructured SAM Surface

Apart from the chemical character of the terminal group or immobilised biological species, surface morphology and nanostructured character can be an essential factor for the resultant performance. This is especially important for wettability, as is demonstrated on a nano- and microstructured gold surface prepared by a galvanic reaction on silicon wafers covered with a derivative with the terminal quaternary ammonium group and anchored via the thiol group (Fig. 6.19). In this case, the rough nano- and microstructured gold surface switches reversibly and rapidly from a superhydrophilic state in the presence of a chloride counter ion to a superhydrophobic state in the presence of a perfluorooctanoate counter ion by simple immersion in the respective salt solution and drying of the surface. Moreover, this type of counter ion-based switch on a plain surface affords a different adsorption of negatively charged citrate-capped Au nanoparticles with high adsorption with chloride counter ion and low adsorption with the perfluorooctanoate counter ion. This decrease in adsorption is due to the long tail of the latter counter ion impeding the due electrostatic attraction of the surface and nanoparticle [33].

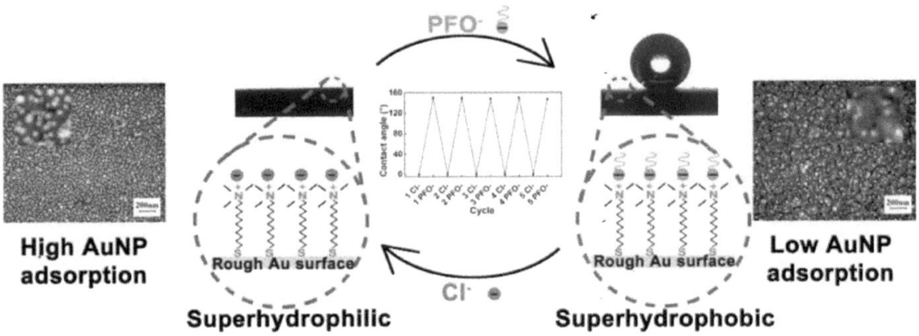

Figure 6.19 Middle: schematic presentation of structured surface with quaternary ammonium derivative on nano-microstructured gold, exhibiting reversible switchable wettability by counter ion exchange and SEM micrographs of plain surface with SAM after incubation with AuNPs with Cl^- (left) and PFO^- (right) as a counter ion. Reprinted with permission from [33]. Copyright 2016 American Chemical Society.

The topology of the surface also determines access to the immobilised biological element. A comparison was made of 2D planar gold and 3D gold nanoparticle-modified SAM with immobilised SNA lectin (Fig. 6.20). It revealed that the 2D planar configuration immobilised a larger amount of lectin of (2.53 ± 0.01) pmol cm^{-2} than the 3D configuration with 20 nm AuNPs (0.94 ± 0.01) pmol cm^{-2}. Despite the higher amount of immobilised SNA lectin, the 3D surface exhibited a lower limit of detection for the analyte glycoproteins fetuin and asialofetuin down to aM level than the 2D with the limit of detection at the fM level. This can be attributed to the better accessibility of lectin on a 3D surface to its analytes, the higher sensitivity of detection and much lower limit of detection [34].

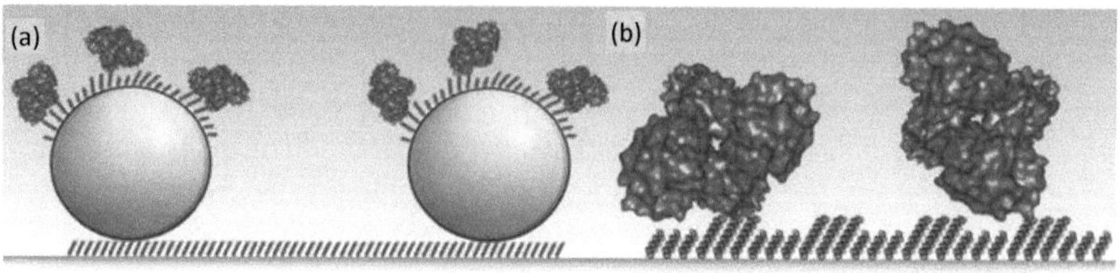

Figure 6.20 Schematic presentation of (a) 3D and (b) 2 D gold architecture. Reprinted with permission from [34]. Copyright 2016 form ESG.

6.6.4 SAM in Nanoelectronics

In nanoelectronics, SAM can be deployed in field effect transistors (FET), as summarised in Fig. 6.21. Such an electronic application with details of the type of monolayers used has been comprehensively summarised in reviews [35, 36].

FETs rely on an electrostatic modulation of charge-carrier mobility in appropriately prepared semiconductors. Monolayers were applied in different parts of the assembly for FET with SAM. SAM can be used as: (a) organic semiconductor, (b) molecular dielectric, (c) gate dielectric primer and (d) electrode modifier (Fig. 6.21).

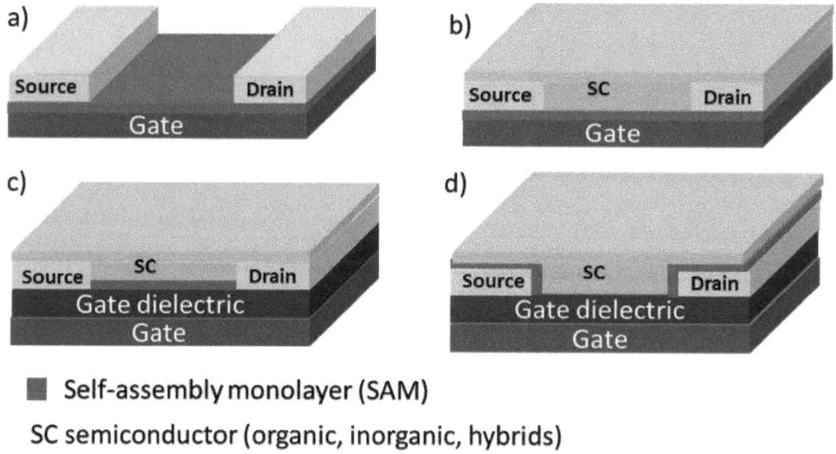

Figure 6.21 Different assemblies for field effect transistor (FET) with SAM. SAM can be used as: (a) organic semiconductor, (b) molecular dielectric, (c) gate dielectric primer and (d) electrode modifier.

As an example, SAM consisting of large self-assembled π-conjugated systems on the gate dielectric can act as a conductive channel between source and drain electrodes and form so-called SAM-based field-effect transistors (SAMFETs) [37]. Typical examples of molecules used in SAM-based field-effect transistors are large aromatic systems from oligocene, oligothiophenes, oligophenyl, phthalocyanine derivatives. The pioneering investigation on devices with oligothiophenes and tetracene derivatives was optimised by design to the application of substituted quinquethiophene as the semiconducting core. In addition, a dimethylchlorosilane derivative as the anchoring group with one binding site is used instead of trichlorosilane and an aliphatic spacer between the semiconducting core and the anchoring group was adjusted to ensure good packing for processability, even on rougher surfaces such as AlOx and SiO_2.

A similar set-up is to use a single nanotube or a network of nanotubes that act as a conductive layer to form nanotube field effect transistor (NTFET) sensors [38]. In this case, non-covalent interactions such as π-π interactions or hydrophobic interactions on engineered polysaccharides, enzymes, antibodies etc. are preferred to avoid defects in NT [39]. Moreover, different types of FET biosensors have been constructed, of which the most typical is modulation of the charge-carrier mobility by employment and formation of SAM on the gate electrode.

A biosensor based on a single-walled nanotube (SWNT) was employed for glycan-lectin interactions with very good detection of 2 nM of the binding bacterial lectins (Fig. 6.22). In this biosensor set-up, the glycan unit is conjugated to a porphyrin derivative based on azido-alkyne click chemistry and this conjugate interacts non-covalently with SWNTs and forms a platform for ligation with lectin molecules [40]. It is worth noting that, for such an application, a higher purity of semiconducting SWNTs is required in order to obtain better signal quality.

6.7 SAM in Glycan-Based Biosensors

The development of biosensors began in the early sixties when L. C. Clark and C. Lyons developed the first enzymatic biosensor

[41]; subsequently, the development has been intensified and commercialised in such applications as glucose tests, pregnancy tests and clinical diagnostics.

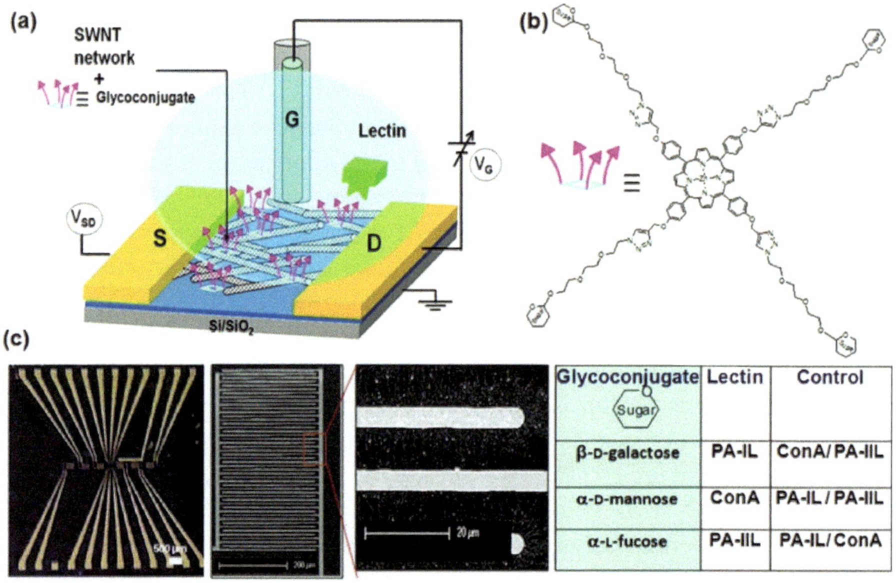

Glycoconjugate (Sugar)	Lectin	Control
β-D-galactose	PA-IL	ConA/PA-IIL
α-D-mannose	ConA	PA-IL / PA-IIL
α-L-fucose	PA-IIL	PA-IL/ConA

Figure 6.22 (a) Schematic illustration of glyco-conjugate-functionalised single-walled carbon nanotubes (SWNTs)-FET for selective detection of lectins. (b) Chemical structure of porphyrin-based glyco-conjugates for non-covalent functionalization of SWNTs. The table shows the selected glyco-conjugates and their specific lectins and controls. (c) Optical image of a Si/SiO_2 chip with micropatterned interdigitated electrodes. SEM image of interdigitated electrodes used for device fabrication. Inset shows the SWNTs deposited by dielectrophoresis technique between the microelectrodes. Reproduced with permission from [40]. Copyright 2011 from American Chemical Society.

In addition, the early detection of disease markers by biosensor technologies has become a crucial factor in the prevention and treatment of diseases and can considerably reduce the costs for health services as well as decrease time involved in treatment leading to the greater comfort and wellbeing of patients. Moreover, a general biosensor platform with robust and efficient clinical detection remains in high demand.

Biosensor is a device that uses specific biochemical reactions mediated by isolated enzymes, immuno-systems, tissues, organelles or whole cells to detect chemical compounds, usually by electrical, thermal or optical signals.

Hence, a biosensor as an analytical device contains an element which makes it possible to recognise a biologically active analyte by means of their interactions. These are converted to measurable signals in a transducer scheme and transferred to a detector, as shown in Fig. 6.23.

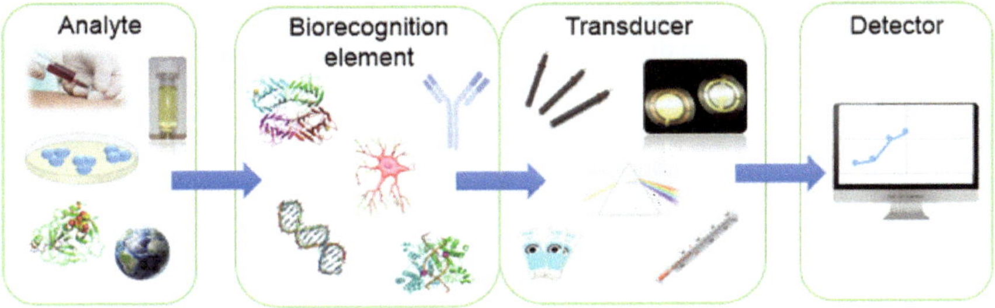

Figure 6.23 Schematic presentation of the biosensor with an analyte, a biorecognition element, a transducer, and a detector. Reproduced with permission from [42].

6.7.1 Advantages and Limitations of SAM in Biosensors

SAM as a part of a biosensing scheme is present at the interface of the transducer and biorecognition elements. As detailed previously, different strategies for linking in SAM have been applied to overcome a number of challenges and the application of SAM affords several significant advantages.

(1) Simple preparation of well-ordered, pinhole-free and robust SAMs.
(2) Robustness over an extended period, affording reliable measurements.
(3) Customised design according to requirement by the terminal group of SAM with different functional groups (differently charged, hydrophobic, hydrophilic...).

(4) SAM surfaces offer cell membrane-like or cellular-like microenvironment for immobilisation of biological element.
(5) Enhanced sensitivity and selectivity due to the presence of terminal group for prevention of non-specific interaction or adhesion.
(6) Small amount of biological element for immobilisation is required (monolayer).
(7) Capacity to provide fundamental molecular-level information on phenomena like protein adsorption, glycan-protein and antigen–antibody interaction, etc.

Despite these many advantages, the SAM surface may also have restricted use due to several factors:

(1) The limited chemical stability of some types of SAM can lead to oxidation or desorption of SAM during measurements.
(2) The electric field applied and possible thermal desorption of SAM is harmful to biosensor applications.
(3) Limited stability and sensitivity of the biological element on SAM (pH, temperature, ionic strengths) can sometimes result in a decrease or loss of biological activity.
(4) Character of SAM (charged hydrophobic) can induce the accumulation of unwanted impurities and contaminants that prevent access to the binding site.

6.7.2 Classification of Glycan-Related Biosensors

There are two main technologies for signal detection in biosensors: label-based technologies, which require a linked electrochemical or optical label probe on the element or analyte for recognition via transducer platform and label-free technologies of detection [43]. A different type of transducer configuration was used for the detection of glycan or glycan unit in the biological element. Most of these applications require SAM formation in order to ensure high stability, reproducibility, sensitivity and selectivity.

Glycan-used biosensors requiring label are predominantly based on glycan microarrays Fig. 6.24. A labelled glycan microarray largely relies on a fluorescent-labelled probe and can be divided into three main groups of this sensorial platform. The first group involves a fluorescent-labelled protein which specifically interacts with the glycan unit (Fig. 6.24a). Another current strategy is binding through a streptavidin-biotin complex (Fig. 6.24b) or by a secondary probe bearing the fluorescent probe in a so-called sandwich configuration (Fig. 6.24C).

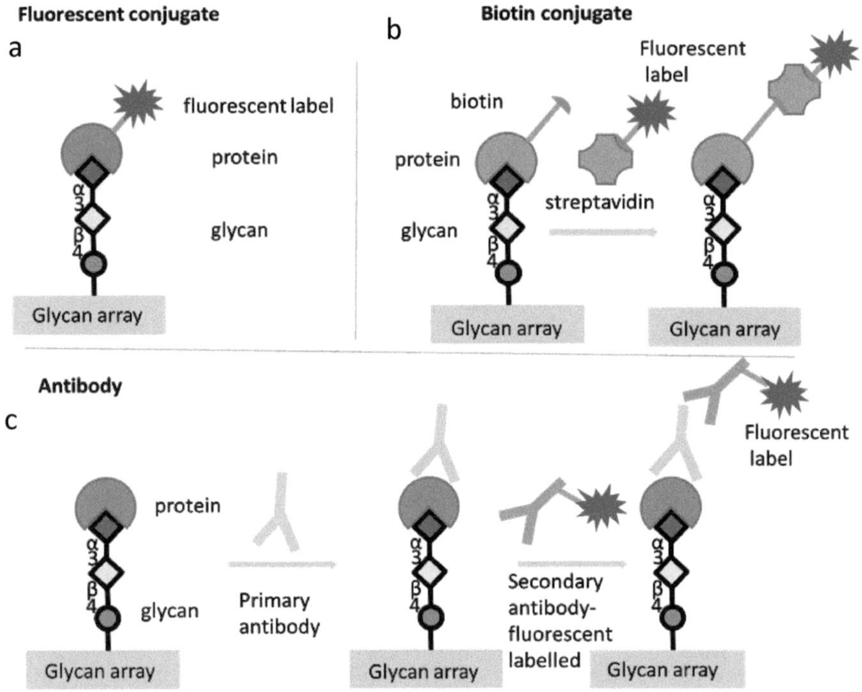

Figure 6.24 Schematic representation of different fluorescent-labelled probes for protein-glycan array configurations: (a) protein conjugate bearing fluorescent probe; (b) protein conjugation to biotin and subsequent complexation with streptavidin-bearing fluorescent probe; (c) binding of secondary fluorescent-labelled antibody to primary antibody bound to protein.

The glycan microarray can be deployed in several applications [44]:

- Rapid analysis of glycan-protein interaction, where the lectin-glycan interaction is most effective (lectins are small proteins from plants, microbes or mammals, which specifically recognise and bind to carbohydrate unit(s) without effecting modifications in the glycan structures present even in biological elements [45]).

- Viral and bacterial proteins glycoprofilation.
- Monoclonal and therapeutic antibodies (for different applications) glycoprofilation.
- Identification of new glycan-binding proteins.

Other widely used applications are profiling of the substrate specificity of enzymes and immune response, the detection of viruses and whole cells and profiling of the immune response and investigations in serodiagnosis and biomarkers.

It should be noted that specific glycan-lectin complexation also provides different types of configuration biosensors based on lectin that detect carbohydrate units in other biological elements, usually glycoprotein [46].

In most glycan biosensor applications, SAM is formed by the immobilised biological element and, at the same time, with an element which prevents non-specific interaction to enhance the limit of detection, selectivity and specificity, as mentioned elsewhere. However, for example, in an EIS biosensor, SAM also acts as a dielectric layer of insulation and this layer should be sufficiently thin to permit measurement of the probe layer capacitance. Defects and imperfect insulation can considerably reduce the sensitivity. The EIS transducer signal in biosensor use relies on changes in the interfacial electron-transfer kinetics between the redox probe in the solution and the surface. An electrochemical change occurs from the interaction between the biological element on SAM and the target analyte from the solution (for example lectin–carbohydrate). Changes in the signal are monitored by the charge-transfer resistance R_{ct} of the system, which increases in proportion to the increase in the amount of the target analyte complex with the biological element (Fig. 6.25a) [47].

Recently, electrochemical capacitance spectroscopy has been applied, using a tethered redox probe on SAM (Fig. 6.25b) [48]. This methodology is largely dependent on the fully capacitive-based response to changes in the density of states occupancy at the interface. The advantages of this method are that it affords good reproducibility and sensitivity and at the same time does not require a redox probe to be added to the solution, as in the EIS method.

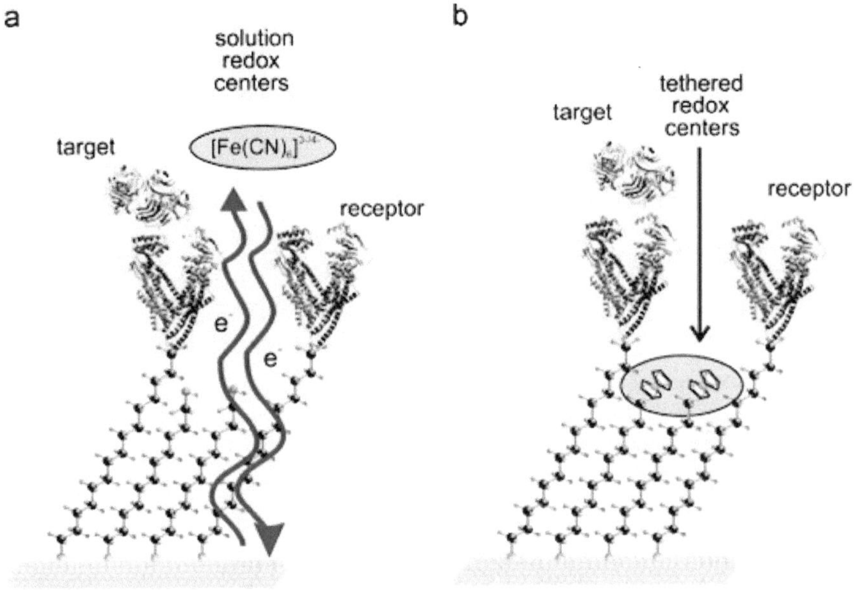

Figure 6.25 Comparative schematic representations of (a) faradaic EIS and (b) capacitive ECS biosensing. (a) The faradaic impedimetric surface engineering is based in the electron transfer resistance (R_{ct}) of the redox probe in solution to metal surface. (b) The capacitive approach does not need a redox probe in solution, but an electrode tethered one. The environmental sensitivity of this charging capability is used as the reporter of neighbouring biorecognition. Reproduced with permission from [48]. Copyright 2014 from Elsevier.

Acknowledgement

This chapter received funding from Qatar University grant # QUUG-CAM-2017-1.

References

1. Whitesides, G. M., and Grzybowski, B. (2002) Self-assembly at all scales, *Science*, **295**, pp. 2418–2421.
2. Ulman, A. (1996) Formation and structure of self-assembled monolayers, *Chem. Rev.*, **96**, pp. 1533–1554.

3. Bigelow, W. C., Pickett, D. L., and Zisman, W. A. (1946) Oleophobic monolayers: I. Films adsorbed from solution in non-polar liquids, *J. Colloid Sci.*, **1**, pp. 513–538.
4. Klukova, L., Filip, J., Belický, S., Vikartovska, A., and Tkac, J. (2016) Graphene oxide-based electrochemical label-free detection of glycoproteins down to aM level using a lectin biosensor, *Analyst*, **141**, pp. 4278–4282.
5. Love, J. C., Estroff, L. A., Kriebel, J. K., Nuzzo, R. G., and Whitesides, G. M. (2005) Self-assembled monolayers of thiolates on metals as a form of nanotechnology, *Chem. Rev.*, **105**, pp. 1103–1170.
6. Chapman, R. G., Ostuni, E., Takayama, S., Holmlin, R. E., Yan, L., and Whitesides, G. M. (2000) Surveying for surfaces that resist the adsorption of proteins, *J. Am. Chem. Soc.*, **122**, pp. 8303–8304.
7. Zhang, Z., Zhang, M., Chen, S., Horbett, T. A., Ratner, B. D., and Jiang, S. (2008) Blood compatibility of surfaces with superlow protein adsorption, *Biomaterials*, **29**, pp. 4285–4291.
8. Ilcikova, M., Tkac, J., and Kasak, P. (2015) Switchable materials containing polyzwitterion moieties, *Polymers*, **7**, pp. 2344–2370.
9. Castner, D. G., and Ratner, B. D. (2002) Biomedical surface science: Foundations to frontiers, *Surf. Sci.*, **500**, pp. 28–60.
10. Putzbach, W., and Ronkainen, N. (2013) Immobilization techniques in the fabrication of nanomaterial-based electrochemical biosensors: A review, *Sensors*, **13**, pp. 4811–4840.
11. Oliverio, M., Perotto, S., Messina, G. C. Lovato, L., and DeAngelis, F. (2017) Chemical functionalization of plasmonic surface biosensors: A tutorial review on issues, strategies, and costs, *ACS Appl. Mater. Interfaces*, **9**, pp. 29394–29411.
12. Fisher, J. M. (2010) Amine coupling through EDC/NHS: Apractical approach. *Methods Mol. Biol.*, **627**, pp. 55–73.
13. Biggs, C.I., Edmondson, S., and Gibson, M.I. (2015) Thiol–ene immobilization of carbohydrates onto glass slides as a simple alternative to gold–thiol monolayers, amines or lipid binding, *Biomater. Sci.*, **3**, pp. 175–181.
14. Fukuda,T., Onogi, S., and Miura, Y. (2009) Dendritic sugar-microarrays by click chemistry, *Thin Solid Films*, **518**, pp. 880–888.
15. Bertok, T., Sediva, A., Filip, J., Ilcikova, M., Kasak, P., Velic, D., Jane, E., Mravcová, M., Rovenský, J., Kunzo, P., Lobotka, P., Smatko, V., Vikartovska, A., and Tkac, J. (2015) Carboxybetaine interface for

electrochemical glycoprofiling of antibodies isolated from human serum, *Langmuir*, **31**, pp. 7148–7157.

16. Houseman, B. T., and Mrksich, M. (2002) Carbohydrate arrays for the evaluation of protein binding and enzymatic modification, *Cell Chem. Biol.*, **9**, pp. 443–454.
17. Bertók, T., Dosekova, E., Belicky, S., Holazova, A., Lorencova, L., Mislovicova, D., Paprckova, D., Vikartovska, A., Plicka, R., Krejci, J., Ilcikova, M., Kasak, P., and Tkac, J. (2016) Mixed zwitterionic-based self-assembled monolayer interface for impedimetric glycomic analyses of human IgG samples in an array format, *Langmuir*, **32**, pp. 7070–7078.
18. Gulino, A. (2013) Structural and electronic characterization of self-assembled molecular nanoarchitectures by X-ray photoelectron spectroscopy, *Anal. Bioanal. Chem.*, **405**, pp. 1479–1495.
19. Wei, Q., Becherer, T., Angioletti-Uberti, S., Dzubiella, J., Wischke, C., Neffe, A. T., Lendlein, A., Ballauff, M., and Haag, R. (2014) Protein interactions with polymer coatings and biomaterials, *Angew. Chem. Int. Ed.*, **53**, pp. 8004–8031.
20. Mendes, P. M., Yeung, C. L., and Preece, J. A. (2007) Bio-nanopatterning of surfaces, *Nanoscale Res. Lett.*, **2**, pp. 373–384.
21. Ginger, D. S., Zhang, H., and Mirkin, C. A. (2004) The evolution of dip-pen nanolithography, *Angew. Chem. Int. Ed.*, **43**, pp. 30–45.
22. Lee, K.-B., Kim, E.-Y., Mirkin, C. A., and Wolinsky, S. M. (2004) The use of nanoarrays for highly sensitive and selective detection of human immunodeficiency virus type 1in plasma, *Nano Lett.*, **4**, pp. 1869–1872.
23. Liu, G. Y., Xu, S., and Qian, Y. (2000) Nanofabrication of self-assembled monolayers using scanning probe lithography, *Acc. Chem. Res.*, **33**, pp. 457–466.
24. Castronovo, M., and Scaini, D. (2011) The atomic force microscopy as a lithographic tool: Nanografting of DNA nanostructures for biosensing applications, *Methods Mol Biol.*, **749**, pp. 209–221.
25. Kolodziej, C. M., and Maynard, H. D. (2012) Electron-beam lithography for patterning biomolecules at the micron and nanometer scale, *Chem. Mater.*, **24**, pp. 774–780.
26. Mendes, P. M., Jacke, S., Critchley, K., Plaza, J., Chen, Y., Nikitin, K., Palmer, R. E., Preece, J. A., Evans, S. D., and Fitzmaurice, D. (2004) Gold nanoparticle patterning of silicon wafers using chemical e-beam lithography, *Langmuir*, **20**, pp. 3766–3768.

27. Nandivada, H., Ross, A. M., and Lahann, J. (2010) Stimuli-responsive monolayers for biotechnology, *Prog. Polym. Sci.,* **35**, pp. 141–154.

28. Liu, M., Wang, S., and Jiang, L. (2017) Nature-inspired superwettability systems, *Nat. Rev. Mater.,* **2**, p. 17036.

29. Mendes, P. M. (2008) Stimuli-responsive surfaces for bio-applications, *Chem. Soc. Rev.*, **37**, pp. 2512–2529.

30. Filip, J., Popelka, A., Bertok, T., Holazova, A., Osicka, J., Kollar, J., Ilčikovaì, M., Tkac, J., and Kasak, P. (2017) pH switchable interaction of carboxybetaine ester based SAM with DNA and gold nanoparticles, *Langmuir*, **33**, pp. 6657–6666.

31. Lahann, J., Mitragotri, S., Tran, T. N., Kaido, H., Sundaram, J., Choi, I. S., Hoffer, S., Somorjai, G. A., and Langer, R. (2003) A reversibly switching surface, *Science,* **299**, pp. 371–374.

32. Peng, D. K., Yu, S. T., Alberts, D. J., Lahann, J. (2007) Switching the electrochemical impedance of low-density self-assembled monolayers, *Langmuir,* **23**, pp. 297–304.

33. Osicka, J., Ilcikova, M., Popelka, A., Filip, J., Bertok, T., Tkac, J., and Kasak, P. (2016) Simple, reversible and fast modulation in superwettability, gradient and adsorption by counterion exchange on self-assembled monolayer, *Langmuir*, **32**, pp. 5491–5499.

34. Bertok, T., Sediva, A., Vikartovska, A., and Tkac, J. (2014) Comparison of the 2D and 3D nanostructured lectin-based biosensors for in situ detection of sialic acid on glycoproteins, *Int. J. Electrochem. Sci.,* **9**, pp. 890–900.

35. Casalini, S., Bortolotti, C. A., Leonardi, F., Biscarini, F. (2017) Self-assembled monolayers in organic electronics, *Chem. Soc. Rev.*, **46**, pp. 40–71.

36. Torsi, L., Magliulo, M., Manoli, K., and Palazzo, G. (2013) Organic field-effect transistor sensors: A tutorial review, *Chem. Soc. Rev.*, **42**, pp. 8612–8628.

37. Smits, E. C. P., Mathijssen, S. G. J., van Hal, P. A., Setayesh, S., Geuns, T. C. T., Mutsaers, K. A. H. A., Cantatore, E., Wondergem, H. J., Werzer, O., Resel, R., Kemerink, M., Kirchmeyer, S., Muzafarov, A. M., Ponomarenko, S. A., deBoer, B., Blom, P. W. M., and de Leeuw, D. M. (2008) Bottom-up organic integrated circuits, *Nature*, **455**, pp. 956–959.

38. Huang, W., Diallo, A. K., Dailey, J. L., Besar, K., and Katz, H. E. (2015) Electrochemical processes and mechanistic aspects of field-effect sensors for biomolecules, *J. Mater. Chem. C*, **3**, pp. 6445–6470.

39. Allen, B. L., Kichambare, P. D., and Star, A. (2007) Carbon nanotube field-effect-transistor-based biosensors, *Adv. Mat.*, **19**, pp. 1439–1451.

40. Vedala, H., Chen, Y. A., Cecioni, S., Imberty, A., Vidal, S., and Star, A. (2011) Nanoelectronic detection of lectin-carbohydrate interactions using carbon nanotubes, *Nano Lett.*, **11**, pp. 170–175.

41. Turner, A. P. F. (2013) Biosensors: Sense and sensibility, *Chem. Soc. Rev.*, **42**, pp. 3184–3196.

42. Pihíková, D., Kasak, P., and Tkac, J. (2015) Glycoprofiling of cancer biomarkers: Label-free electrochemical lectin-based biosensors, *Open Chem.*, **13**, pp. 636–655.

43. Paleček, E., Tkac, J., Bartosik, M., Bertoìk, T., Ostatnaì, V., and Paleček, J. (2015) Electrochemistry of nonconjugated proteins and glycoproteins. Toward sensors for biomedicine and glycomics, *Chem. Rev.*, **115**, pp. 2045–2108.

44. Park, S., Gildersleeve, J. C., Blixt, O., and Shin, I. (2013) Carbohydrate microarrays, *Chem. Soc. Rev.*, **42**, pp. 4310—4326.

45. See website with lectin binding specificity to selected carbohydrate units, www.functionalglycomics.org.

46. Klukova, L., Bertok, T., Kasak, P., and Tkac, J. (2014) Nanoscale controlled architecture for development of ultrasensitive lectin biosensors applicable in glycomics, *Anal. Methods*, **6**, pp. 4922–4931.

47. Fernandes, F. C. B., Santos, A., Martins, D. C., Góes, M. S., and Bueno, P. R. (2014) Comparing label free electrochemical impedimetric and capacitive biosensing architectures, *Biosens. Bioelectron.*, **57**, pp. 96–102.

48. Fernandes, F. C. B., Góes, M. S., Davis, J. J., and Bueno, P. R. (2013) Comparing label free electrochemical impedimetric and capacitive biosensing architectures, *Biosens. Bioelectron.*, **50**, pp. 437–440.

Chapter 7

Introduction to Glycan Enrichment Using Nanomaterials

Jan Tkac

Glycanostics, Ltd., Department of Glycobiotechnology, Institute of Chemistry, Slovak Academy of Sciences, Dubravska cesta 9, Bratislava, 84538, Slovak Republic

Jan.Tkac@savba.sk

There are few main reasons behind necessity of glycan enrichment *prior* MS analysis: (1) glycans are present in samples at the level well below limit of detection of a MS instrument; (2) glycans/glycopeptides have to be separated from more abundant peptides and (3) ionisation of glycans/glycopeptides during MS analysis in the presence of peptides is less efficient. Glycans, glycopeptides or glycoproteins can be enriched implementing various strategies described here. The main aim of the chapter is to convince a reader that especially nanotechnology revolutionised glycan enrichment strategies.

Glyconanotechnology: Nanoscale Approach for Novel Glycan Analysis and Their Medical Use
Edited by Tomáš Bertók

ISBN 978-981-4800-16-7 (Hardcover), 978-0-429-44849-2 (eBook)
www.jennystanford.com

7.1 Glycan Enrichment Using Nanoparticles

7.1.1 Interactions behind Glycan Enrichment

Glycan enrichment can be performed using various ways, i.e. using a diverse range of physical (hydrophobic, hydrophilic and ionic interactions) or chemical (boronic acid and hydrazide chemistry) interactions, their combination (molecularly imprinted polymers-MIPs) and finally using biospecific/affinity-based recognitions using lectins (Fig. 7.1) [1]. Figure 7.2 summarises the main advantages of the most popular glycan enrichment strategies.

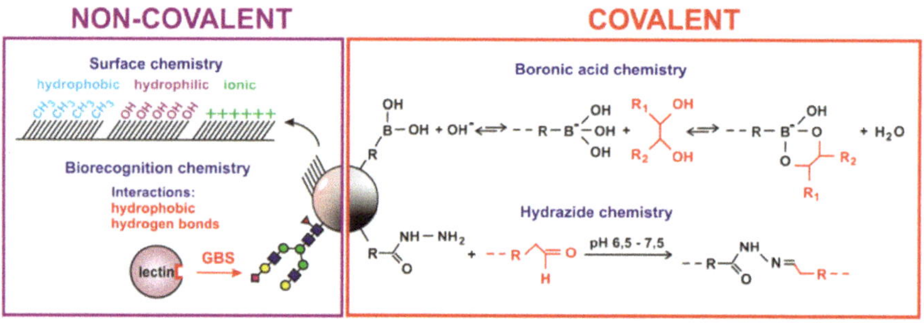

Figure 7.1 Various ways glycans can be enriched using hydrophobic, hydrophilic or ionic interactions (upper left part) or biorecognition interactions involving glycan binding proteins (lectins) (lower left part). Alternatively for glycan capture/release particles modified by boronate functional groups or hydrazine can be applied, as well. GBS–glycan binding surface of a lectin. Image taken from [2] with permission from Wiley.

7.1.1.1 Boronate-based enrichment

Boronate-based materials are the most frequently applied for glycan enrichment due to strong interactions between boronate functional groups and *cis*-diols present in glycans (Fig. 7.1), requiring basic conditions [3]. The pH during glycan enrichment should be above pK_a of a boronate moiety (i.e. above pH 8.9 for frequently applied phenylboronic acid having pK_a = 8.9). The exception from this rule is enrichment of sialic acid containing glycans occurring with boronate-modified material at pH lower than pK_a of the boronate functional group. Steric factors and buffer components play a significant role during boronate-based interactions, as well [4]. After enrichment process was completed, glycans are released by decreasing pH of the eluting buffer.

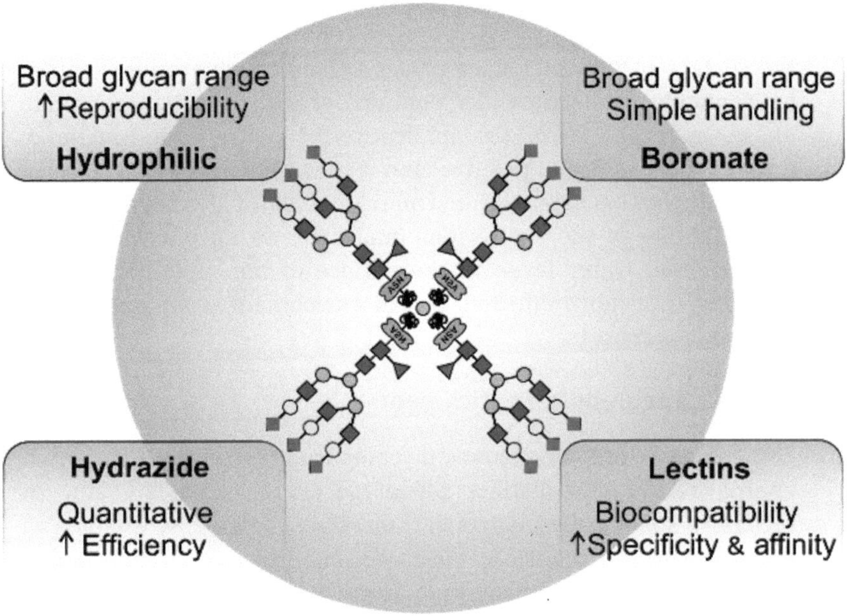

Figure 7.2 Advantages of using various enrichment strategies.

Some biomolecules like glycoproteins can be unstable under basic conditions, so unmodified boronate material is not an ideal enrichment matrix. To solve this problem, synthesis of boronate materials having electron withdrawing groups in the proximity to boronate functional groups was successfully applied [5].

7.1.1.2 Hydrazide-based enrichment

Application of hydrazide-modified materials is another popular choice for glycan enrichment. There is, however, a need to oxidise glycans first to form aldehydes from *cis-diol* groups. Such aldehydes will in turn chemically react with hydrazide forming hydrazone (Fig. 7.1). Such modified glycans are stable under basic conditions and glycan release can be triggered by application of acidic pH.

7.1.1.3 Hydrophilic interactions

Hydrophilic interaction-based method for glycan enrichment is also a popular pre-concentration method involving use of a polar stationary phase and a nonpolar mobile phase for separation. Glycans strongly interact with a hydrophilic water-rich layer on the surface of the stationary phase. Other less hydrophilic (more hydrophobic) molecules like peptides are dissolved in a mobile phase. In recent years, the application of zwitterionic functional groups (i.e. having a positive and a negative charge in a close proximity) has become more and more popular for glycan enrichment due to ability of the zwitterionic molecules to form a very stable and oriented water layer on the surface of the separation phase resulting in a more effective glycan enrichment compared to the traditional supports.

7.1.1.4 Lectin-based enrichment

Lectin-based pre-concentration compared to above-mentioned approaches have a distinct advantage since lectins are able to recognise specifically distinct carbohydrate moieties within glycan structures (Fig. 7.3 left). At the same time, it is worth mentioning that due to different recognition strategy compared to antibodies (interaction via a shallow groove rather than by a deep pocket), lectins are able to recognise glycans with lower affinity and specificity. Low specificity means that lectins can recognise glycans with different composition, when having the same or similar core structure like mannose units in case of Concanavalin A lectin binding (Fig. 7.3 right). On the other hand, when a minor modification of glycan has a significant physiological consequence, lectins are able to detect such perturbation within glycan structure. A typical example is *Sambucus nigra* agglutinin (SNA) recognising sialic acid bound to galactose via α2-6 bond, while *Maackia amurensis* agglutinin binds to α2-3 linked sialic acid (Fig. 7.4). Finally, it is important to point out to the fact that lectins are an effective tool to fractionate glycans into different groups (i.e. containing α2-3 linked sialic acid, α2-6 linked sialic acid high mannose ones, etc.). Glycans can be released from lectins using free carbohydrates or by application of an acidic elution buffer.

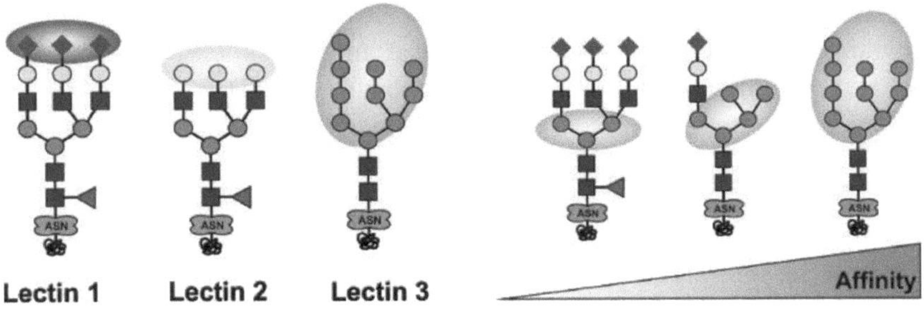

Figure 7.3 Binding of three lectins to different carbohydrates present in glycans (left) and affinity of Concanavalin A lectin towards glycans containing different content of mannose units (right).

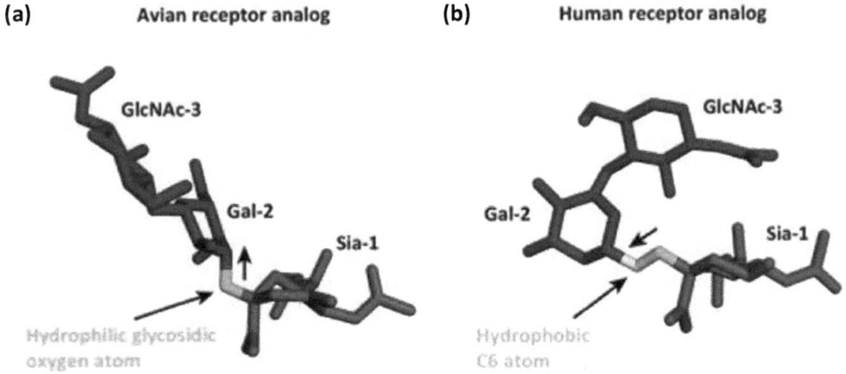

Figure 7.4 Configurations of avian and human receptor analogues. (a) The avian receptor analogue. The Sia-1 (sialic acid, SA) is linked to Gal-2 (galactose) via an α2-3-glycosidic bond. The glycans extend forward and the hydrophilic glycosidic oxygen atom is exposed to the receptor-binding site. (b) The human receptor analogue. The Sia-1 is linked to Gal-2 via an α2-6-glycosidic bond. The glycans fold back and the hydrophobic C6 atom is exposed to the receptor-binding site. In both (a) and (b) *N*-acetylglucosamine is abbreviated as GlcNAc-3; black arrows indicate the orientations of the glycosidic linkage. Reprinted from [8] with permission from Elsevier.

7.1.1.5 Other ways for enrichment

There are other options for glycan enrichment such as a reverse phase mode, an electrostatic mode (for glycans containing negatively charged sialic acids), an electrostatic repulsion hydrophilic interactions and an affinity capture based on protein A, which is applicable for selective enrichment of antibodies with subsequent glycoprofiling of glycans present especially in the Fc fragment [2].

7.1.2 Nanoparticles

Nanoparticles (NPs) by definition are any objects with at least one dimension below 100 nm with unique physico-chemical properties including large surface to volume ratio. Large surface area can address problems with mass transfer kinetics and allows interfacial modifications with high density of active functional groups, what are prerequisites for time effective glycan enrichment process [6]. During synthesis of NPs we can control size, morphology, surface chemistry and composition of NPs with a possibility to prepare hybrid nanoparticles with unique characteristics (i.e. magnetic NPs with magnetic core and a silica shell) [7].

7.1.2.1 Silica NPs (SiNPs)

Silica NPs (SiNPs), which are frequently applied for glycan enrichment, are mechanically and chemically stable over a wide pH, a feature important for column packing material. Nanoporous silica materials with pore size spanning range 2–150 nm are the first choice material, but other NPs made of metal oxides can be applied as a packing/enrichment material either in its pristine or modified state. SiNPs or NPs composed of metal oxides could be easily patterned via a silane chemistry forming a thin self-assembled monolayer (SAM). The initial step of silanisation is hydrolysis of silane molecule with subsequent condensation reaction on the surface making covalent Si-O bond (Fig. 7.5) [9]. By proper choice of the silane molecule it is possible to deliver to the surface functional groups of interest. It is worth mentioning that a thin silica shell is very often applied to cover various types of NPs to make them more stable forming hybrid NPs with a shell made of silica and the core made of the other material.

For example SiNPs (90 nm) with 3D polymer brush containing zwitterionic molecules (thickness of 5 nm) showed effective glycan enrichment properties. Ultralow level of glycans (10 fmol) could be enriched with high recovery index of 88%, performing much better compared to commercial matrices [10].

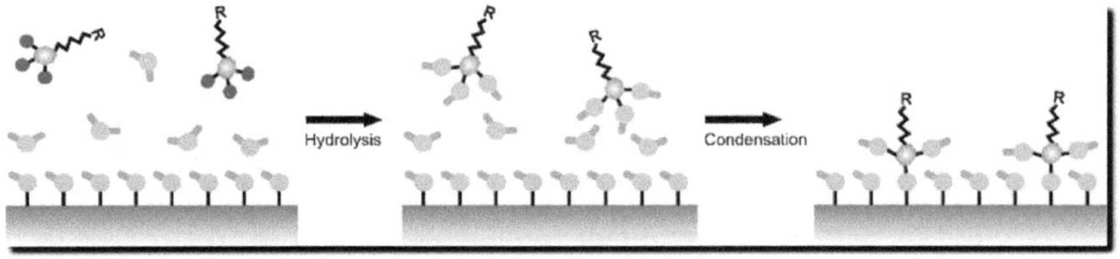

Figure 7.5 Simplified mechanism describing formation of silane-based SAMs (atoms: H, blue; O, green; Cl, red; and Si, pink). Image taken from [9] with permission from Wiley.

7.1.2.2 Gold NPs (AuNPs)

AuNPs like any gold surface can be easily patterned at nanoscale using formation of a SAM by its incubation with thiols molecules. By a proper choice of ω-functional groups of thiols it is feasible to tune interfacial properties (density and type of functional groups) of gold in a simple and highly reproducible way [2].

A simplified glycan enrichment procedure within a pipette tip for example was based on application of 20 nm AuNPs attached to the polymer backbone to enhance surface area. SAM formed on AuNPs was utilised for covalent immobilisation of a lectin. Such a lectin-modified monolith selectively enriched galactosylated glycans with a high recovery of 95%. Finally, the device was used for glycan profile analysis in the lysate of *E. coli* [11].

Hydrazide functionalised ultrasmall AuNPs with the size of 1.2 nm were applied for very selective capture of periodate oxidised glycopeptides by Tran et al., when as much as 97% of all peptides captured from rat kidney tissue were glycopeptides [12]. This highly selective capture of glycopeptides was possible due to extremely high density of hydrazide on AuNPs, i.e. 630 nmol mg^{-1}, what is 79-fold higher density compared to hydrazide density on magnetic particles having size of 200–500 nm [12].

7.1.2.3 Carbonaceous nanoparticles

Graphene and graphene oxide (GO)

Graphene as two-dimensional (2D) crystalline material consisting of a single layer of carbon atoms bound together via sp^2 hybridisation with unique properties is attracting a considerable attraction since its discovery in 2004 [13]. Graphene with its enormous specific surface area of 2,630 m^2 g^{-1} predetermines its application in various enrichment strategies. The main problem with using pristine graphene is its very low solubility, low reproducibility of handling and the surface which is prone to hydrophobic rather than hydrophilic interactions needed for glycan enrichment. This is why an oxidised form of graphene—graphene oxide (GO)—is preferentially applied in pre-concentration of glycans. It is not only presence of a wide range of hydrophilic oxygen containing functional groups, what makes this nanomaterial so interesting, but it is also a simple and cost-effective route behind preparation of GO, what matters a lot. GO is produced from graphite by oxidation using strong acids with a final sonication applied to release individual GO flakes from graphite oxide. GO can be modified in many different ways depending on the density of oxygenated functional groups. At their low-density π–π stacking interactions between pristine graphene areas and pyrene derivatives could be applied. Since pyrene derivatives are highly toxic compounds, other alternative ways are applied. For example, when the density of oxygen-containing groups on GO is moderate or high we can apply silane chemistry (interaction with –OH groups of GO) or covalent amine coupling by activation of –COOH groups of GO. Alternatively, GO could be integrated with a diverse range of (bio)polymers or other NPs, which could be directly employed for glycan enrichment or applied for glycan pre-concentration after subsequent functionalisation. Typically, GO are integrated with magnetic NPs for simple single-step glycan enrichment using an external magnetic field (i.e. magnet).

For example, GO was non-covalently modified by pyrenebutyric acid via π–π stacking interaction with activation of –COOH group by $SOCl_2$ to prepare an active GO derivative used for selective glycan binding (Fig. 7.6). Glycan enrichment can be monitored by naked eye since modified GO was crosslinked in presence of glycans. Finally, glycans were released from GO by sonication in an acidic environment at 60°C. GO-based glycan enrichment was very effective in presence of excessive protein level [14].

Single-walled carbon nanotubes (SWCNTs)

SWCNTs are in essence graphene flakes rolled into the form of a tube with diameter of few nm. Due to curvature of SWCNTs, modification of its surface by π-π stacking interactions is problematic a due to much lower specific surface area of 1,315 $m^2\ g^{-1}$ compared to graphene, this nanomaterial is not that often applied for glycan enrichment. Typically, SWCNTs are integrated with magnetic NPs for simple single-step glycan enrichment using an external magnetic field.

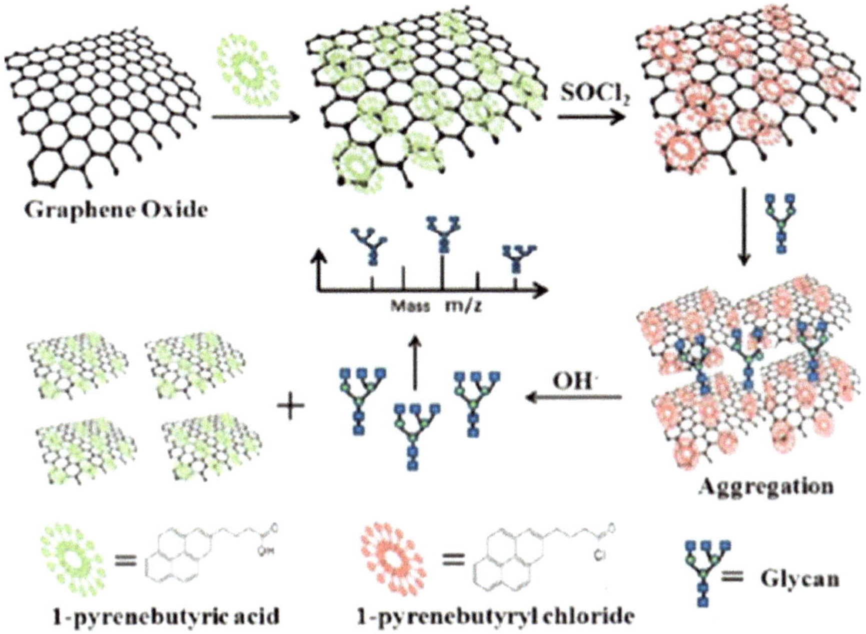

Figure 7.6 Graphene oxide modified by 1-pyrenebutyric acid with subsequent activation of –COOH group by $SOCl_2$ for effective glycan enrichment. Reprinted with permission from [14] Copyright 2013 American Chemical Society.

Strano's group modified SWCNTs with derivatives of phenylboronic acids containing –COOH, $-NO_2$ and $-NH_2$ functional groups with various substitutions generating 144 distinct interfacial modifications with some of them exhibited remarkable binding affinity to certain monosaccharides [15].

Nanodiamonds

Nanodiamond particles contrary to GO and SWCNT are not conductive since carbon atoms within the diamond structure are bound with three neighbouring C atoms making sp^3 hybridisation without presence of double bonds. At the same time nanodiamond particles are able to effectively transfer heat. This is why such type of NPs was not used for glycan enrichment, but rather for efficient ionisation of glycan molecules by Matrix assisted laser desorption ionisation time-of-flight mass spectrometry (MALDI-TOF MS) [16].

7.1.2.4 Magnetic nanoparticles (MNPs)

MNPs are becoming more and more popular since a simple application of an external magnetic field can separate them from the solution. There is no need to apply any traditional separation techniques such as centrifugation and/or filtration and the magnetic separation can be very quick. The most often applied material to make MNPs is ferrite (Fe_3O_4). In order to modify MNPs, silane chemistry can be effectively applied to deliver on the surface functional groups necessary for glycan enrichment. MNPs are very often combined with other nanoparticles making hybrid NPs (see later) to apply a synergetic effect (i.e. combination of a large surface area of carbonaceous material and magnetic properties of MNPs) for a pre-concentration procedure. Glycan enrichment can be effectively done using magnetic nanoparticles with different glycan recognising functional groups present on the surface of such nanoparticles.

MNPs modified by a thin layer of SiO_2 further patterned by zwitterionic molecules were effective in glycan enrichment with high binding capacity of 100 mg g^{-1} and high enrichment recovery of 74% with a rapid magnetic separation. Moreover, when combined with MS analysis glycans from just 0.1 fmol of IgG could be detected [17].

MNPs modified by a branched and ordered polymer (dendrimer) containing boronate functional groups exhibit 3–4 orders of magnitude higher affinity constant for glycoproteins compared to single boronate moieties. Moreover, glycan enrichment could be completed within 1 min with limit of detection (LOD) of 180 amol [18].

7.1.2.5 Hybrid nanoparticles

Hybrid NPs are composed of at least two different NPs with the most popular option being integration of various types of NPs with MNPs. Magnetic carbon nanotubes were composed of carbon nanotubes with d = 40–60 nm modified by in situ formed MNPs with d = 10–15 nm from Fe^{3+} ions [19]. Another example is the composite consisted of graphene accommodated MNPs (100 nm). Such composite was effectively employed for enrichment of as low as 1 fmol of glycopeptides from as little as 1 uL of a sample [20].

7.1.2.6 Molecularly imprinted polymers (MIPs)

MIPs can be effectively applied in a selective capture and release of glycoproteins, since the analyte molecule is imprinted and after formation of a thin layer of a polymer around the template (analyte), the analyte is removed from the MIP. Thus, such MIP is then able to selectively recognise the analyte over other molecules. MIPs are considered as affinity-like matrices with properties (i.e. binding pockets for size exclusion-based interactions with functional groups applied for effective interactions with glycans), which can be tuned by the preparation process. Such MIPs can finally exhibit lectin-like properties [6].

MIPs could be prepared with a polymer thickness controlled at the nanoscale for the selective affinity-like enrichment of specific glycoproteins. In order to study this phenomenon, three glycoproteins (ribonuclease B, HRP and glucose oxidase) were imprinted within a thin film. The thickness of the film (5.7; 10.2; 16.3 and 25.3 nm) was controlled by polymerisation time (1; 3; 6 and 15 h). The higher the size of the glycoprotein (15–80 kDa) the thicker film (5.7–16.3 nm) was needed to achieve maximal binding capacity (16.7–19.5 mg g^{-1}) [22].

The most sophisticated MIP-based approach for selective enrichment of a glycoprotein combines traditional MIP strategy for imprinting a protein backbone with an additional glycan interacting feature (Fig. 7.7). Glycan of the glycoprotein (prostate specific antigen [PSA]) interacts with boronate functional groups imprinted within the polymer. Since boronate-glycan interaction can be disrupted by low pH, simple pH lowering was applied to remove the PSA template from the surface. Such approach exhibited 30-fold higher affinity for its target compared to other (glyco)proteins [21].

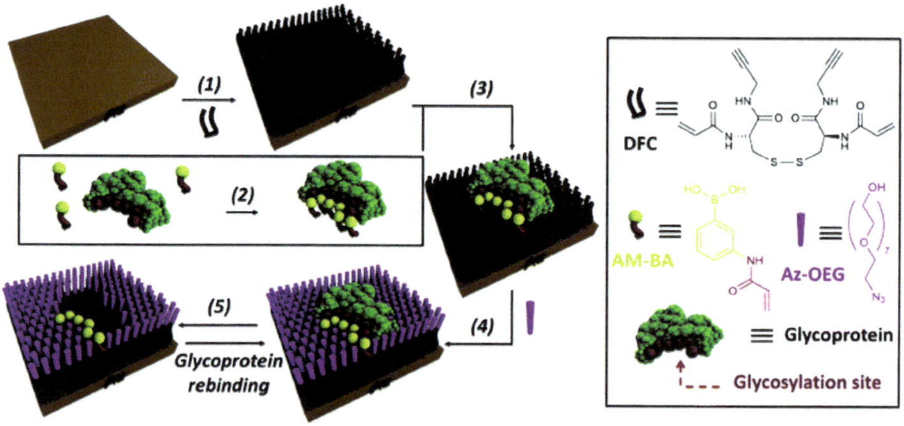

Figure 7.7 Experimental design for the formation of surface-restricted click-imprinted binding sites for glycoproteins. Disulphide dimer (**DFC**) SAMs were prepared by immersing clean gold substrates in 0.1 mM methanolic solutions of **DFC** for 24 h (step 1). In step 2, BA receptor units are introduced via (3-acrylamidophenyl)boronic acid (**AM-BA**) that is incubated for 30 min at an optimised pH (8.5) with a template target glycoprotein. Multiple boronate esters are formed reversibly between the **AM-BAs** and the carbohydrate structures of the glycoprotein template. The pre-assembled glycoprotein–**AM-BA** complex is then grafted on the **DFC** SAM via acrylamide co-polymerisation, affording the creation of spatially arranged sets of BAs on the surface that are specific for the target glycoprotein (step 3). In order to provide complimentary allosteric specificity, a mould or imprint is created around the glycoprotein template at the surface by so-called click chemistry functionalization of the alkynes of the **DFC** on the SAM by reacting azide-terminated heptaethylene glycol (**Az-OEG**) moieties with the terminal alkynes on the **DFC** SAM via a copper-catalysed alkyne–azide cycloaddition (CuCAAC) reaction (step 4). The glycoprotein targets are removed by washing under acidic conditions (step 5). Reproduced from [21] with permission of The Royal Society of Chemistry.

Figure 7.8 shows examples of three different types of material applied in glycan enrichment including mesoporous silica material, magnetic particle covered by a thin SiO_2 shell and magnetic particles attached to graphene.

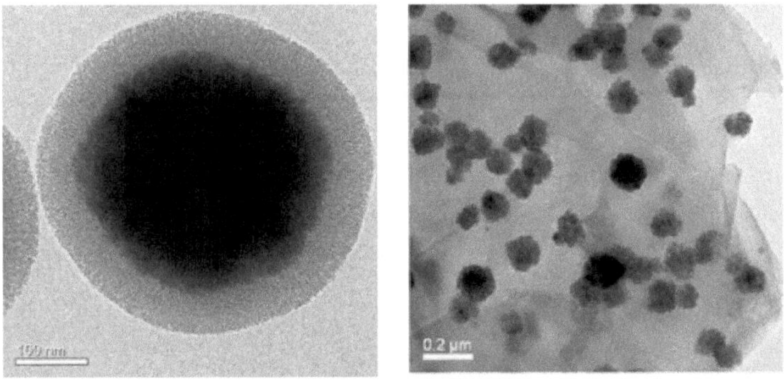

Figure 7.8 Transmission electron microscopy (TEM) image of magnetic particle covered by a thin SiO_2 shell. Reprinted from [24], Copyright 2014, with permission from Elsevier (left image). TEM image of graphene modified by magnetic particles. Reprinted from [25], Copyright 2014, with permission from Elsevier (right image).

7.1.3 Nanoporous Materials

Nanoporous materials are according to IUPAC definition only those materials having the pore size below 1,000 nm [26]. In the scientific literature, however, nanoporous materials are divided into three subcategories according to the pore size: macroporous (above 50 nm), mesoporous (2–50 nm) and microporous (below 2 nm). There are other parameters besides pore size, which are important for nanoporous materials, i.e. pore shapes (cylindrical, slit shaped, funnel shaped, etc.) (Fig. 7.9) and degree of pore interconnections. We can characterise porosity of the material by the total pore volume and a specific surface area. Size and interfacial modification of pores play an important role in interactions with ligands due to a confinement effect [27].

For example, when glycoprotein (ribonuclease B) interacted with a mesoporous material, an enhancement factor (i.e. affinity of interaction with mesoporous silica compared to non-porous silica) increased from 45 to 900, when besides boronate affinity, also electrostatic interactions were involved [28]. Upon decrease of pore size of mesoporous silica from 2.6 nm to 2.1 nm, further increase of an enhancement factor up to 2,100 was observed.

This indicates that for effective ligand capture it is important to carefully design surface chemistry and morphology of porous materials [28]. An interesting material prepared by polymerisation contains mesopores (3.9 nm) and macropores (1.4 μm) to mimic function of protein A to bind antibodies [29]. Boronate modified mesopores are utilised for accommodation of IgG's glycan, while the IgG molecules could not enter the mesopores (Fig. 7.10). IgG binding capacity of the material was comparable to a number of protein A mimics. The monolith was quite stable and cost-effective alternative to use of a protein A with the price of 40 US dollars per 1 g [29].

Figure 7.9 Common types of pore geometries: cylindrical, ink bottle, funnel shaped, pore open on both ends of the material or particle and also branched, and a pore of high tortuosity.

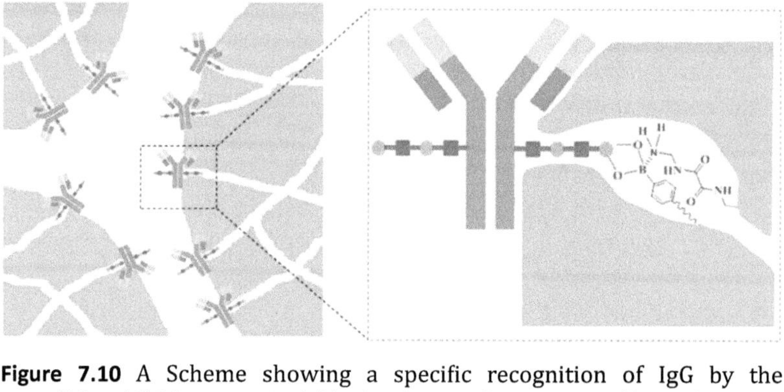

Figure 7.10 A Scheme showing a specific recognition of IgG by the monolith exhibiting protein A-like binding with presence of bimodal pores (macropores and mesopores) modified by boronate functional groups. Reproduced from [29] with permission of The Royal Society of Chemistry.

7.1.4 Active Glycan Enrichment by Microengines

Nanotechnology helped to develop novel devices, which can be applied for active glycoprofiling of various samples. Such devices propelled by different means, when modified by glycan binding agents (boronate functional groups or lectins) can selectively pick, transport and release glycan cargo (or cells) upon demand. The device could be then controlled magnetically and the device was propelled by formation of oxygen bubbles generated by catalytic decomposition of H_2O_2 (a fuel) by platinum (Fig. 7.11). The device selectively picked up *E. coli* cells even in presence of yeast cells, transported them, and release of the cargo was triggered change of pH. The device could travel with speed of 33 μm s^{-1} [30]. The topic was recently reviewed discussing many novel approaches [31].

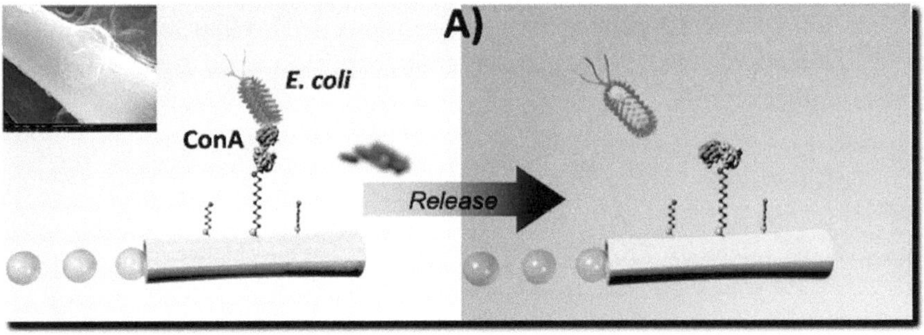

Figure 7.11 Lectin-modified microengines for bacteria isolation. Schemes depicting the selective pick-up, transport, and release of the target bacteria by a lectin-modified microengine. Upon encountering the cells, the lectin-functionalised microengines recognise the *E. coli* cell walls by O-antigen structure binding—allowing for selective pick-up and transport. Inset (in Scheme A, top left side): a SEM image of a portion of a lectin-modified microengine loaded with an *E. coli* cell. Scheme A, right side: Release of the capture bacteria by navigation in a 10 mM glycine solution, pH 2.5. Reprinted with permission from [30]. Copyright 2012 American Chemical Society.

7.1.5 Applications

We can conclude that application of functionalised NPs or mesoporous materials can be applied in a time effective way. While typically glycan enrichment can last from 60 min to 160 min [32],

10 min is sufficient for boronate-modified mesoporous silica material [23]. Using hybrid NPs integrating MNPs glycan enrichment can be completed within 1 min [18], while separation time using an external magnet can be as short as 15 [33]. The highest binding capacity of 882 mg g^{-1} towards glycoproteins was achieved using modified magnetic particles [34].

While using commercially available approaches there is a need to have at least 10 pmol [19] or 40 fmol [35] of glycans as a starting material, for the most sensitive enrichment procedures as low as 100 amol of glycan could be sufficient [17]. High effectivity of NPs-based glycan enrichment also means that ultralow amount of often precious sample is required, i.e. 50 nL of human serum with an application of carbonised mesoporous silica composite [36].

MIPs with controlled thickness of a polymer over imprinted analyte within nanometre scale can offer quite high selectivity for its analyte of 30 over other glycoproteins [21], what is attractive to selectively fish out a particular glycoprotein also from a quite complex sample.

References

1. Palaniappan KK, Bertozzi CR (2016) Chemical glycoproteomics. *Chemical Reviews* 116(23): 14277–14306.
2. Dosekova E, Filip J, Bertok T, Both P, Kasak P, Tkac J (2017) Nanotechnology in glycomics: Applications in diagnostics, therapy, imaging, and separation processes. *Medicinal Research Reviews* 37(3): 514–626.
3. Li D, Chen Y, Liu Z (2015) Boronate affinity materials for separation and molecular recognition: Structure, properties and applications. *Chem Society Reviews* 44(22): 8097–8123.
4. Stine KJ (2015) Glycans in mesoporous and nanoporous meterials. In: Stine KJ (ed) *Carbohydrate Nanotechnology*. John Wiley & Sons, Inc, pp. 233–266.
5. Wu X, Li Z, Chen X-X, Fossey JS, James TD, Jiang Y-B (2013) Selective sensing of saccharides using simple boronic acids and their aggregates. *Chemical Society Reviews* 42(20): 8032–8048.
6. Huang B-Y, Yang C-K, Liu C-P, Liu C-Y (2014) Stationary phases for the enrichment of glycoproteins and glycopeptides. *Electrophoresis* 35(15): 2091–2107.

7. Zhao H, Li Y, Hu Y (2014) Nanotechnologies in glycoproteomics. *Clinical Proteomics* 11(1): 21.

8. Wu Y, Wu Y, Tefsen B, Shi Y, Gao GF (2014) Bat-derived influenza-like viruses H17N10 and H18N11. *Trends in Microbiology* 22(4): 183–191.

9. Hushegyi A, Klukova L, Bertok T, Tkac J (2015) Carbohydrate nanotechnology and its application to biosensor development. In: Stine KJ (ed) *Carbohydrate Nanotechnology*. John Wiley & Sons, Inc, pp. 387–421. doi:10.1002/9781118860212.ch15.

10. Huang G, Xiong Z, Qin H, Zhu J, Sun Z, Zhang Y, Peng X, ou J, Zou H (2014) Synthesis of zwitterionic polymer brushes hybrid silica nanoparticles via controlled polymerization for highly efficient enrichment of glycopeptides. *Analytica Chimica Acta* 809: 61–68.

11. Alwael H, Connolly D, Clarke P, Thompson R, Twamley B, O'Connor B, Paull B (2011) Pipette-tip selective extraction of glycoproteins with lectin modified gold nano-particles on a polymer monolithic phase. *Analyst* 136(12): 2619–2628.

12. Tran T, Park S, Lee H, Park S, Kim B, Kim O-H, Oh B-C, Lee D, Lee H (2012) Ultrasmall gold nanoparticles for highly specific isolation/enrichment of N-linked glycosylated peptides. *Analyst* 137(4): 991–998.

13. Novoselov K, Geim A, Morozov S, Jiang D, Zhang Y, Dubonos S, Grigorieva I, Firsov A (2004) Electric field effect in atomically thin carbon films. *Science* 306(5696): 666–669.

14. Zhang WJ, Han HH, Bai HH, Tong W, Zhang YJ, Ying WT, Qin WJ, Qian XH (2013) A highly efficient and visualized method for glycan enrichment by self-assembling pyrene derivative functionalized free graphene oxide. *Analytical Chemistry* 85(5): 2703–2709.

15. Reuel NF, Grassbaugh B, Kruss S, Mundy JZ, Opel C, Ogunniyi AO, Egodage K, Wahl R, Helk B, Zhang JQ, Kalcioglu ZI, Tvrdy K, Bellisario DO, Mu B, Blake SS, Van Vliet KJ, Love JC, Wittrup KD, Strano MS (2013) Emergent properties of nanosensor arrays: Applications for monitoring igg affinity distributions, weakly affined hypermannosylation, and colony selection for biomanufacturing. *Acs Nano* 7(9): 7472–7482.

16. Wu C-L, Wang C-C, Lai Y-H, Lee H, Lin J-D, Lee YT, Wang Y-S (2013) Selective enhancement of carbohydrate ion abundances by diamond nanoparticles for mass spectrometric analysis. *Analytical Chemistry* 85(8): 3836–3841.

17. Chen YJ, Xiong ZC, Zhang LY, Zhao JY, Zhang QQ, Peng L, Zhang WB, Ye ML, Zou HF (2015) Facile synthesis of zwitterionic polymer-coated

core-shell magnetic nanoparticles for highly specific capture of N-linked glycopeptides. *Nanoscale* 7(7): 3100–3108.

18. Wang H, Bie Z, Lu C, Liu Z (2013) Magnetic nanoparticles with dendrimer-assisted boronate avidity for the selective enrichment of trace glycoproteins. *Chemical Science* 4(11): 4298–4303.
19. Ma RN, Hu JJ, Cai ZW, Ju HX (2014) Facile synthesis of boronic acid-functionalized magnetic carbon nanotubes for highly specific enrichment of glycopeptides. *Nanoscale* 6(6): 3150–3156.
20. Wang JX, Wang YN, Gao MX, Zhang XM, Yang PY (2015) Multi layer hydrophilic poly(phenol-formaldehyde resin)-coated magnetic graphene for boronic acid immobilization as a novel matrix for glycoproteome analysis. *Acs Applied Materials & Interfaces* 7(29): 16011–16017.
21. Stephenson-Brown A, Acton AL, Preece JA, Fossey JS, Mendes PM (2015) Selective glycoprotein detection through covalent templating and allosteric click-imprinting. *Chemical Science* 6(9): 5114–5119.
22. Wu G, Li JY, Qu X, Zhang YX, Hong H, Liu CS (2015) Template size matched film thickness for effectively in situ surface imprinting: A model study of glycoprotein imprints. *RSC Advances* 5(58): 47010–47021.
23. Yan Y, Deng C, Zheng Z, Zhang X, Yang P (2014) Synthesis of polyboronic acid functionalized hierarchically ordered macro-/mesoporous silica for selective enrichment of glycopeptides for mass spectrometric analysis. *ChemPlusChem* 79(1): 31–34.
24. Zhao M, Deng C (2016) Fluorous modified magnetic mesoporous silica composites incorporated fluorous solid-phase extraction for the specific enrichment of N-linked glycans with simultaneous exclusion of proteins. *Talanta* 159: 111–116.
25. Wang J, Wang Y, Gao M, Zhang X, Yang P (2016) Versatile metal-organic framework-functionalized magnetic graphene nanoporous composites: As deft matrix for high-effective extraction and purification of the N-linked glycans. *Analytica Chimica Acta* 932: 41–48.
26. Rouquerol J, Avnir D, Fairbridge CW, Everett DH, Haynes JM, Pernicone N, Ramsay JDF, Sing KSW, Unger KK (1994) Recommendations for the characterization of porous solids (Technical Report). *Pure and Applied Chemistry*, vol 66. doi:10.1351/pac199466081739.
27. Li QJ, Tu XY, Ye J, Bie ZJ, Bi XD, Liu Z (2014) Nanoconfining affinity materials for pH-mediated protein capture-release. *Chemical Science* 5(10): 4065–4069.

28. Chen Y, Wang SS, Ye J, Li DJ, Liu Z, Wu XC (2014) Insights into the effect of nanoconfinement on molecular interactions. *Nanoscale* 6(16): 9563–9567.

29. Liu Y, Lu Y, Liu Z (2012) Restricted access boronate affinity porous monolith as a protein A mimetic for the specific capture of immunoglobulin G. *Chemical Science* 3(5): 1467–1471.

30. Campuzano S, Orozco J, Kagan D, Guix M, Gao W, Sattayasamitsathit S, Claussen JC, Merkoci A, Wang J (2012) Bacterial isolation by lectin-modified microengines. *Nano Letters* 12(1): 396–401.

31. Katuri J, Ma X, Stanton MM, Sanchez S (2017) Designing micro- and nanoswimmers for specific applications. *Accounts Chemical Research* 50(1): 2–11.

32. Kolarich D, Jensen PH, Altmann F, Packer NH (2012) Determination of site-specific glycan heterogeneity on glycoproteins. *Nature Protocols* 7(7): 1285–1298.

33. Dong LP, Feng S, Li SS, Song PP, Wang JD (2015) Preparation of concanavalin a-chelating magnetic nanoparticles for selective enrichment of glycoproteins. *Analytical Chemistry* 87(13): 6849–6853.

34. Zhang XH, He XW, Chen LX, Zhang YH (2014) A combination of distillation-precipitation polymerization and click chemistry: Fabrication of boronic acid functionalized Fe_3O_4 hybrid composites for enrichment of glycoproteins. *Journal of Materials Chemistry B* 2(21): 3254–3262.

35. Jin T, Xiong ZC, Zhu X, Mehio N, Chen YJ, Hu J, Zhang WB, Zou HF, Liu HL, Dai S (2015) Template-free synthesis of mesoporous polymers for highly selective enrichment of glycopeptides. *Acs Macro Letters* 4(5): 570–574.

36. Qin H, Hu Z, Wang F, Zhang Y, Zhao L, Xu G, Wu Ra, Zou H (2013) Facile preparation of ordered mesoporous silica-carbon composite nanoparticles for glycan enrichment. *Chemical Communications* 49(45): 5162–5164.

Chapter 8

Glyconanobiotechnology for Medical Applications

Jaroslav Filip

Department of Environment Protection Engineering, Tomas Bata University, Vavreckova 275, Zlin, 76272, Czech Republic

jfilip@utb.cz

For some two decades, a solid scientific base for recent impressive advance in nanotechnology has been developing. Parallel with that, investigation of roles of glycans in living organisms has brought intriguing news. Therefore, medical applications of glyconanotechnology have been anticipated and many scientific papers reported on glycan-based nanotherapeutics with great effect towards cancer as well as on precise imaging of malignant tissues. These reports assume that glyconanotechnology approach in medicine is extremely potent and promising.

Glyconanotechnology: Nanoscale Approach for Novel Glycan Analysis and Their Medical Use
Edited by Tomáš Bertók

ISBN 978-981-4800-16-7 (Hardcover), 978-0-429-44849-2 (eBook)
www.jennystanford.com

This chapter will provide a concise introduction to glyconanobiotechnology applications in medicine. The reader will learn about glyco-nano-vaccines, basic nano-therapeutic principles based on glycomics and imaging methods using nanotechnology and glycomics.

The chapter will end with a section on smart nano-drugs integrating diagnostic and therapeutic functions as the hallmark of the current research in this extremely intriguing field with strong promises for future development.

8.1 Introduction

There is no doubt that medicine is a field anticipating commercial-scale employment of nanotechnology, though with equally high hopes and concerns. Amazing progress has been already achieved in nanotechnology applications for DNA sequencing, in diagnosis and combination of immunology and nanotechnology, to name the highlights. These successes have fueled a further development of nanovaccines and novel therapeutic approaches, for example, immunotherapy of cancer. Furthermore, a deeper insight into glycomics has been achieved, quite separately from nanotechnology in the beginning, but recent years have evidenced their merging into extremely interesting field of glyconanotechnology applied in medical diagnosis, imaging and therapy. Glycans—sugars on cell or protein surface—were initially considered as molecules without any significant function, but their necessity for inter- and intra-cellular communication started to be recognized during 1980s. It is known now that the enormous diversity of glycans constitutes extremely complex communication and "executive" frame tool on molecular level. Some of the basic glycan functions in a single cell are outlined in Fig. 8.1.

Information in "glycocommunication" is delivered by recognition and binding event between a ligand (glycan) and a specific receptor. This fact has been employed in four main areas: (i) glycan-based antigens (glycan nanovaccines); (ii) smart drugs; (iii) cell imaging and (iv) other treatments, e.g., blocking of glycans receptors responsible for entering viruses or bacteria into the body. These areas will be discussed in dedicated sections to introduce what glyconanotechnology offers in medicine.

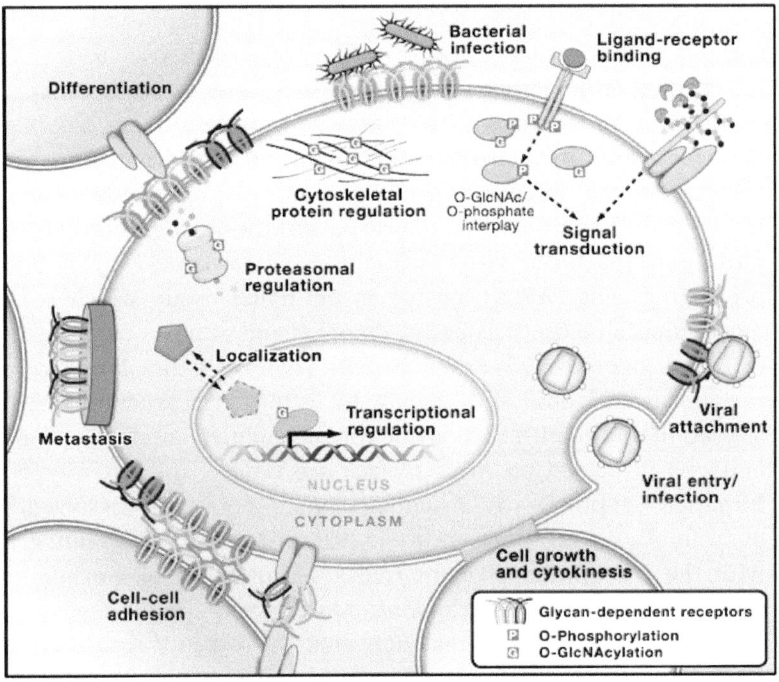

Figure 8.1 Depiction of different interactions provided by glycans. Adapted from Hart and Copland, 2010.

For preparation of novel therapeutic and/or diagnostic glyconanoparticles, a large pool of routine nanotechnology techniques is now available. This knowledge base is, nevertheless, equally useful for discovering new glycan-based markers and obtaining deeper insight into glycomics necessary for further glyconanotechnology development.

8.2 Glycan Nanovaccines

To describe glycan nanovaccines, one should start with the introduction of vaccines in general, nanovaccines and, finally, glycans employable in this area.

8.2.1 Vaccines

The principle of vaccination has been discovered long before any idea about functioning of immune system has appeared. At the end of 18th century Edward Jenner observed that exposition to cowpox (not dangerous) prevented later development of (more deadly) smallpox. Now, human **immune system** has been described and its main functions and two major components recognized—an information/recognition and an executive one. In so-called **cellular** immunoresponse the recognition of the **pathogen** (bacteria, virus, parasite...) is performed by antigen presenting cells (**APCs**) located in epithelial tissue which is the most vulnerable towards pathogen enter and attack. Consequently, in blood, lymphocytes called **B-cells** (type of white blood cells) recognize antigens and signal molecules released by APCs' activation and maturation. While for the cellular immunoresponse antigens displayed on APCs surface are typical, the latter type—**humoral response**—is based on signal molecules dissolved in body fluids. As can be seen in Fig. 8.2, encounter of immunogens with the recognition element releases a plethora of biochemical signal molecules (interleukins, interferons and—the most important—**antibodies**) that activates the executive part of the immune system. It is important to note that the intracellular communication occurring during immunoresponse relies on selective "fitting" of signaling molecules into specific receptor on a cell surface.

Intriguing on immune system is that the same "signal" chemicals (i.e., interferons and interleukins) are responsible also for attenuation of the immune response after the pathogens are eliminated. Unattenuated immune response is a serious problem for living organism—it causes allergies and **autoimmune disorders** like rheumatoid arthritis and diabetes.

To understand the effect of vaccination, it should be added that after the immune response attenuation to "stand-by" level, most immune cells undergo apoptosis and are "cleaned out"—not all of them, actually. The small surviving portion still bears information about the pathogen (more precisely about the pathogen's antigen) and is ready to set on an immune reaction in case of repeated attack of the same pathogen. This is called

acquired (or adapted) immunity and this is the fundament of vaccination principle.

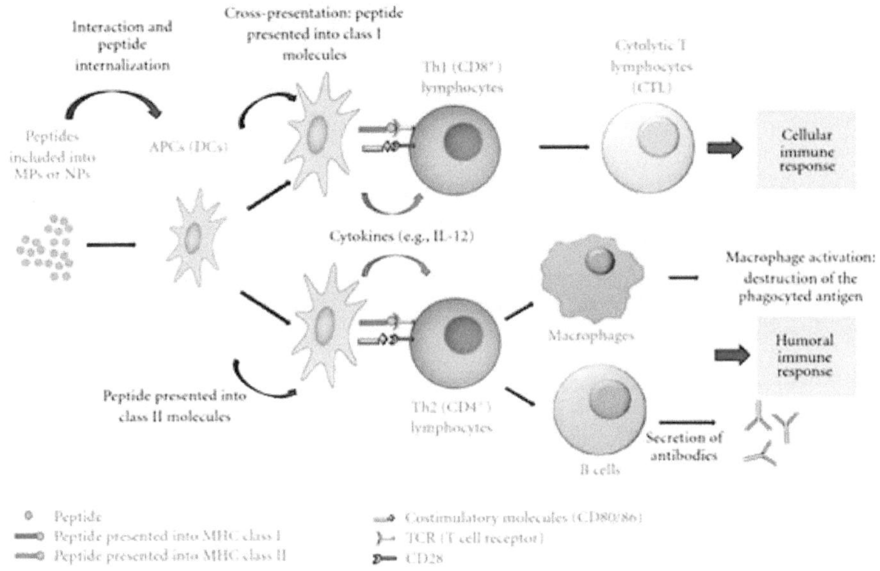

Figure 8.2 Schematic illustration of immunoresponses: macrophages internalize the pathogen, digest it, and display its antigens on their surface. Activated macrophages activate Helper T-cells via specific bonding between MHC/presented antigen complex and T-cell receptors. Further actions of humoral immunoresponse (B-cell activation) and cell-induced immunoresponse (T-cell maturation into cytotoxic cells) are described in the figure. Creative Commons Attribution License.

8.2.2 Nanovaccines

The antigen is the most important—"active"—part of any vaccine and its type determines the type of consequent immune reaction. First vaccines containing dead or inactive microbial cells or virus particles, both covered with variety of complex antigens, eliciting complex immune reaction. On the other side, this complexity means risk especially for persons with less developed or deficient immunity. To avoid this, vaccines containing only a selected immunogenic part of the pathogen were developed, for example, inactivated bacterial toxins called toxoids (e.g., toxin from *Corynebacterium diphtheria* treated with formalin (Glenny and

Hopkins, 1923). The research went on, bringing, in the 1980s and later, development of conjugated and subcellular[1] vaccines and, with a huge help from nanotechnology and without a strict scission from a field of conjugation vaccines, **nanovaccines**.

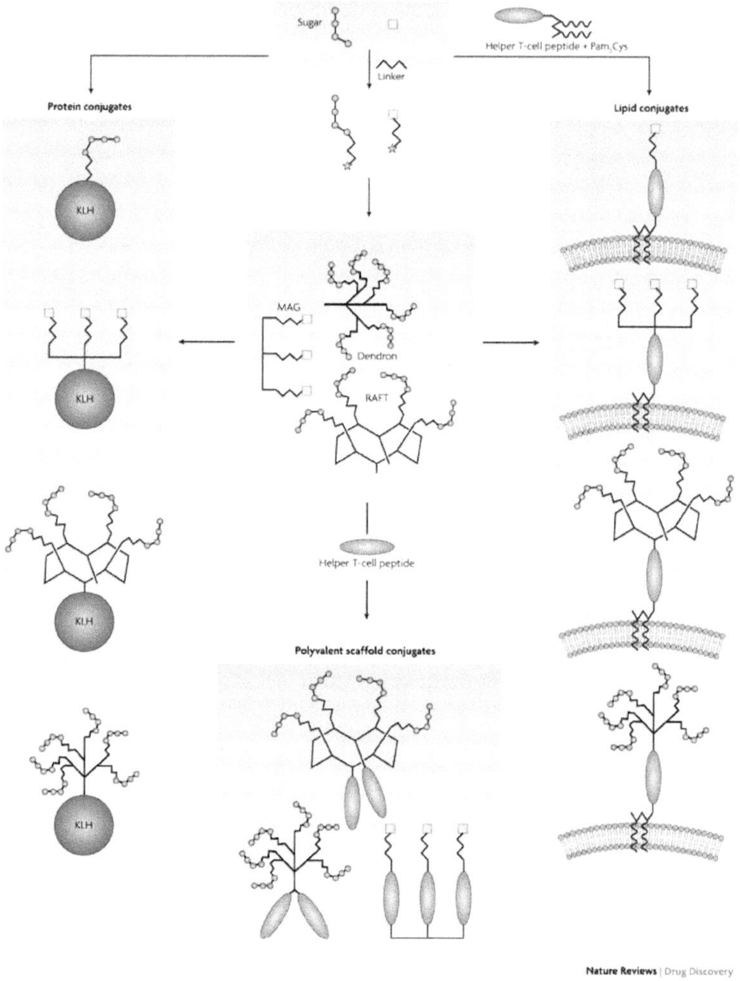

Figure 8.3 Different types of nanovaccines prepared by conjugation of glycans to carrier nanoparticle via a linker. Adapted from Astronomo et al., 2010.

[1]The term "subcellular vaccines" refers to vaccines prepared from isolated parts of pathogenic cells while "conjugated vaccines" denotes complex of subcellular immunogen with an adjuvant and/or carrier particle.

The latter consist of a desired antigen integrated with inorganic nanoparticles and/or auxiliary moieties. The antigens can be synthesized de novo, instead of being isolated from pathogens. These achievements allow nanovaccines to be prepared more precisely and with higher reproducibility, in comparison to conventional vaccines. It also means that the immune response elicited by nanovaccines is more predictable and well characterized, but it is usually weaker compared to conventional vaccines because typical bacterial cell or virus particle contain more than one immunogen on their surface. The weak response is actually one of the toughest tasks to be solved in nanovaccine development. One of the most common ways to tackle this issue is to conjugate less active antigen to an adjuvant—a molecule or a particle with better immunogenic activity. In many conjugated vaccines the carrier particle plays also a role of an adjuvant.

The typical example of carrier adjuvant is tetanus toxoid protein conjugated with bacterial lipopolysaccharide or keyhole limpet hemocyanin (KLH). From the scheme in Fig. 8.3 it can be seen that nanotechnology allow us to synthesize nanovaccine particles with a high precision reaching, for example, exact number of immunogenic units per one nanoparticle and their exact location.

8.2.3 Glycans Employed in Nanovaccines

A decision about which antigen should be used is a key question in nanovaccine preparation. Traditionally, protein-based antigens were employed consisting of protein parts with peptide moieties called epitopes.[2] Isolated epitopes were also employed, as well as molecular constructs prepared from nucleic acids. In recent years, however, great potential of glycan-based immunogens have been discovered, even thought it has been known for quite a long time that some bacterial ligands responsible for immunoresponse contain oligo- or polysaccharides. For example, dendritic cells and other leucocytes are abundantly covered with danger-recognition and signaling receptors of "TLR" family, some of them being selective towards bacterial glycoproteins or glycopeptides. Furthermore, the activity of many proteins, including immunoglobulins, is

[2]Each protein antigen bears several epitopes, which is the reason for complex immune response triggered by this kind of immunogens.

modified by changes in glycosylation pattern. More information about glycan-based receptors and ligands and their roles in infection development and treatment can be found, e.g., in recent review by Baum and Cobb (2017).

Pristine glycans, although necessary for the recognition, scarcely show any significant immunogenicity when detached from their protein or lipid backbone, they have to be conjugated with adjuvant protein. An example is a vaccine against *Burkholderia mallei* (bacteria causing glanders) consisting of gold nanoparticles coated with adjuvant protein and lipopolysaccharide molecules from non-virulent strain *B. thailandesis*. This nanovaccine has efficiently protected non-human primates against glanders and it is reasonable to anticipate the same effect also for people. Immunoresponse was also elicited by gold nanoparticles conjugated with glycan units mimicking capsular polysaccharides of *Neisseria meningitidis*—bacteria responsible for life-threatening meningitis—and by specific glycans conjugated to diphtheria toxoid resulting in antistreptococcus vaccine.[3]

Besides the known fact that glycans modulate "natural" immunoresponse there are also studies suggesting that immunogenic effect can be induced also by glycans conjugated only with inorganic carrier nanoparticles. Their immunoresponse, however, strongly depends on an exact number of glycan units and their exact placement on the surface of the nanoparticle. The request on nanotechnology skills needed for glycan nanovaccine development can be illustrated by a slight change in glycosylation that determines the species specificity of influenza. The particles are covered with hemagglutinin—proteins with specific affinity to sialic acid molecules. While avian influenza is capable to bind to glycans containing α2,3-linked sialic acid, the flu infecting mammals recognizes only α2,6-linked sialic acid. The difference in the structure of both glycans is shown in Fig. 8.4.

It is also known that polysaccharides (e.g., hyaluronic acid) are more immunogenic than their monomeric or oligomeric counterparts. Bearing this in mind, it is no surprise that multivalent glycan units displayed on inorganic nanoparticles imitate, to

[3]Note that these results are only experimental; it is just a first step of a several-year-long journey to possible applications of the nanovaccines in conventional human medicine.

some level, natural polysaccharides which makes some glycan nanovaccines working even without adjuvant proteins.

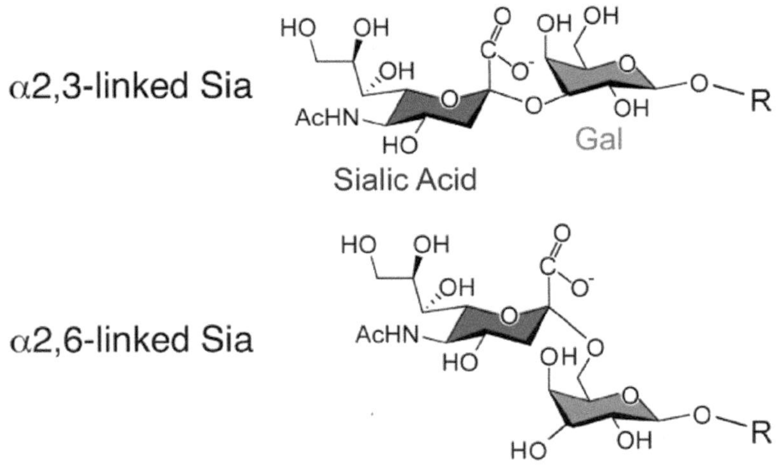

Figure 8.4 α2,3-linked (above) and a 2,6-linked sialic acid. Reprinted from Byrd-Leotis et al., 2017. Creative Commons Attribution License.

8.2.4 Immunotherapy

Even though antibacterial vaccines are very important, glycan-based antigens are more intensively investigated for anticancer therapy. Until recently, it was thought that immune system is incapable to destroy cancer cells. Even though they are covered by numerous receptors and ligands, they are not recognized as pathogenic; therefore cancer cells are safe from depletion by immune system. But progress in biochemistry, molecular biology, and nanotechnology resulted in development of nanovaccines capable to elicit immunoreaction also against cancer cells. It was achieved especially by close investigation of potential cancer cells antigens called tumor-associated (TACA) or Tumor-specific carbohydrate antigens (TSCA). While the former are aberrant variations of carbohydrates displayed on health cells, the latter are more specific to cancer cells. This topic was targeted, for example, in a recent review by Soliman et al. (2017) where known TACA antigens have been described along with antibodies that binds to them. For example, "Thomsen–Friedenreich antigen" (TF; see Fig. 8.5) and "Thomsen nouvelle monosaccharide" (Tn) antigen are known products of cancer-induced aberration of surface glycans, as well as "Lewis system" antigens (denoted as Le^A, Le^B, Le^X, and Le^Y regarding to their exact composition, see Fig. 8.6), Blood group ABO(H) and surface glycosphingolipids with sialic acid ending denoted to as GD2.

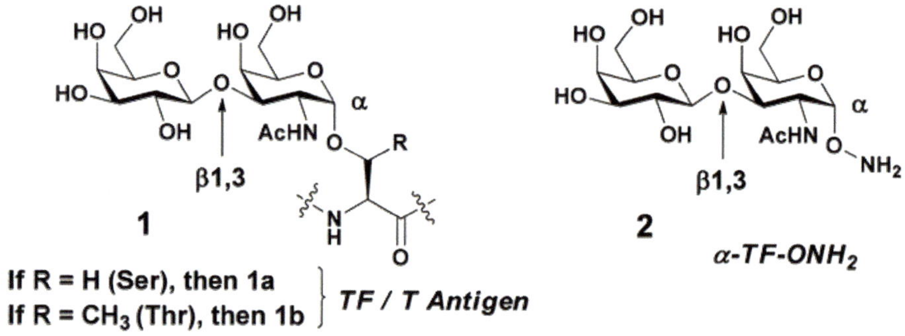

Figure 8.5 Structure of TF antigen. Reprinted from Bourgault et al., 2017.

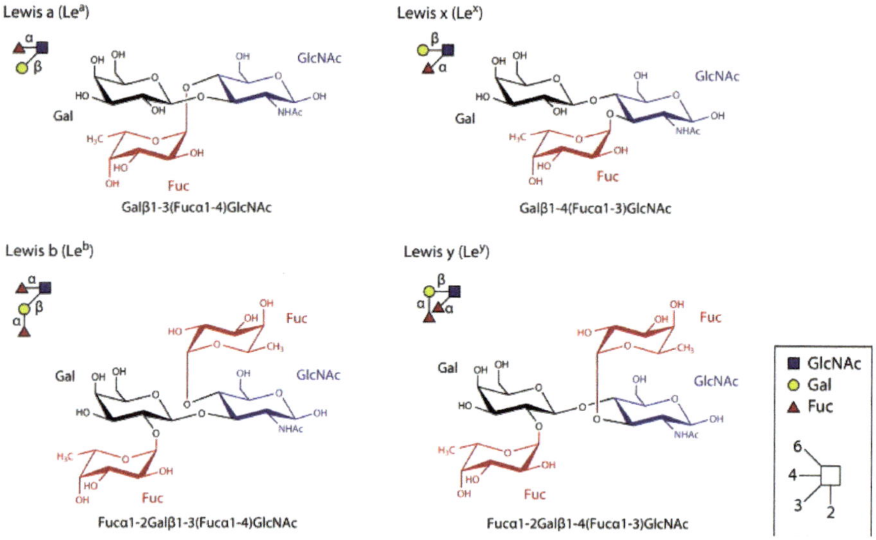

Figure 8.6 Structures of Lewis-type antigens. Adapted from Dingjan, et al., 2015.

Due to the low immunogenicity of cancer cell antigens, efficient anticancer immunotherapeutic drugs and vaccines are more difficult to prepare. However, it was found that conjugation of tumor-associated antigens with adjuvant proteins (e.g., KLH) could induce production of the desired antibodies. Such approach was tested for different antigens, first in "one-to-one" molecular ratio and, later, multivalent and clustered vaccines were prepared where one adjuvant protein molecule born as much as five different carbohydrate antigens. Efficiency of these conjugated vaccines was confirmed during the most preclinical and even phase II clinical trials. Unfortunately, in 2013, phase III clinical trial of conjugate vaccine consisting of KLH protein, GD2 antigen and synthetic QS-21 adjuvant failed[4] to prove any benefit for stage II melanoma patients. Recently, there are only few drugs based on antibodies approved for cancer treatment, for example, alemtuzumab (trade name Campath®)—a "naked" monoclonal antibody targeting specific glycoproteins overexpressed on leukemic cells surface which makes them vulnerable to destruction by the immune system.

Regardless of a seemingly insignificant outcome of the cancer immunotherapy methods, glycan-based or -containing cancer immunogens mentioned in this section are a subject of ever more intensive research and many clinical trials are currently under way. It should be also noted that to sale any novel therapeutic is always preceded by about a decade of experiments and trials.

Much more work has to be done in this field, for example, to employ the fact that glycans are known to modulate immune cells also indirectly, which will help to prepare more efficient anticancer vaccines.

8.3 Other Glycan Nanotherapeutics

Apart from vaccination, glycans can be also applied in following types of treatments: (i) "saturation" of glycan receptors on cells surfaces to block pathogen adhesion; (ii) restoring functions of impaired enzymes; (iii) specific inactivation of enzymes.

[4]This conjugate vaccine has been developed as soon as in mid 1990s while the phase III trial was finished in 2013! This is a particular and very significant reason why there is still not commercially available pool of anticancer immunotherapeutic drugs.

8.3.1 Treatment Effect of Competitive Bonding

As was stated earlier, every microbial or viral infection starts with a recognition and binding of pathogen by specific cell receptors which will not occur once these receptors are blocked. Since most pathogen ligands in question contain glycans as the recognition unit, the therapeutics employing such blocking effect will be nanoparticles bearing specific glycan units on their surface and designed so as to attach to the desired receptors with high specificity and high efficiency. These particles can be taken as analogues of pathogen particles with maximized amount of recognition-responsible glycans on their surface.

Efficiency of these nanoparticles is often correlated to valency and structural arrangement of the surface-displayed glycan units. In general, these constructs consist of a carrier nanoparticle and glycan moieties often separated by spacers with selected length.

For carrier nanoparticles different materials can be used. Metal nanoparticles (Ag, Au, ...) can be prepared with high reproducibility and in desired range of properties, which is crucial for any kind of potential nanodrug. Furthermore, they can be easily modified by thiol chemistry—molecules with the anchoring SH-group on one end possess the other end accessible to further modification. In a more simple approach, Ag and Au nanoparticles were conjugated with synthetic derivative of natural polysaccharide chitosan forming antimicrobial agents towards *E. coli*[5] due to their partial negative electric charge. Other sulfated polysaccharides (carrageenans, chondroitin sulfate, heparan sulfate, and dextran sulfate, to name just a few) are important compounds produced by all types of cells (i.e., plant, bacterial and mammal cells) and their applications in glyco-nano technology have been reviewed, for example, by Raveendran et al. (2013).

Carbon-based molecules can be modified with molecules bearing pyrene moiety on one side. Though carbon nanoparticles are known for their low reproducibility, it can be improved when giving up their low price. Of all carbon nanomaterials fullerenes—

[5]Gold and silver nanoparticles themselves exhibit intriguingly complex bactericidal effect by triggering non-specific oxidation stress, production of oxygen radicals and binding to sulfuric protein side chains. This complexity and low specificity prevent from development of any resistance. Due to such effect many researchers are talking about a new class of antibiotics (nanoantibiotics).

stable spherical clusters of typically less than 100 highly ordered carbon atoms discovered in 1980s—provide probably the highest reproducibility of preparation and properties. Further, methods have been developed for their chemical modification, including synthesis of "glycoballs"—fullerene particles covered with glycans (see Fig. 8.7). For example, "fucofullerenes" have attached 36 fucose units on their surface via spacers could block cell lectins (receptors) responsible for recognition of pathogens including Ebola virus. It is agreed that high efficiency of lectin blocking is caused mainly by multivalency effect—glycofullerenes can bear dozens of repeating glycan units on their surface which dramatically increases their chance to bind to the desired lectin. Further increase of glycan valency was achieved by the synthesis of fullerenes modified with other glycofullerenes.

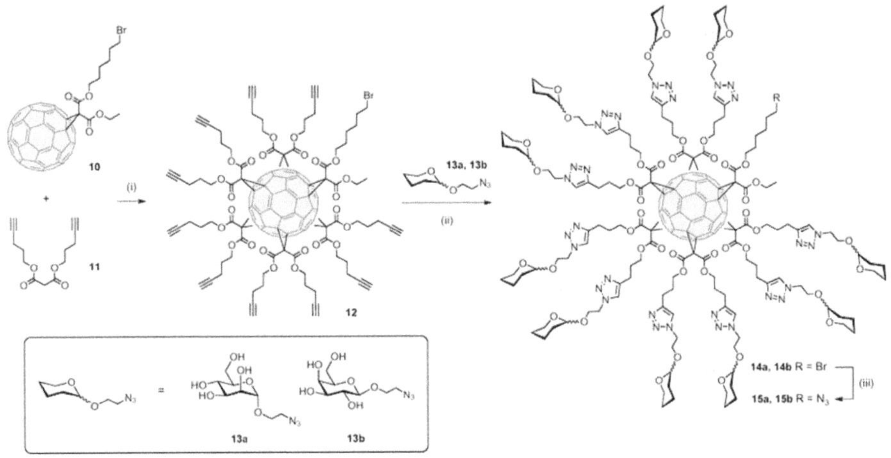

Figure 8.7 Synthesis and structure of fucofullerene. Adapted from Muñoz 2015.

Nevertheless, the most promising materials in this field are polymers. Both synthetic and natural (e.g., proteins) macromolecules have been conjugated with appropriate glycans and tested towards their capability to block recognition of pathogens. Similarly, also supramolecular constructs—dendrimers—have been employed. These fractal-like large molecular scaffolds can be prepared with striking precision and very high reproducibility. Due to their versatility they represent a perfect tool for investigation of influence of glycan valency and structural arrangement. Poly(amidoamine) (PAMAM) dendrimers can be also synthesized with high precision and reproducibility. These molecular constructs with globular shape of several nm in diameter and many accessible moieties allow for precise surface modification (i.e., binding of glycan units). It was employed, for example, in fabrication of 6-sialic acid-PAMAM dendrimer supramolecular hybrid inhibiting efficiently adhesion of influenza virus in the same manner as glycofullerenes described above. Since size is the matter here, it should be noted that glycofullerenes and dendrimers are typically several nanometers in diameter.

8.3.2 Pharmacochaperons and Protein Inhibitors

Another glycans' feature employable for treatment is modulation of activity of some enzymes. This function was investigated especially in context of lysosomal storage diseases (LSDs) caused by genetically induced malfunction of enzymes responsible for hydrolysis of "waste" molecules (e.g., glycoproteins, lipids). The well-known example of LSD is Gaucher disease caused by impaired cerebral β-glucocerebrosidase. Iminoglycan derivatives termed nojirimycin are known to efficiently bond to the active site of the impaired enzyme forming enzyme-glycan complex with recovered activity. Conjugates of nojirimycin with nanoparticles were found to partially enhance the activity of glucocerebrosidases in vitro but it was not accompanied with the expected significant clinical effect. Notably, multivalency of iminoglycan units also plays role here, but in a different manner as compared to glycan/lectin interactions.

Quite recent studies suggested that biomolecules called "heparan sulfate proteoglycans" play significant role in development of neurodegenerative diseases, e.g., in protection of aggregated proteins. The protective conjugation between β-amyloid proteins and heparan sulfate in proteoglycans can be cleaved by certain glucuronidases (cf. treatment of Gaucher disease) namely heparanase, which promises certain therapeutic effect of this enzyme. Regardless of the recent lack of knowledge, it can be anticipated that some nanotherapeutics will be developed that will modulate the heparan sulfate proteoglycans action.

8.4 Carbohydrates and Targeting and Therapies

"Smart drugs" idea emerged many decades ago but has been brought to life only quite recently. Unlike their less smart relatives, smart drugs are capable of a targeted transportation of a pharmacologic substance into a selected tissue/organ which significantly reduces potential adverse side effects. The targeted nanoparticles can be loaded also with some markers allowing us to perform precise diagnostics of, e.g., tissue damage or disease progress. Similarly, this technique can be used for imaging of the selected cells, for example, microscopic tumor metastases which are difficult to detect in conventional ways.

8.4.1 Targeted Delivery of Chemicals

While in the beginning of the 20th century, a German doctor (and 1908 Nobel Prize recipient) Paul Ehrlich formulated an idea of targeted medicine ("magic bullet"),[6] its second half witnessed publications reporting on encapsulation and a controlled release of different compounds from liposomes and polymeric delivery systems developed, among others, by Prof. Robert Langer. The work of the latter was recognized worldwide, reaching more than hundreds of thousands citations and rewards, including Queen Elizabeth Prize.

Even though polymeric nanoparticles, liposomes, and similar structures have been known since 1950s–1960s, only in the early 1980s first reports on conjugation of anticancer drugs with polymers appeared and as soon as in the 1990s first commercial nanodrugs based on this principle reached markets. Besides simple conjugates of polymer nanoparticles with drugs, cytostatics were also encapsulated into liposomes. Later, reports on targeted accumulation of specially designed nanoparticles loaded with the drugs have appeared (see Fig. 8.8) and, in 2013, the Nobel Prize in physiology and medicine was awarded to Randy Schekman, James Rothman, and Thomas C. Südhof for their discovery of molecular machinery behind cellular transportation system via natural nanoparticles—vesicles.

---

[6]Inspiration is anecdotally believed to come from von Veber's opera about magic archer, but tissue staining and specific affinity of different tissues was Ehrilch's whole-life scientific passion and surely had not emerged after one opera watching.

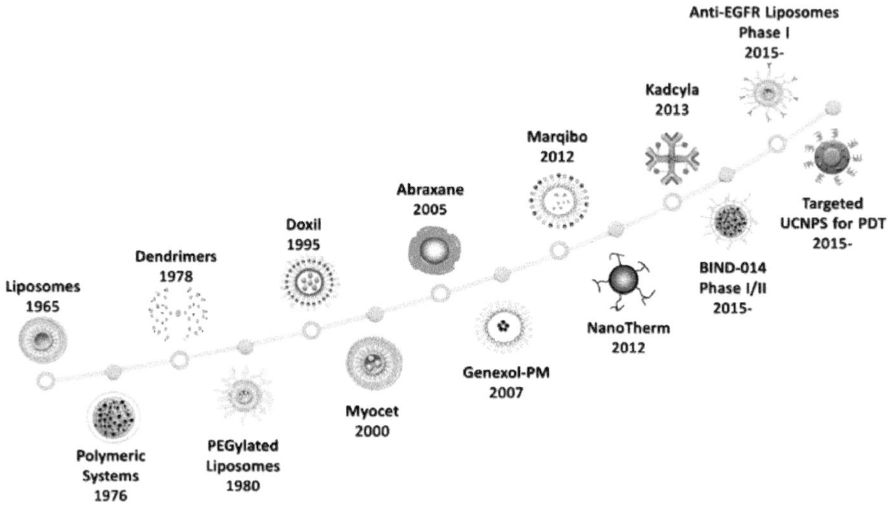

Figure 8.8 Timeline of development of drug nanocarrier systems and commercially available formulations. Creative Commons Attribution License.

For targeted drug delivery two main options are recognized—either passive or active targeting. The following sections will introduce both of them.

8.4.1.1 Carbohydrates in passive targeting

Passive targeting employs higher permeability of less dense endothelium of vessels in tumors. It is caused by tumor's rapid development and allows smaller particles to leak from veins to the tumor cells more readily than into a healthy tissue. This preference is called the enhanced permeability and retention (EPR) effect. While "leaky" veins allow nanoparticles to get closer to the tumor cells, deficient lymphatic drainage (another feature typical for tumor tissues) is responsible for the enhanced retention of the nanoparticles (see Fig. 8.9).

Even though this kind of targeting is passive, it can be modulated, e.g., by nanoparticle size adjusting. It was found that particles as large as 400 nm are capable of leakage from the tumor veins, but much efficient leakage occurs when the size drops below 200 nm.

It is also important to make such passively targeting nanoparticles "invisible" to prevent them from possible recognition by macrophages and consequent releasing from the body without any observable effect. Polysaccharides hyaluronic acid, dextrans, and pullulans could prolong retention of modified liposomes in the body, which is the crucial parameter for nanoparticle accumulation in tumor and release their cytotoxic cargo here (see Fig. 8.10).

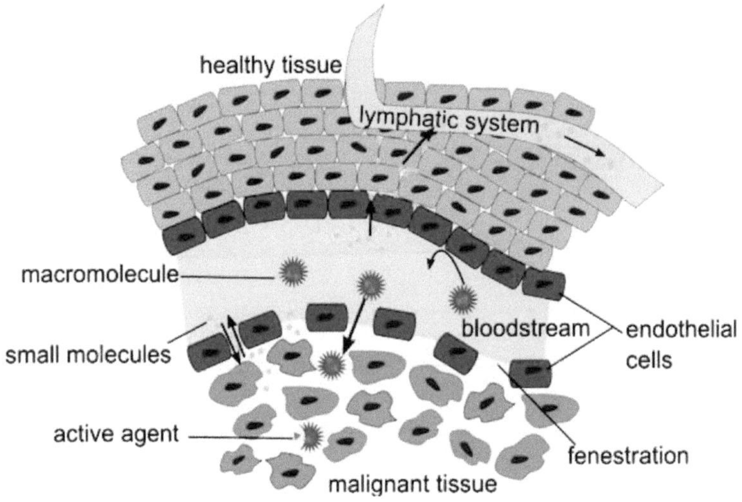

Figure 8.9 Schematic illustration of the "EPR" effect. Creative Commons Attribution License.

Another crucial parameter is that the payload must be released not before reaching the targeted tissue where, ideally, chemical or physical properties differ from the rest of the organism. It triggers degradation of the nanocarrier or disjunction of the drug molecules from the carrier. Polysaccharide nanocarriers, for example, degrade in the organism slowly and when the degradation rate is exactly tuned and related with the retention time, quasi-selective drug release can be achieved.

Even though the passive targeting suffers from some shortcomings (especially the hot topic is how to improve intratumoral transportation of liposomes), it is agreed (and clinically tested and also FDA-approved) that conjugation of cytotoxic drugs like doxorubicin with polymer nanoparticles improves their pharmacokinetics. For example, doxorubicin conjugated with dextran entered clinical trial stage I, similar to dextran derivatives conjugated with cytostatic drug camptothecin. Chitosan and hyaluronic acid were also investigated in this field and approximately fourfold higher drug accumulation in tumor cells was achieved when the carrier liposome contained aliphatic glycolipid. It should be noted that all named polysaccharides exhibited good biocompatibility, biodegradability, and low toxicity making them promising candidates for passive-targeted cytostatic nanocarriers.

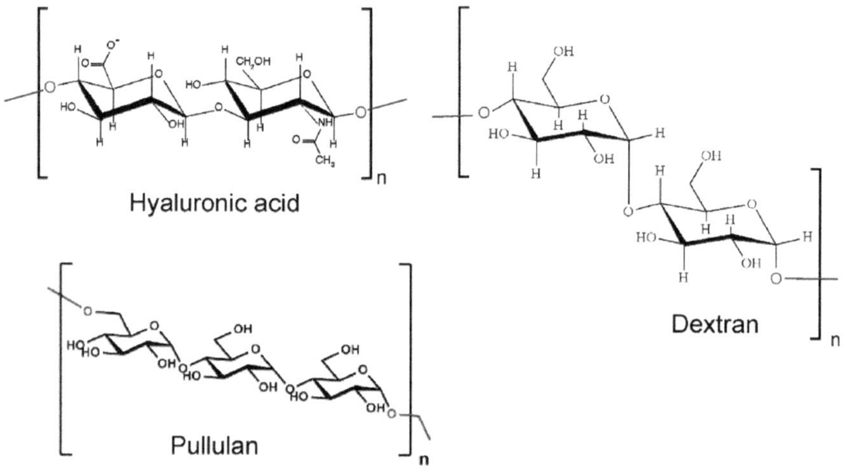

Figure 8.10 Chemical structure of natural polysaccharides hyaluronic acid, dextran, and pullulan.

It should be also noted that the EPR effect is quite often the prerequisite for effective active targeting since the target is most often located in cancer tissue and intravenous injection of smart drugs is also one of the most convenient way of their application. It is illustrated in Fig. 8.11.

8.4.1.2 Carbohydrate-based active targeting

The **active targeted delivery** employs special ligands displayed on the nanocarrier's surface with specific affinity only to the cells bearing the appropriate receptors (see Fig. 8.11). This significantly improves selectivity of the nanocarriers assuming further decrease of the chemotherapeutic dose and harmful side effects. The nanocarriers' ligands can also bind to receptors triggering their internalization which further improves the drug distribution.

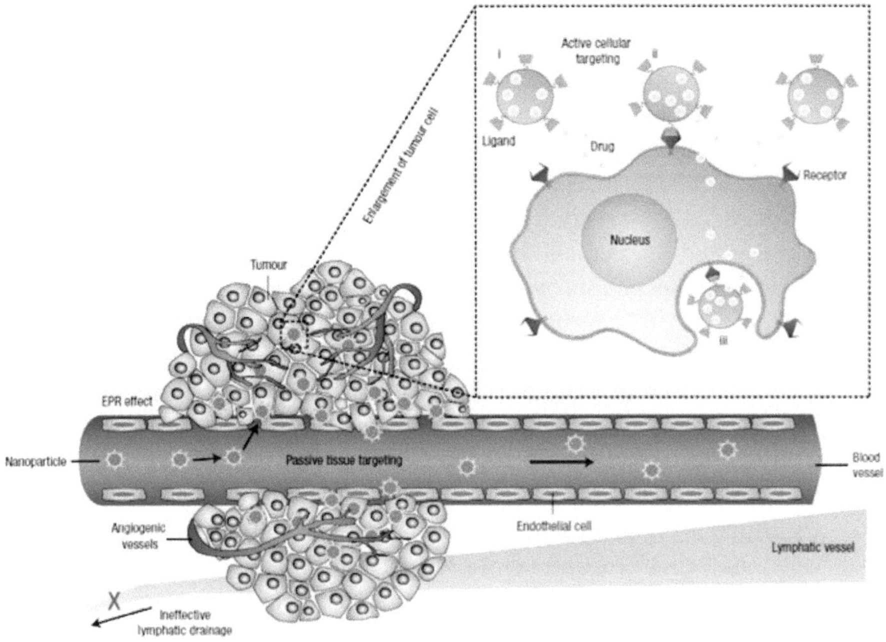

Figure 8.11 Illustration of the EPR effect followed by an active targeting. The inset shows nanocarriers releasing their cargo either outside or inside the targeted cell. Adapted from Peer et al., 2007.

This type of drug targeting emerged no more than two decades ago with nanocarriers grafted with antibodies against specific "cancer" receptors. Targeting with antibodies (and, later, with their smaller parts or with peptides) prevailed for relatively long time. For example, in a 2007 *Nature Nanotechnology* review article on nanocarriers in cancer therapy, employment of targeting ligands other than antibodies-derived seems to be in obvious minority. More recently, however, the increased importance and usefulness of carbohydrates in this field has been recognized. Now, many kinds of nanocarriers are available with displayed targeting glycans.

As for the nanocarrier type, the first choice are polymeric scaffolds and polymer nanoparticles due to their biocompatibility and/or biodegradability. While some polysaccharide nanocarriers are passively targeted, micelles with an additional surface glycosylation were found to dramatically enhance distribution of non-soluble drugs sorafenib and darunavir. Enhanced drug delivery was possible also with glycosylated liposomes or simple systems like polysaccharide molecules (e.g., hyaluronic acid or pullulan) complexed with the cytostatic drug.

Hyaluronic acid is actually very attractive polysaccharide because of its biocompatibility and relative inexpensiveness, and, above all, it can be recognized by the so-called CD-44 receptor and cleaved by enzymes hyaluronidases abundant on surface and in cytosol of many cancer cells, respectively. HA-covered nanocarriers will be then selectively internalized by cancer cells and consequently enzymatically degraded, releasing their cargo. Similarly, polysaccharide pullulan is recognized by asialoglycoprotein receptors, also typical for cancer cells. The carriers can be also marked with glucose because of increased glucose uptake by cancer cells necessary for their rapid growth.[7] Similarly, the rapidly growing cancer cells have increased need for iron; therefore a higher amount (up to 100 fold compared to normal cells) of "TfRs" receptors are displayed on their surface. These receptors can be rather easily targeted by glycoproteins called transferrins.[8]

Smart drugs were prepared also from inorganic nanomaterials including mainly carbon nanomaterials, metal nanoparticles of different shapes, and mesoporous silica nanoparticles. Graphene sheets seem to be convenient for this purpose because of their surface chemistry, high surface area, and flexibility. For example, graphene-based nanosheets/doxorubicin nanoparticles with good stability were prepared and further "wrapped" into hyaluronic acid or chitosan to impose selective intake of the nanoparticles by the cancer cells and higher stability.

Another example is PEG-grafted carbon nanotubes "glued" into bundles by cyclodextrins—molecular "cups" of several saccharide molecules arranged into ring (see Fig. 8.12). These nanoparticles were found to be a versatile chemotherapeutics delivery tool,

[7]The so-called Warburg effect was discovered by Otto Heinrich Warburg and appreciated by the Nobel Prize in physiology or medicine in 1931. According to this hypothesis, cancer cells switch to anaerobic glucose fermentation from normal aerobic pyruvate oxidation to obtain energy for their rapid growth. The cancer then can be seen as mitochondrial disorder.

[8]Conjugation of toxins, cytotoxic drugs, genes, and other molecules with transferrins for targeted delivery into different kinds of cells began to be investigated some three decades ago, as stated by Wagner et al. (1994).

especially for hydrophobic drugs which can be engulfed into cyclodextrin cavities, while the outer side can be either chemically modified by targeting ligand or attached to some larger nanocarrier (see Fig. 8.13).

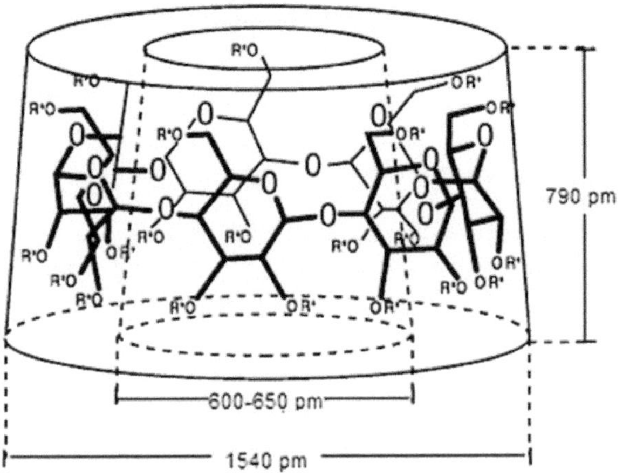

Figure 8.12 Structure of β-cyclodextrin. Creative Commons Attribution License.

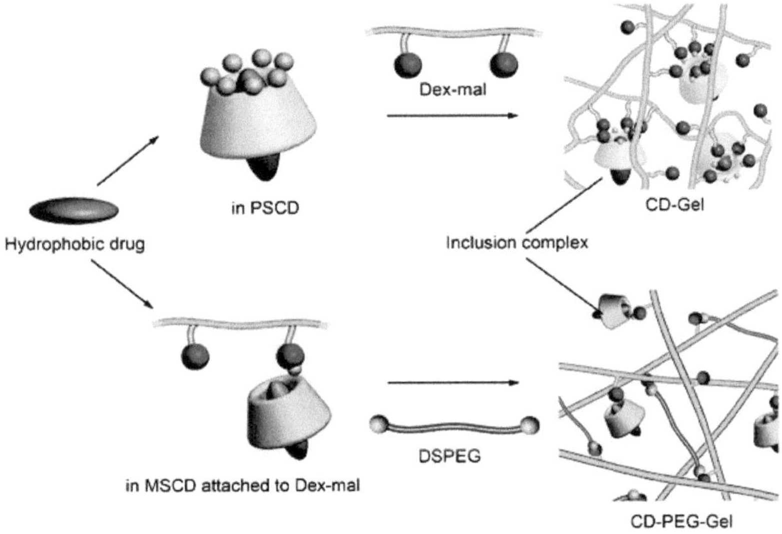

Figure 8.13 Cyclodextrin(CD)–drug complexes integrated into gel matrixes with modified CDs (top) used as crosslinking agent or (bottom) CDs attached to the polymer backbone. Adapted from Peng et al., 2010.

Another option is to load the drug molecules into pores of mesoporous carbon nanoparticles, which are spherical and rather irregular. Of greater interest here, however, are mesoporous silica nanoparticles. Their surface can be easily modified by carbohydrate moieties (polyvalent glucose units or hyaluronic acid) while the drug to be delivered is adsorbed into the pores. Besides the targeting effect of glycans on nanoparticle surface, their role can be also in selective drug release, as exemplified in Box 8.1.

Box 8.1 How the cargo can be released at the right site?

After the internalization, nanoparticles get into so-called lysosomes where naturally a lower pH exists. This different acidity can unfold micelles and degrade liposomes releasing the transported drugs. Another way of targeted unload is hypoxic reductive environment with abundant presence of reduction agents like glutathion in cytosol. This environment can break, for example, disulfide bond -S-S-. Hence, the drug molecule conjugated by this type of covalent bond will be released selectively in tumor cells. Another option is an enzymatic destruction of the nanocarrier by hyaluronidases, as mentioned above. Systems based on de-capping of drug-containing pores and triggered by specific carbohydrate recognition were also introduced. For example, mesoporous silica nanoparticles with drug-loaded pores was modified by synthetic lectin—phenylboronic acid (PBA). This complex was incubated with dextran which, due to its affinity to PBA, formed additional, pore-capping surface layer preventing the cargo from release. In a glucose-rich environment, glucose replaced large dextran molecules; therefore, the drug was free to diffuse from the pores. Polymer nanoparticles (including liposomes, micelles, vesicles, ...) can also have incorporated in their structure molecules sensitive to external stimuli. In this way, for example, liposomes were fabricated that loosened their composition after irradiation by electromagnetic signal with certain wavelength (from radiofrequency wave, infrared (heat) to UV) or as a response to ultrasound waves.

Crossing the barriers

The Box 8.1 above is only very brief description of basic molecular mechanisms employable in smart drugs. More sophisticated approaches employed, for example, combination of glycans for selective targeting and lipids for improved penetration through biological barriers. Out of all biological barriers, the blood–brain barrier (BBB) represents one of the toughest obstacles for smart drugs to treat brain diseases. Under the normal conditions, nanoparticles will hardly get from a vein to the brain cells. BBB is "impermeable for 98% of small-molecule drugs and 100% of macromolecular drugs". However, there are receptors for transportation of small molecules—for example, the above-mentioned transferrins can cross BBB via active "TfR" receptors. It was employed for selective delivery of nanoparticles to brain by i.v. dosage. Lactoferrin is another glycoprotein for the delivery of metal atoms to neural tissue which was also employed as targeting ligand for smart drugs to cross BBB. It has been reported that modification of nanoparticles by transferrin or lactoferrin increased the uptake of the delivered drugs 2–4-fold.

Besides receptors, BBB contains also so-called hexose transporters which improved administration of liposomes coated with mannose or its analogues. Furthermore, solid brain tumors also exhibit less tight endothelium; hence the EPR effect can be also employed here, at a certain rate.

Alternative delivery via nasal mucosa has been also investigated, again with the recognized importance of glycans. For example, chitosan-decorated nanoparticles can travel through nasal epithelium to CNS and further to the brain because of their good affinity to nasal mucosa. A "reversed" approach employed therapeutic proteins conjugated with a wheat germ agglutinin (WGA) lectin binding specifically to glycosyls contained in the nasal membrane. The same effect was observed also for "odorranalectin," very small protein with lectin-like activity isolated from a skin of Asian frog *Odorrana grahami*.

Other diseases than cancer

Although the cancer treatment is the major area for smart drugs, other diseases were also investigated. Levostatin, a drug for atherosclerosis treatment, was successfully loaded into artificial liposomes covered by hyaluronic acid. Amphotericin B, an antiparasitic drug with severe adverse side effect, was integrated with mannosylated dendrimer which improved its bioavailability and effective therapeutic dose decreased approximately five-fold compared to the unconjugated drug.

Recent studies also suggested that small magnetic Fe_2O_3 nanoparticles covered with molecules bearing both positive and negative electric charge and trisaccharide trehalose units could migrate through BBB to the brain where they inhibit the aggregation of proteins responsible for neurodegenerative diseases (Huntington disease; see Fig. 8.14). Similar inhibition in case of Alzheimer disease was observed for antibodies towards the aggregating proteins (anti-β amyloid antibodies) conjugated with above-mentioned glycoprotein transferrin.

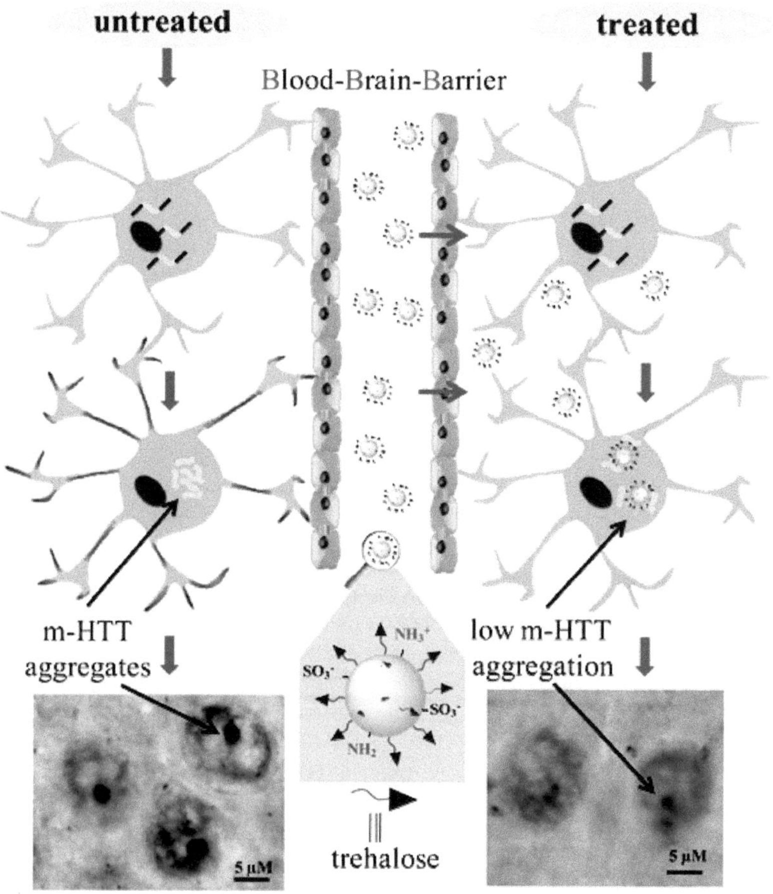

Figure 8.14 Effect of trehalose-modified iron oxide nanoparticles on inhibition of protein aggregates formation in Huntington disease. Adapted from Debnath et al., 2017.

8.4.1.3 Other ways of treatment based on targeted delivery

Besides the chemical treatment there are other ways to deplete cancer cells. Back in the 1950s British scientist R. F. Spiers revealed that on the surface of particles of metal with higher atomic number (Z) the dose of the applied ionization radiation is amplified. It is because the interaction of the X-ray photons with high-Z metal atoms generates secondary radiation. A few decades later, gold nanoparticles began to be used as radiotherapy sensitizer, especially for brain tumor treatment. Nanoparticle accumulation in the targeted cancerous tissue has to be reached, e.g., by active targeting using glycans.

Besides gold, materials like hafnium oxide nanocrystals have been successfully marketed as radiosensitizers (trade name NBTX or NBTXR) for the cancer treatment. It is not clear, however, what the composition of HfO nanocrystals' biocompatibilizing surface coating is.

Delivering boron

Another treatment is "**Boron neutron capture therapy**" relying on targeted delivery of boron ^{10}B atoms into chemo- and radio-resistant cancer cells (e.g., brain tumors called glioblastoma). After exposure to neutron radiation, ^{10}B atoms undergo α-decay and emit α-radiation, which is lethal for cancer cells and very local because of its extremely low penetration depth (less than 10 μm) (see Fig. 8.15). Selective allocation of boron was achieved, for example, by fucosyl conjugates preferentially internalized by cancer cells exhibiting enhanced activity of enzyme fucosyl transferase. Other glycans have been also employed, for example, glucose employing the Warburg effect (see footnote 7). Glycans can be either conjugated with boron-containing molecules (e.g., carborane, molecules of 10 ^{10}B atoms and 2 C atoms; Fig. 8.15) or with larger nanoparticles where ^{10}B atoms are loaded in other ways. An example is mesoporous silica nanoparticles conjugated with carborane and surface coated with carbohydrates.

Photothermal and photodynamic therapy

So-called photothermal and photodynamic therapy are both based on an efficient absorption of near infrared (NIR) light by nanoparticles or molecules. For the former, the absorbed energy causes heating of nanoparticles to such extent, that the targeted cancer cells can be "cooked from inside". In photodynamic therapy the energy of NIR is absorbed by molecule of so-called photosensitizer which, in turn, reacts with adjacent water molecules and produce locally cytotoxic reactive oxygen species.

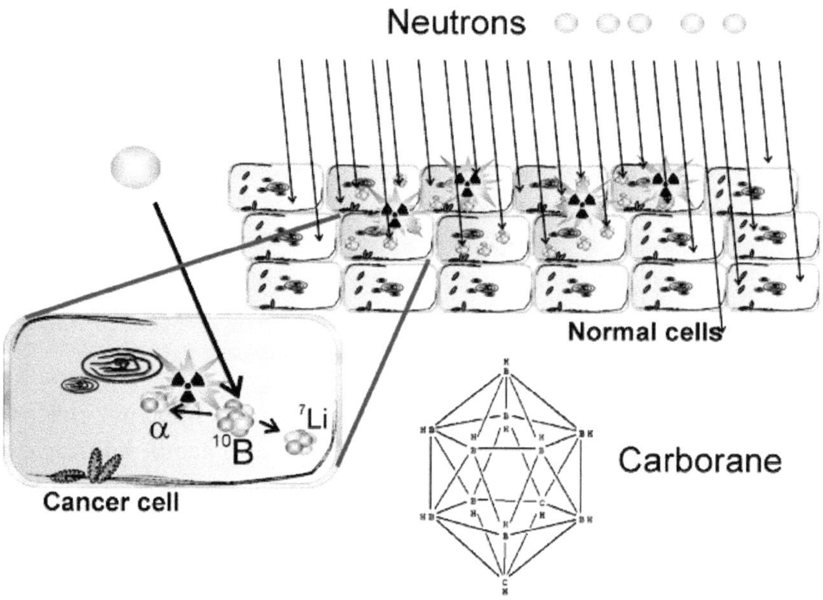

Figure 8.15 Schematic illustration of boron neutron capture therapy. (Right bottom) Structure of carborane molecule.

For **photothermal therapy** typically various gold nanoparticles are used, coated with moieties enhancing their selective accumulation in the cancer tissue. Both passive and active targeting can be employed here, with less possibilities to tailor efficiency of EPR effect with carbohydrates.[9] On the other

[9]The most frequent way of biocompatibilization of nanoparticles in passive targeting remains the modification of their surface with polyethyleneglycole, which is not glycan. It prevents proteins in body fluids to adsorb on nanoparticles and mark them as pathogens to be cleared out of the body. It was found, however, that modification with polysaccharide dextran exhibits similar effect.

side, coating of gold nanoparticles with polysaccharides like hyaluronic acid is very simple and effective procedure applicable for this kind of nanoparticles.

Some organic dyes could also induce photothermal effect, e.g., a compound "IR820" assembled with hyaluronic acid and chitosan molecules to form stable and biocompatible nanoparticles internalized by cancer cells and with hyperthermia induction by NIR irradiation.

Furthermore, carbon nanoparticles including graphene-based nanosheets and spherical mesoporous carbon nanoparticles were also found to induce hyperthermia after NIR absorption. It was investigated in few studies, but mostly without an employment of carbohydrate targeting.

Photodynamic therapy is most often performed by nanoparticles containing molecules of a photosensitizer, typically a porphyrine derivative. These molecules can be attached to inorganic, e.g., gold, nanoparticles and further modified by targeting glycan moieties, for example, hyaluronic acid. Photosensitizers have been also selectively delivered in scaffolds of glucose-containing polymer or encapsulated into hyaluronic acid-coated PLA-based copolymer.

Another option is to use inorganic sensitizers, e.g., CuS or ZnO nanoparticles deposited on reduced graphene oxide (rGO) sheet surfaces further decorated by hyaluronic acid (HA). Interestingly, in vivo investigation of rGO/ZnO/HA nanoparticles revealed dual effect of such smart drug—while rGO exhibited photothermal effect, ZnO nanocrystals were acting as a photosensitizer.

Nucleic acid delivery

Another type of cargo delivered by nanocarriers targeted by glycans is genetic material. "Gene vectors"—a tool for efficient transport of the selected genes into desired cells, co called **transfection**—have been originally developed by molecular biologists for recombinant organisms creation. Later, it was found that DNA or RNA chain fractions impose also therapeutic effect (see Box 8.2).

Box 8.2 Therapeutic effect of nucleic acids

Nucleic acids can help fight cancer, for example, in these ways: (i) delivery of DNA plasmids into cancerous "M2" type of macrophages. After a successful transfection with the plasmid, M2 is switched to "M1" type macrophage which, in contrary, recognizes cancer cells as dangerous and destroys them. (ii) "Small interfering" RNA (siRNA) is delivered into cancer cells where it inhibits translation of certain protein, e.g., "EFG receptors, EFGR". EFGR is important part of growth factor-induced tumor growth and its inhibition slows spreading of tumor.

Since DNA and RNA are negatively charged molecules, they are mostly complexed with oppositely charge polymer chains, most often poly(ethyleneimine) (PEI). For in vivo applications, however, active targeting of the PEI/D(R)NA conjugates should be achieved. Glycans can be employed, for example, via complexation of PEI and DNA with hyaluronic acid, as can be seen in Fig. 8.16. Another way is to replace PEI with polymer containing glycan moieties, for example, methacrylamide backbone bearing glucose, α-D-mannopyranosyl or trehalose units were successfully complexed with DNA plasmids or siRNA and delivered their cargo selectively into cancer cells.

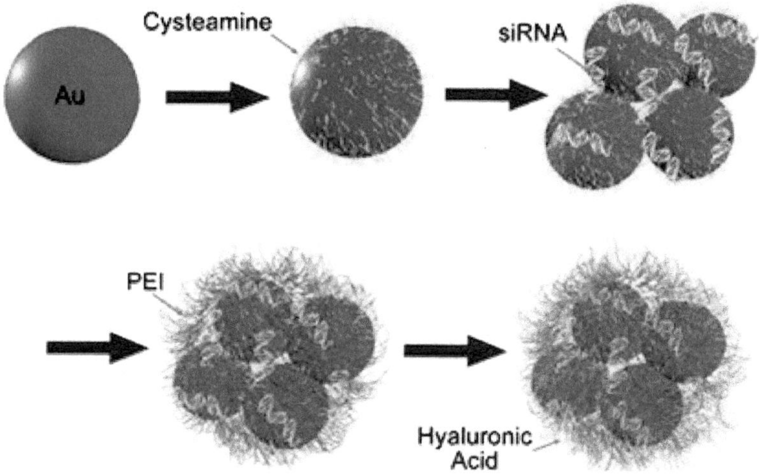

Figure 8.16 Clusters of gold nanoparticles modified by cysteamine and covered with complex of siRNA, PEI and hyaluronic acid. Adapted from Ariga 2012.

8.4.1.4 Multieffect nanotherapeutics

So far, nanoparticles with one therapeutic mode were discussed. But they can be designed so as to provide multiple therapeutic effects in powerful synergy, for example, photothermal nanoparticles loaded with cytostatic drug. To achieve this, e.g., gold nanoparticles were employed, "wrapped" in graphene nanosheets which bonded cytostatic drug doxorubicin and the resulted nanoparticles were further covered with hyaluronic acid for active targeting. These nanoparticles exhibited better performance compared to just one-function configurations.

A nice example of the combined chemo-photothermal therapy is gold nanoparticles coated by spacers ended with cations imposing partial positive charge and spacers ended with 3-bromopyruvate. Such nanoparticles were targeted on subcellular level—the positive charge caused their accumulation in mitochondria. There the conjugated bromopyruvate bound to the enzyme hexokinase 2—a crucial part of unique cancer cells metabolic cycle (see footnote 7—Warburg effect) which hampers the cancer spreading. Furthermore, physical damage to cancer cells' mitochondria was achieved by hyperthermia induced by NIR laser irradiation.

Some treatments require synergy of more chemicals, for example, insulin distribution was found to be more efficient when accompanied with supply of cyclic adenosinemonophosphate (cAMP). For this purpose, silica nanoparticles with pores loaded with cAMP molecules and surface modified with tagged insulin molecules were developed allowing for glucose-triggered insulin release and consequent pore decapping releasing the desired amount of cAMP (see Fig. 8.17).

8.5 Carbohydrates and Cell Imaging

8.5.1 Glyco-Nano Imaging with Organic Fluorescent Probes

While radiolabels were commonly used in cell imaging, now their role has been mostly overtaken by fluorescent UV-VIS probes allowing noninvasive tracking of living cells. In combination with advanced digital data processing they are a powerful imaging tool. It is mostly performed by fluorophores with absorption maxima in NIR range of wavelength which penetrates quite well into the tissue[10] without damaging it and without induction of a background autofluorescence. The employed fluorophores also need to remain stable in physiological environment and non-toxic for health cells. Examples include cyanine dyes and others, reviewed, for example, by Miksa et al. (2016).

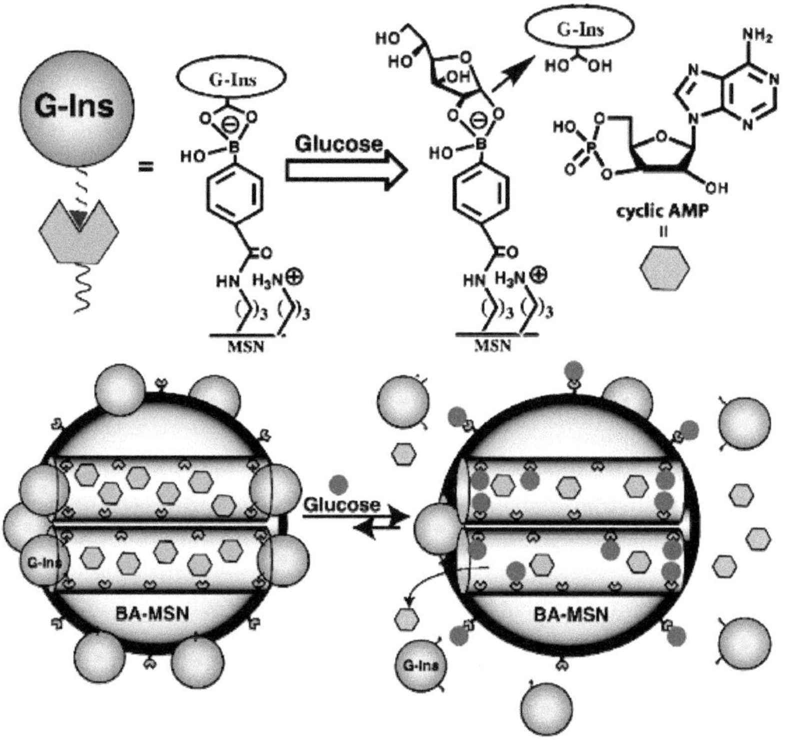

Figure 8.17 Schematic illustration of mesoporous silica (MSN) with pores loaded by cAMP and surface covered by phenylboronic acid (BA) binding labeled insulin (G-Ins) or glucose, once the concentration of latter increases. Glucose binding results in insulin release along with decapping of the pores from which cAMP molecules diffuse. Adapted from Zhao et al., 2009.

[10]For λ = 500–600 nm the penetration depth is approximately 4 mm, above 800 nm the penetration depth doubles to approx. 8 mm. It is deep enough to penetrate epidermis and dermis (the first two layers of skin) and to reach and sometimes also penetrate (its thickness varies between 1 and 6 mm) the subcutaneous fat tissue.

Some fluorophores are designed to "switch on" in cancer cell microenvironment, for example, 4-amino-1,8-naphthalimide conjugated with a single galactose unite (targeting hepatocytes) by a disulfide bond. This bond is cleaved in biologic thiol-rich hepatocyte cytosol turning the probe emission from yellow to green. "Switching on" the probe in the targeted site is then secondary information making this kind of imaging even more precision.

Some chemotherapeutics (e.g., doxorubicin) exhibit also a sufficient fluorescence by themselves allowing for their tracking without any additional labeling. They can be also delivered in an optically inactive form that switches on after the desired target is reached and the site-specific disjunction, e.g., disulfide bond cleaved by glutathione, occurs.

Fluorescent probes for tumor imaging can be delivered in liposomes employing either EPR effect or the glycan-based active targeting, mostly by hyaluronic acid, glucose, or glycoproteins including transferrin. Also a conjugation of the organic fluorophore with the simple glycan unite was possible and led to efficient imaging of the selected tissue.

Fluorophores can be also loaded into polymer nanoparticles, e.g., by "layer-by-layer" architecture where layers of fluorophore molecules intercalate molecules of "structural" polymer. It is also possible to covalently attach the fluorophore molecules on polymer backbone or on inorganic nanoparticles surface.

8.5.2 Glyco-Nano Imaging with Inorganic Probes

Inorganic imaging probes can be used either for optical imaging because many of them are fluorophores or as so-called contrasting agents for imaging by magnetic resonance or ionization radiation-based imaging techniques.

8.5.2.1 Optical imaging

In the simplest approach, gold nanoparticles coated with glycans are used for fast control of aggregation induced by some investigated lectins. Here, gold nanoparticles' capability of "localized surface plasmon resonance" is why some optically detected change is induced upon aggregation of nanoparticles by lectin (or any other external stimuli).

Certain rate of fluorescence is inherent also to silica nanoparticles as well as to carbon nanotubes and graphene. Their optical properties, however, are inferior to synthetically prepared flourophores which can be attributed to a lower level of reproducibility. But there are reports, e.g., on efficient fluorescent imaging by graphene oxide conjugated with gold nanoparticles coated with transferrin molecules (see Fig. 8.18).

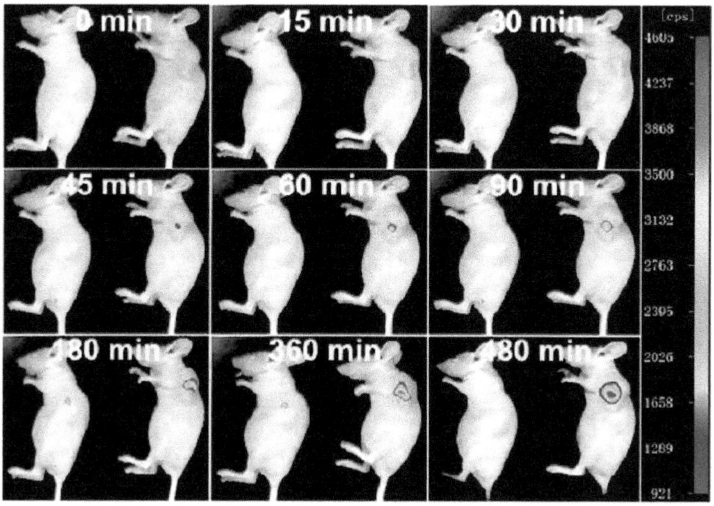

Figure 8.18 Fluorescent detection of induced tumor in mice injected with blank solution (left) and graphene-gold nanoparticles decorated with transferrin. Nanoparticle accumulation only in the tumor site can be observed very well as the detected fluorescence signal. Adapted from Wang et al., 2013.

Fluorescence quenching

When employing graphene-based nanosheets or gold nanoparticles, a phenomenon termed fluorescence quenching must be taken into account. It occurs when energy from the excited fluorophore molecule is not emitted as the light photon(s), but it is transferred to an adjacent molecule which does not turn it into visible emission. This effect lowers the detected fluorescence intensity, but it can be also employed in switch-on systems and selective detection. For example, when changed microenvironment of the quencher/fluorophore conjugate cause their disjunction or at least such rearrangement of some spacer between them that they get further from each other. Energy transfer to such further distance is not possible and the original fluorophore fluorescence can be detected.

Besides these applications, metallic, metal oxide and carbonaceous nanoparticles have been employed as the optical labels nanocarriers with features very similar to those mentioned in targeted delivery of chemotherapeutics.

Semiconductor nanocrystals termed **quantum nanodots** are known to be excellent inorganic fluorophores with the emission wavelength depending only on their size. Since this parameter can be effectively controlled by synthesis condition setup, wide scale of fluorophores can be prepared. However, quantum nanodots mostly contain cadmium (CdS, CdZn, ...); therefore their toxicity must be tackled by biocompatibilization, i.e., fabrication of organic film on their surface. Furthermore, their conjugation with glucose, lactose, or combination of glycan moieties was achieved targeting them into cancer cells with increased glucose uptake. Quantum dots-like fluorescence probes were prepared also from carbon, for example, graphene-like quantum nanodots were employed for selective imaging after their conjugation with hyaluronic acid.

Upconversion luminescence is another phenomenon employable for optical cell imaging. It occurs when nanophosphor[11] particles absorbs photon and transfer its energy to an adjacent rare-earth nanocrystal, which then emits the detectable radiation. The **upconversion nanoparticles** typically consist of inorganic nanocrystals doped with lanthanoid atoms and their emitted wavelength is not determined by the crystal size, but only by the lanthanoid species, which makes the preparation more controllable. Furthermore, methods for their successful surface modification have emerged recently, e.g., modification by α-cyclodextrin for biocompatibilization or glycans for active targeting.

8.5.2.2 Contrasting agents

In today's clinical practice imaging methods employ most often different susceptibility of tissues to ionizing radiation (X-ray computed tomography [CT] or positron emission tomography

[11]Regardless of the term, phosphor is general label for a luminescent compound, including both phosphorescence and luminescence. Some of these materials contain also phosphorus atoms, but others are for example Cu-doped zinc sulfide or mixture of magnesium and potassium fluoride doped with manganese atoms.

[PET]) or different magnetic properties of atomic nuclei (magnetic resonance [MR]). These imaging methods are also used together with direct detection of radiolabeled molecules/particles.

Though "traditional" **radiolabeling** is not so frequent now, it is still sometimes used. An example is γ-tilmanocept, short dextran chains decorated with mannose units and chelating "holder" of ^{99m}Tc atom. It is an FDA-approved compound for mapping of sentinel lymph nodes (SLNs)—the most probable gateways for spreading of metastases from a primary tumor. SLN cells have overexpressed mannose receptors therefore accumulate tilmanocept very well. The γ-tilmanocept-marked lymph node is then identified by a gamma probe (see Fig. 8.19), but tilmanocept can bear also a fluorescent label. For example, in a study by Azad et al. (2015), both γ- and Cy3-modified tilmanocept (Cy3 is a cyanine-based fluorescent dye) was used to investigate their affinity to different types of mannose receptors. Multiple probe conjugation with tilmanocept is also possible, e.g., a fluorescent one and two probes for PET imaging and quantitative radiometric evaluation of the nodes after excision (see Fig. 8.20).

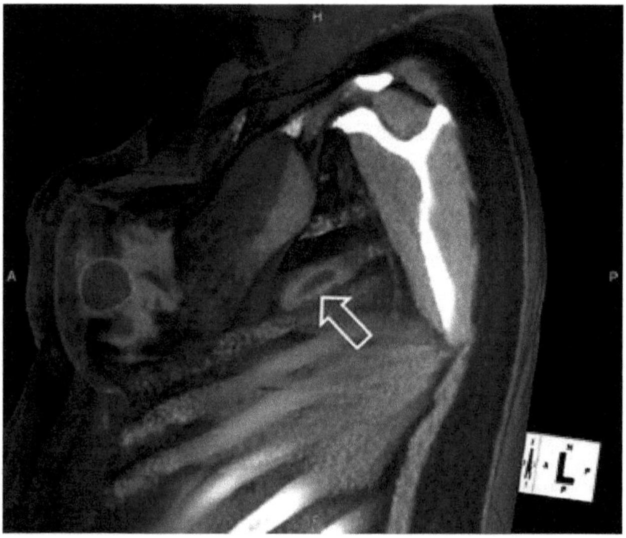

Figure 8.19 Imaging of sentinel lymph nodes (SLN) in breast cancer patient marked with γ-tilmanocept. Arrow points to SLN, the other detected spot is injection site. Adapted from Wallace et al., 2013.

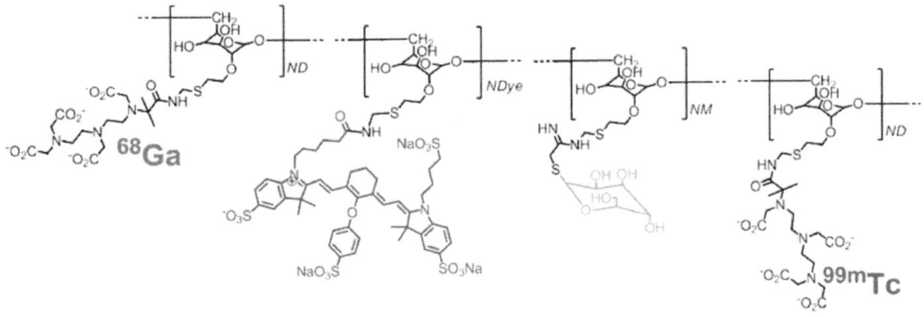

Figure 8.20 Structure of tilmanocept molecule modified with one fluorescent and two radiometric (^{68}Ga and ^{99m}Tc) probes. Adapted from Qin et al., 2015.

In PET, positron-emitting β^+ radioisotope is injected into scoped body. Each positron annihilate while two identical gamma photons are emitted in opposite directions. PET imaging then uses pairs of oppositely located gamma detectors with a body to be scanned located between them. This allows for 3D imaging, similar to CT.

For **PET tumor imaging**, an FDA-approved standard is ^{18}F-FDG (2-deoxy-2-fluoro-glucose) accumulating in cells overexpressing glucose reporter I on their surface. PET imaging was also used in research, for example, to record distribution of radiolabeled (^{68}Ga) siRNA oligonucleotides conjugated with galactose (see Fig. 8.21).

Computed tomography (**CT**) and single-photon emission computed tomography (**SPECT**) are imaging methods with lower resolution than PET. While the former relies on different susceptibility of tissues to X-rays, the latter employs detection of gamma photons emitted by gamma radioisotopes injected into the investigated body. Both methods harness also from glyconanotechnology. For example, gold nanoparticles have been found to be efficient contrast enhancers for CT, even after their coating with glucose and PEG for improved intake by tumor cells and increased retention in the body, respectively. CT image of these nanoparticles after 24 h can be seen in Fig. 8.22 Other examples include gold nanoparticles grafted with lactobionic acid for distribution into asialoglycoprotein receptors abundant on hepatocarcinoma cells surface, or glucose-grafted gold nanoparticles for enhanced CT imaging of tumors.

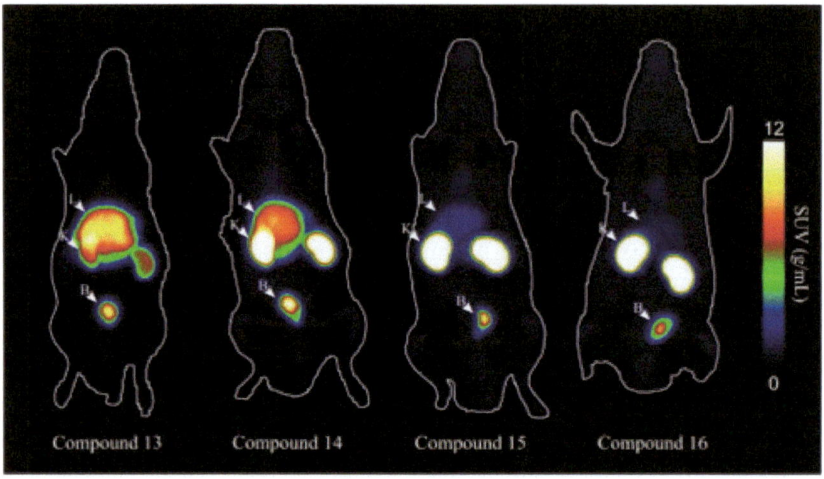

Figure 8.21 PET recording of distribution of radiolabeled siRNA conjugated with 7 (compound 13), 3 (compound 14), 1 (compound 15) and no galactose units. While conjugate with highest amount of glucose accumulated preferentially in liver (L in the figure), the non-glycosylated siRNA accumulate in kidney (K in the figure). Adapted from Mäkilä et al., 2014.

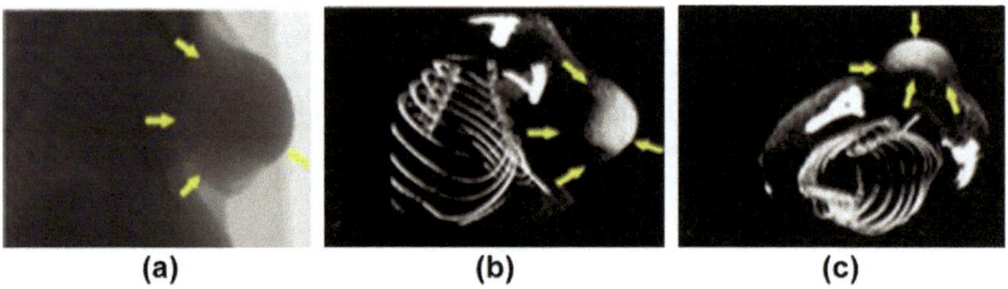

Figure 8.22 CT image of tumor–object marked with arrows. The investigated animals were injected with gold nanoparticles modified with glucose and PEG to enhance the imaging. Adapted from Feng et al., 2014.

8.5.2.3 Contrasting agents for MR

Magnetic resonance is a noninvasive imaging method where nonionizing radiofrequency radiation and strong magnetic field is used. Nuclear properties of protons in different tissues are used to compute the 3D image of the whole living organism/body. To improve the image output, contrasting agents with good paramagnetic properties are used, mostly iron oxide magnetic nanoparticles or molecular scaffolds with Gd^{3+} ions. The latter can be conjugated with carbohydrate-based targeting units, e.g., phenylboronic acid derivative with selective affinity to sialic acid abundant on cancer cell surface. This enabled of selective imaging of tumors in mice.

There are also "switch-on" MR probes, for example, a Gd^{3+} chelate associated with galactose-containing molecular shielding. In a presence of active β-galactosidase—the enzyme used as the gene expression marker—the galactose is hydrolyzed disabling the shielding and permits water molecules to access the Gd^{3+} ions. By this mechanism a significant MR signal can be obtained only in the presence of active β-galactosidase.

Liposomes loaded with magnetic iron oxide nanoparticles ("magnetoliposomes") can aim specific organs similarly to liposomes described in previous sections.

Iron oxide nanoparticles can be also grafted with glycan units, e.g., with hyaluronic acid promoting the specificity to cancer or inflammation cells overexpressing CD-44 receptors, sialic acid for affinity to β-amyloids associated with early stage of Alzheimer's disease, Le^X-type glycans applicable for selective MR imaging of lesions in brains or mannose-containing polymer delivering the nanoparticles specifically to lung cancer cells. In Fig. 8.23 typical output of MRI scanning can be seen with high contrast reached for vasculature imaging. The images were acquired after i.v. application of biocompatibilized iron oxide nanoclusters to examined animals.

In this field the nanotechnology reached the market level quite successfully—there are few FDA-approved MR sensitizers based on iron oxide nanoparticles with dextran coating, e.g., Feridex by AMAG Pharmaceuticals or Combidex nanoparticles in phase III clinical trials. It should be noted that the dextran capping is generally taken as the mean to increase nanoparticles stability in solution.

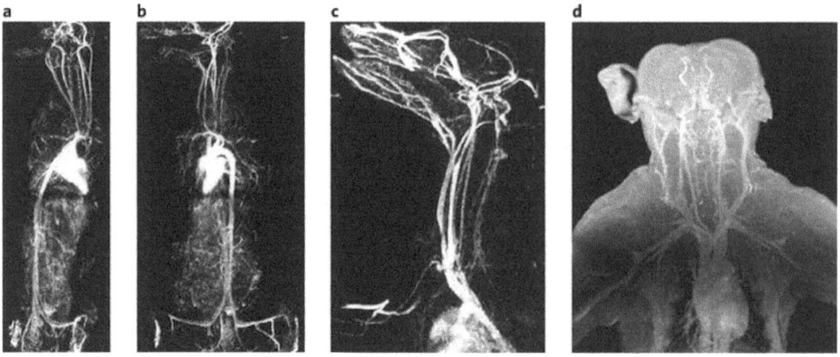

Figure 8.23 MR imaging of test animals injected with PEGylated iron oxide nanoparticles. Adapted from Lu et al., 2017.

8.6 One for All, All for the Future: Theranostic Nanoparticles

Regardless of achievements in smart drugs development, the true miracle of glyco-nano-technology relies in the nanoparticles with integrated multifunctions. Two different therapy modes have been already mentioned and there are also reports on multimodal diagnostics nanoparticles integrating, e.g., fluorescence probe with MR contrast agent. Recently, nanoparticles integrating both therapeutic and diagnostic/imaging functions are of the greatest interest. They have been termed "theranostic"—term coined in 2002 by John Funkhouser.

First theranostic-like approaches can be recognized in the beginning of cancer radiotherapy; in 1948 ^{131}I was found to ablate certain types of metastases in which it preferentially accumulate and, at the same time, it can be detected by gamma cameras. Later, experiments on radiolabeled misonidazole (a nitrogen-containing molecule used as a sensitizer for radiotherapy) were performed and allowed for successful tracking of the drug after its application.[12] This approach was employed more often to assess the fate of drugs after their applications in individuals. It appeared to be very helpful especially in cancer therapy because

[12]Conjugation of misonidazole-like radiosensitizers with ribosyl or fucosyl unit enhanced several times a selective accumulation of the drug in hypoxic parts of tumors (Mannan et al., 1991).

of large inter-individual differences in specific markers or receptors/ligands occurring even within the same type of cancer.

Recent achievements in this field in fact reflect the progress in nanocarrier-performed targeted delivery of drugs, labels, or sensitizers introduced above. There are few basic formulations of theranostic nanoparticles. First mode is to conjugate the drug and targeting moieties covalently to iron oxide or gold nanoparticles which possess innate imaging capabilities as contrasting agents (see the Section 8.5.2). Notably, gold nanoparticles themselves are also radiosensitizers and photothermal agents and small iron oxide nanoparticles are also innately theranostic since they induce hyperthermia in an external high-frequency altering magnetic field. These features are must-have for any candidate for multimodal theranostic nanoparticles. Similarly, graphene and carbon nanotubes exhibit fluorescence, drug loading capability and NIR light-induced cytotoxic effect.

Another option is to use nanocarrier without a labeling capability and conjugate it with both optical probe and a drug. In many studies silica-based mesoporous or hollow nanoparticles have been employed for this purpose, e.g., silica-based nanoparticles with pores loaded with doxorubicin and capped by cyclodextrin conjugated with fluorescence probe. Upon a cleavage by glutathione the probe changes its optical properties and, at the same time, the pores are de-capped which releases the chemotherapeutic from the pores (see Fig. 8.24). Theranostic agents including fluorophores and chemotherapeutics can be also encapsulated into polymer nanoparticles including liposomes, vesicles, or polysaccharide scaffolds.

Theranostic nanoparticles are supposed to differentiate between health and ill tissue, and mark the latter, hence they should perform a pre-treatment diagnosis of disease as well as a monitoring of the therapy progress. These data can be further employed for selection of less (or non-) responsive individuals in therapy and replace it by more appropriate therapeutic approach for them. All these tools can extremely boost efficiency of cancer (and some other diseases) treatment. Hence, the theranostic can be considered as a step toward a personalized medicine, an approach supposed to push the cure efficiency of the conventional medicine ever further.

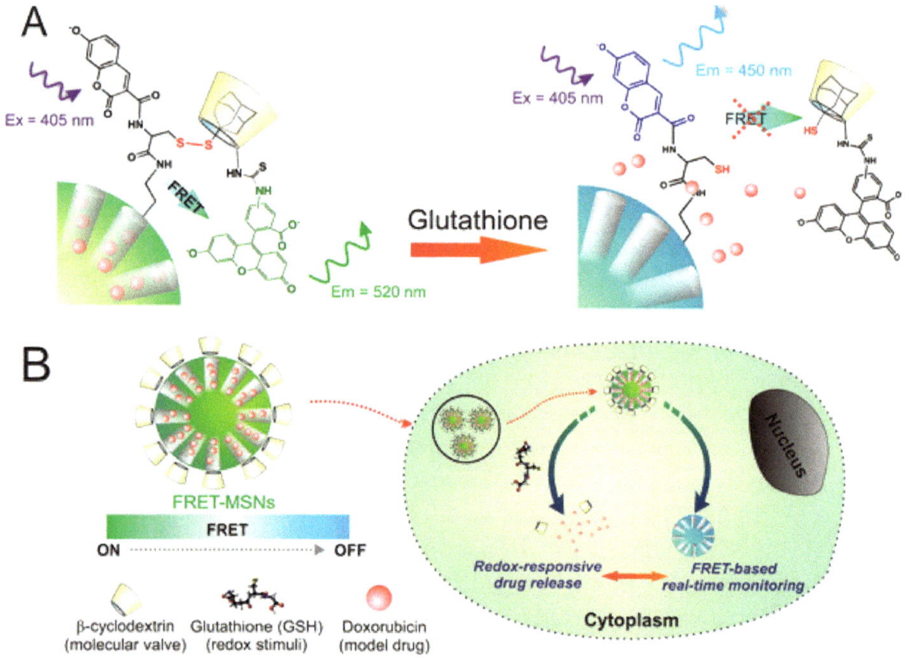

Figure 8.24 Theranostic nanoparticle prepared from mesoporous silica loaded with doxorubicin and modified with glutathione-activated optical probe conjugated with cyclodextrins for capping the probes with doxorubicin. Adapted from Lai et al., 2013.

In this chapter only a very few examples of theranostic nanoparticles employing glyconanotechnology are given. Much more studies have been conducted to develop such smart multieffect drugs with all of them exhibiting good performance in in vitro experiments as well as in in vivo models. Also, the given examples were chosen so as to illustrate the nanotechnology progress in this field. It would be not an exaggeration to claim that nanotechnology is now routinely used in glycobiotechnology. There are though a few more years to get theranostic drugs into common medicinal practice. However, the necessary heading to the new nano-theranostic paradigm in pharmacologic treatment has been already set.

References

Ariga, K., Ji, Q., Hill, J. P., Bando, Y., and Aono, M. (2012). Forming nanomaterials as layered functional structures toward materials nanoarchitectonics, *Npg Asia Materials* **4**, p. e17.

Astronomo, R. D., and Burton, D. R. (2010). Carbohydrate vaccines: Developing sweet solutions to sticky situations?, *Nature Reviews Drug Discovery* **9**, p. 308.

Azad, A. K., Rajaram, M. V. S., Metz, W. L., Cope, F. O., Blue, M. S., Vera, D. R., and Schlesinger, L. S. (2015). γ-tilmanocept, a new radiopharmaceutical tracer for cancer sentinel lymph nodes, binds to the mannose receptor (CD206), *Journal of Immunology (Baltimore, Md.: 1950)* **195**, pp. 2019–2029.

Baum, L. G., and Cobb, B. A. (2017). The direct and indirect effects of glycans on immune function, *Glycobiology* **27**, pp. 619–624.

Bourgault, J. P., Trabbic, K. R., Shi, M., and Andreana, P. R. (2014). Synthesis of the tumor associative [small alpha]-aminooxy disaccharide of the TF antigen and its conjugation to a polysaccharide immune stimulant, *Organic & Biomolecular Chemistry* **12**, pp. 1699–1702.

Byrd-Leotis, L., Cummings, R. D., and Steinhauer, D. A. (2017). The interplay between the host receptor and influenza virus hemagglutinin and neuraminidase, *International Journal of Molecular Sciences* **18**, p. 1541.

Choi, C. H. J., Alabi, C. A., Webster, P., and Davis, M. E. (2010). Mechanism of active targeting in solid tumors with transferrin-containing gold nanoparticles, *Proceedings of the National Academy of Sciences* **107**, pp. 1235–1240.

Debnath, K., Pradhan, N., Singh, B. K., Jana, N. R., and Jana, N. R. (2017). Poly(trehalose) nanoparticles prevent amyloid aggregation and suppress polyglutamine aggregation in a Huntington's disease model mouse, *ACS Applied Materials & Interfaces* **9**, pp. 24126–24139.

Dingjan, T., Spendlove, I., Durrant, L. G., Scott, A. M., Yuriev, E., and Ramsland, P. A. (2015). Structural biology of antibody recognition of carbohydrate epitopes and potential uses for targeted cancer immunotherapies, *Molecular Immunology* **67**, pp. 75–88.

Eggermont, A. M. M., Suciu, S., Rutkowski, P., Marsden, J., Santinami, M., Corrie, P., Aamdal, S., Ascierto, P. A., Patel, P. M., Kruit, W. H., Bastholt,

L., Borgognoni, L., Bernengo, M. G., Davidson, N., Polders, L., Praet, M., and Spatz, A. (2013). Adjuvant ganglioside GM2-KLH/QS-21 vaccination versus observation after resection of primary tumor > 1.5 mm in patients with stage II melanoma: Results of the EORTC 18961 randomized phase III trial, *Journal of Clinical Oncology* **31**, pp. 3831–3837.

Feng, G., Kong, B., Xing, J., and Chen, J. (2014). Enhancing multimodality functional and molecular imaging using glucose-coated gold nanoparticles, *Clinical Radiology* **69**, pp. 1105–1111.

Frullano, L., Rohovec, J., Aime, S., Maschmeyer, T., Isabel Prata, M., Pedroso De Lima, J. J., Geraldes, C. F. G. C., and Peters, J. A. (2004). Towards targeted MRI: New MRI contrast agents for sialic acid detection, *Chemistry: A European Journal* **10**, pp. 5205–5217.

Hart, G. W., and Copeland, R. J. (2010). Glycomics hits the big time, *Cell* **143**, pp. 672–676.

Jiang, B.-P., Zhang, L., Zhu, Y., Shen, X.-C., Ji, S.-C., Tan, X.-Y., Cheng, L., and Liang, H. (2015). Water-soluble hyaluronic acid-hybridized polyaniline nanoparticles for effectively targeted photothermal therapy, *Journal of Materials Chemistry B* **3**, pp. 3767–3776.

Kwatra, D., Venugopal, A., and Anant, S. (2013). Nanoparticles in radiation therapy: A summary of various approaches to enhance radiosensitization in cancer, *Translational Cancer Research* **2**, pp. 330–342.

Li, Z., Tan, S., Li, S., Shen, Q., and Wang, K. (2017). Cancer drug delivery in the nano era: An overview and perspectives (review), *Oncology Reports* **38**, pp. 611–624.

Lorente-Gea, L., García, B., Martín, C., Quirós, L. M., and Fernández-Vega, I. (2017). Heparan sulfate proteoglycans and heparanases in Alzheimer's disease: Current outlook and potential therapeutic targets, *Neural Regeneration Research* **12**, pp. 914–915.

Lu, Y., Xu, Y.-J., Zhang, G.-b., Ling, D., Wang, M.-Q., Zhou, Y., Wu, Y.-D., Wu, T., Hackett, M. J., Hyo Kim, B., Chang, H., Kim, J., Hu, X.-T., Dong, L., Lee, N., Li, F., He, J.-C., Zhang, L., Wen, H.-Q., Yang, B., Hong Choi, S., Hyeon, T., and Zou, D.-H. (2017). Iron oxide nanoclusters for T 1 magnetic resonance imaging of non-human primates, *Nature Biomedical Engineering* **1**, pp. 637–643.

Mäkilä, J., Jadhav, S., Kiviniemi, A., Käkelä, M., Liljenbäck, H., Poijärvi-Virta, P., Laitala-Leinonen, T., Lönnberg, H., Roivainen, A., and Virta, P. (2014). Synthesis of multi-galactose-conjugated 2′-O-methyl oligoribonucleotides and their in vivo imaging with positron emission tomography, *Bioorganic & Medicinal Chemistry* **22**, pp. 6806–6813.

Marrache, S., and Dhar, S. (2015). The energy blocker inside the power house: Mitochondria targeted delivery of 3-bromopyruvate, *Chemical Science* **6**, pp. 1832–1845.

Miksa, B. (2016). Fluorescent dyes used in polymer carriers as imaging agents in anticancer therapy, *Med Chem (Los Angeles)* **6**, pp. 611–639.

Pardridge, W. M. (2007). Drug targeting to the brain, *Pharmaceutical Research* **24**, pp. 1733–1744.

Peer, D., Karp, J. M., Hong, S., Farokhzad, O. C., Margalit, R., and Langer, R. (2007). Nanocarriers as an emerging platform for cancer therapy, *Nature Nanotechnology* **2**, p. 751.

Peng, K., Cui, C., Tomatsu, I., Porta, F., Meijer, A. H., Spaink, H. P., and Kros, A. (2010). Cyclodextrin/dextran based drug carriers for a controlled release of hydrophobic drugs in zebrafish embryos, *Soft Matter* **6**, pp. 3778–3783.

Plotkin, S. A., and Plotkin, S. L. (2011). The development of vaccines: How the past led to the future, *Nature Reviews Microbiology* **9**, p. 889.

Qin, Z., Hoh, C. K., Hall, D. J., and Vera, D. R. (2015). A tri-modal molecular imaging agent for sentinel lymph node mapping, *Nuclear Medicine and Biology* **42**, pp. 917–922.

Raveendran, S., Yoshida, Y., Maekawa, T., and Kumar, D. S. (2013). Pharmaceutically versatile sulfated polysaccharide based bionano platforms, *Nanomedicine: Nanotechnology, Biology and Medicine* **9**, pp. 605–626.

Soliman, C., Yuriev, E., and Ramsland, P. A. (2017). Antibody recognition of aberrant glycosylation on the surface of cancer cells, *Current Opinion in Structural Biology* **44**, pp. 1–8.

Stockhofe, K., Postema, J., Schieferstein, H., and Ross, T. (2014). Radiolabeling of nanoparticles and polymers for PET imaging, *Pharmaceuticals* **7**, p. 392.

Varki, A. (1993). Biological roles of oligosaccharides: All of the theories are correct, *Glycobiology* **3**, pp. 97–130.

Veldman, R. J., Koning, G. A., van Hell, A., Zerp, S., Vink, S. R., Storm, G., Verheij, M., and van Blitterswijk, W. J. (2005). Coformulated *N*-octanoyl-glucosylceramide improves cellular delivery and cytotoxicity of liposomal doxorubicin, *Journal of Pharmacology and Experimental Therapeutics* **315**, pp. 704–710.

Vidal, S. (2015). Sweet fullerenes vanquish viruses, *Nature Chemistry* **8**, p. 4.

Wagner, E., Curiel, D., and Cotten, M. (1994). Delivery of drugs, proteins and genes into cells using transferrin as a ligand for receptor-mediated endocytosis, *Advanced Drug Delivery Reviews* **14**, pp. 113–135.

Wallace, A. M., Han, L. K., Povoski, S. P., Deck, K., Schneebaum, S., Hall, N. C., Hoh, C. K., Limmer, K. K., Krontiras, H., Frazier, T. G., Cox, C., Avisar, E., Faries, M., King, D. W., Christman, L., and Vera, D. R. (2013). Comparative evaluation of [99mTc]tilmanocept for sentinel lymph node mapping in breast cancer patients: Results of two phase 3 trials, *Annals of Surgical Oncology* **20**, pp. 2590–2599.

Wang, Y., Chen, J.-T., and Yan, X.-P. (2013). Fabrication of transferrin functionalized gold nanoclusters/graphene oxide nanocomposite for turn-on near-infrared fluorescent bioimaging of cancer cells and small animals, *Analytical Chemistry* **85**, pp. 2529–2535.

Zhao, Y., Trewyn, B. G., Slowing, I. I., and Lin, V. S. Y. (2009). Mesoporous silica nanoparticle-based double drug delivery system for glucose-responsive controlled release of insulin and cyclic AMP, *Journal of the American Chemical Society* **131**, pp. 8398–8400.

Index